Molecular Biology in Crop Protection

Edited by

George Marshall

and

Dale Walters

Department of Plant Science,
The Scottish Agricultural College,
Ayr, UK

CHAPMAN & HALL

London · Glasgow · Weinheim · New York · Tokyo · Melbourne · Madras

Published by Chapman & Hall, 2–6 Boundary Row, London SE1 8HN, UK

Chapman & Hall, 2–6 Boundary Row, London SE1 8HN, UK

Blackie Academic & Professional, Wester Cleddens Road, Bishopbriggs, Glasgow G64 2NZ, UK

Chapman & Hall GmbH, Pappelallee 3, 69469 Weinheim, Germany

Chapman & Hall Inc., One Penn Plaza, 41st Floor, New York NY 10119, USA

Chapman & Hall Japan, Thomson Publishing Japan, Hirakawacho Nemoto Building, 6F, 1–7–11 Hirakawa-cho, Chiyoda-ku, Tokyo 102, Japan

Chapman & Hall Australia, Thomas Nelson Australia, 102 Dodds Street, South Melbourne, Victoria 3205, Australia

Chapman & Hall India, R. Seshadri, 32 Second Main Road, CIT East, Madras 600 035, India

First edition 1994

Typeset in 10/12pt Palatino by ROM-Data Corporation Ltd, Falmouth, Cornwall.

Printed in Great Britain at the University Press, Cambridge

ISBN 0 412 54390 7 (HB) 0 412 54400 8 (PB)

A catalogue record for this book is available from the British Library

Library of Congress Cataloging-in-Publication data: 94-70977

∞ Printed on permanent acid-free text paper, manufactured in accordance with the proposed ANSI/NISO Z39.48-1992 and ANSI/NISO Z39.48-1984 (Permanence of Paper)

Contents

Contributors

J.A. Butters, Department of Agricultural Sciences, University of Bristol, AFRC Institute of Arable Crops Research, Long Ashton Research Station, Bristol BS18 9AF, UK

D.J. Cole, Rhone-Poulenc Agriculture Ltd, Fyfield Road, Ongar, Essex CM5 0HW, UK

R.P. Finch, Department of Plant Science, Scottish Agricultural College, Auchincruive, Ayr KA6 5HW, UK

S.G. Foster, Biochemistry Section, Agr. Ev. UK Ltd, Chesterford Park Research Station, Saffron Walden, Essex CB10 1XL, UK

R.L. Fuchs, Crop Protection Department, Monsanto Company, 700 Chesterfield Parkway North, St Louis, MO 63198, USA

A.M.R. Gatehouse, Department of Biological Sciences, University of Durham, South Road, Durham DH1 3LE, UK

L.A. Glover, Department of Molecular and Cell Biology, University of Aberdeen, Marischal College, Broad Street, Aberdeen AB9 1AS, UK

V.A. Hilder, Department of Biological Sciences, University of Durham, South Road, Durham DH1 3LE, UK

D.W. Hollomon, Department of Agricultural Sciences, University of Bristol, AFRC Institute of Arable Crops Research, Long Ashton Research Station, Bristol BS18 9AF, UK

L.A. King, c/o Dr Louis Miller, Dept of Genetics, 101B Life Science Building, University of Georgia, Athens, Georgia 30602, USA

A. McHughen, Crop Development Centre, University of Saskatchewan, Saskatoon S7N 0W0, Canada

M. Miller, Potato Marketing Board, Broad Field House, 4 Between Towns Road, Cowley, Oxford OX4 3NA, UK

T.E. Nickson, Crop Protection Department, Monsanto Company, 700 Chesterfield Parkway North, St Louis, MO 63198, USA

J.B. Pillmoor, Biochemistry Section, Agr. Ev. UK Ltd, Chesterford Park Research Station, Saffron Walden, Essex CB10 1XL, UK

R.D. Possee, Natural Environment Research Council, Institute of Virology and Environmental Microbiology, Mansfield Road, Oxford OX1 3SR, UK

W. Powell, c/o Dr Antoni Rafalski, E.I. DuPont de Nemours & Co., DuPont Agricultural Products, Biotechnology Research Exp. St., Wilmington, Delaware 19880-0402, USA

J. Ryals, Agricultural Biotechnology Research Unit, CIBA-GEIGY Corporation, 3054 Cornwallis Road, PO Box 12257, Research Triangle Park, NC 27709-2257, USA

S. Uknes, Agricultural Biotechnology Research Unit, CIBA-GEIGY Corporation, 3054 Cornwallis Road, PO Box 12257, Research Triangle Park, NC 27709-2257, USA

E. Ward, Agricultural Biotechnology Research Unit, CIBA-GEIGY Corporation, 3054 Cornwallis Road, PO Box 12257, Research Triangle Park, NC 27709-2257, USA

Preface

Few individuals can be unmoved by the impact of molecular biology. Advances in the discipline over four decades have progressed at a rate unrivalled in other scientific areas. In its formative years, molecular biology examined the chemical and physical structures of biological molecules, subsequently elucidated the nature and function of DNA and evolved into molecular genetics. From this exponential growth of scientific knowledge, tremendous opportunities were created for the application of molecular approaches to solve problems in applied biology.

This book describes the new productive association between novel state-of-the-art molecular biology and crop protection, a discipline with a sound heritage in traditional applied biology and chemistry. Never before has crop protection faced such diverse challenges. It is charged with improving global food supplies and with the pressure of population increases of one billion in the next decade. But to consider protection of crops simply in terms of weed, pest and disease control would be a gross oversimplification of the mission. Rather, crop protectionists must develop measures which will maintain crop yield and quality without harm to the environment. Chemical, cultural and biological approaches to crop protection must also fulfil evolving legislative demands and address the issues which confer public acceptability.

To meet these challenges it is not suprising that crop protection research has embraced molecular biology. The practical benefits are evident in this book. We now have extensive programmes of research which are truly multidisciplinary in their approaches, while collaboration between scientists in academia and industry provides a synergy of output and improved innovation.

We have prepared this book for crop protectionists who wish to become familiar with the benefits which new molecular techniques can bring, but students of molecular biology will also find crop protection a discipline full of challenges which can test their ingenuity. This book begins by laying a sound foundation in the terminology and techniques used in molecular biology. The following six chapters cover the application of molecular biology in solving a range of technological problems in crop protection. This section begins with three chapters devoted to aspects of control via biological and chemical agents, including an assessment of molecular determinants for resistance to crop protection chemicals. The next three chapters examine the methods which are applied to the genetic engineering of crops for improved resistance to diseases, herbicides and pests. There follows coverage of the assessment methods for genetically engineered plants in the laboratory and field. Issues of Plant Breeders' Rights, patents and legal matters are discussed, with practical examples from around the world. In addition, the environmental impact and legislative issues which surround the release of genetically modified plants and micro-organisms are described, with reference to international case studies.

We are grateful to the authors, who quite apart from providing their chapters on time, also managed to make them exciting reading. We hope that you, the reader will find the following pages equally stimulating.

George Marshall
Dale Walters

PART ONE

Introduction

An introduction to molecular technology

R.P. Finch

1.1 INTRODUCTION

Straightforward explanations of the theory and methods associated with the now widely used molecular/recombinant technologies are not easy to find in the literature. Current research articles in the field rarely explain the theory behind the techniques utilized, or quote the original publication, and this makes life difficult for the student or researcher who is new to the subject. In this chapter, an attempt is made to explain, in relatively simple terms, how nucleic acids can be isolated, analysed and utilized for the genetic manipulation of plants. This serves to provide explanations of the terminology used in the following chapters and the reader is encouraged to refer back to this chapter as necessary. Only a few key papers relating to specific techniques are quoted in Chapter 1, and these citations, along with the suggestions for further reading, are for readers who require more than what is, essentially, an outline and overview of the more widely used tools and techniques.

1.2 NUCLEIC ACID STRUCTURE AND ANALYSIS

1.2.1 DNA structure

Deoxyribonucleic acid (DNA) stores and transmits genetic information and, as such, is the principal molecule of life. DNA normally exists as two linear polymeric strands, each composed of subunits called nucleotides that

Molecular Biology in Crop Protection. Edited by George Marshall and Dale Walters. Published in 1994 by Chapman & Hall, London. ISBN 0 412 54390 7(HB) 0 412 54400 8(PB).

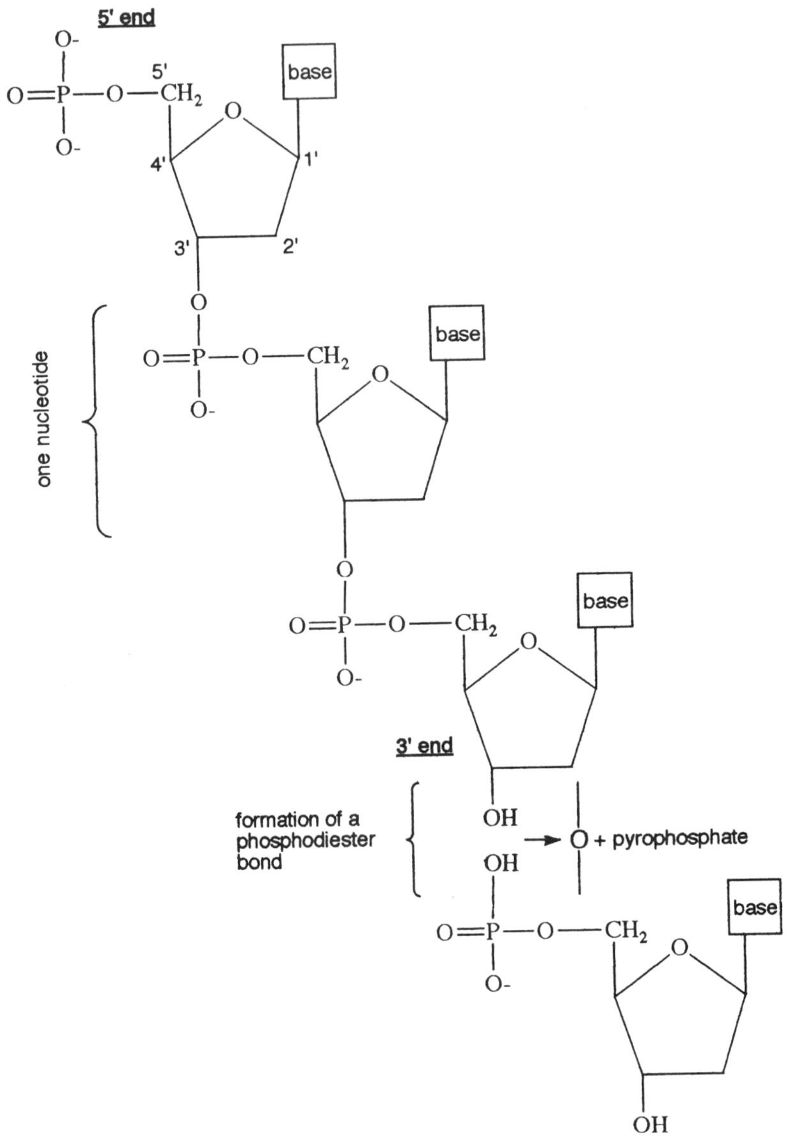

Figure 1.1 A polynucleotide sequence showing the 5′ and 3′ orientation that corresponds to the free phosphate group at carbon position 5 (top) and the free OH group at carbon position 3 of the deoxyribose ring (bottom), respectively. Under the influence of DNA ligase, nucleoside triphosphates can form a phosphodiester bond with the 3′ end of the molecule.

comprise a pentose sugar (deoxyribose), a phosphate group and one of four characteristic organic 'base' molecules, two of which are 'purines', i.e. adenine [A] and guanine [G], and two of which are 'pyrimidines', i.e. cytosine [C] and thymine [T]. Nucleotides are linked by 'phosphodiester bonds' in a characteristic sequence and in a way that infers a specific orientation (Figure 1.1). The two strands of the DNA molecule pair in an antiparallel manner (reverse orientation) according to a chemical complementarity between nucleotide bases such that adenine always pairs with thymine, and guanine with cytosine, by hydrogen bonding (Figure 1.2). The two DNA strands form the familiar double helix with the 'base pairs' stacked between the intertwined sugar–phosphate spines. The double helical structure of DNA is very stable and *in vivo* the DNA strands only unwind and separate in a localized manner during replication and transcription. *In vitro*, however, complete separation of linear molecules can be achieved by heating a solution of the DNA to its melting point of approximately 95°C. This 'denaturation' can be reversed by lowering the temperature; the strands rehybridize in their normal, antiparallel, complementary fashion.

1.2.2 Isolating DNA in the laboratory

The presence of nucleases in bacteria on the skin and in some grades of commercial chemicals means that a number of important precautions must be observed when working with naked DNA. These include wearing surgical gloves as a matter of course when handling all reagents. The reagents themselves should normally be autoclaved and stored in siliconized or acid-washed, heat-sterilized bottles.

A diversity of methods exist for the isolation of DNA from cells and tissues. Generally, however, there are three common objectives for most protocols. Firstly, a cell extract is prepared in a suitable aqueous buffer. Depending on the source material, this may require simple cell lysis or a vigorous tissue homogenization. The extract is then treated with organic solvents, such as chloroform and/or phenol, at alkaline pH to precipitate protein. The DNA is retained in the aqueous phase. Finally the DNA is precipitated from the partially purified aqueous solution by adding alcohol and incubating at low temperature in the presence of a salt (e.g. sodium acetate), and resuspending in water or buffer at the desired concentration (determined spectrophotometrically). Specific modifications to this outline protocol are noted, where appropriate, in the following sections.

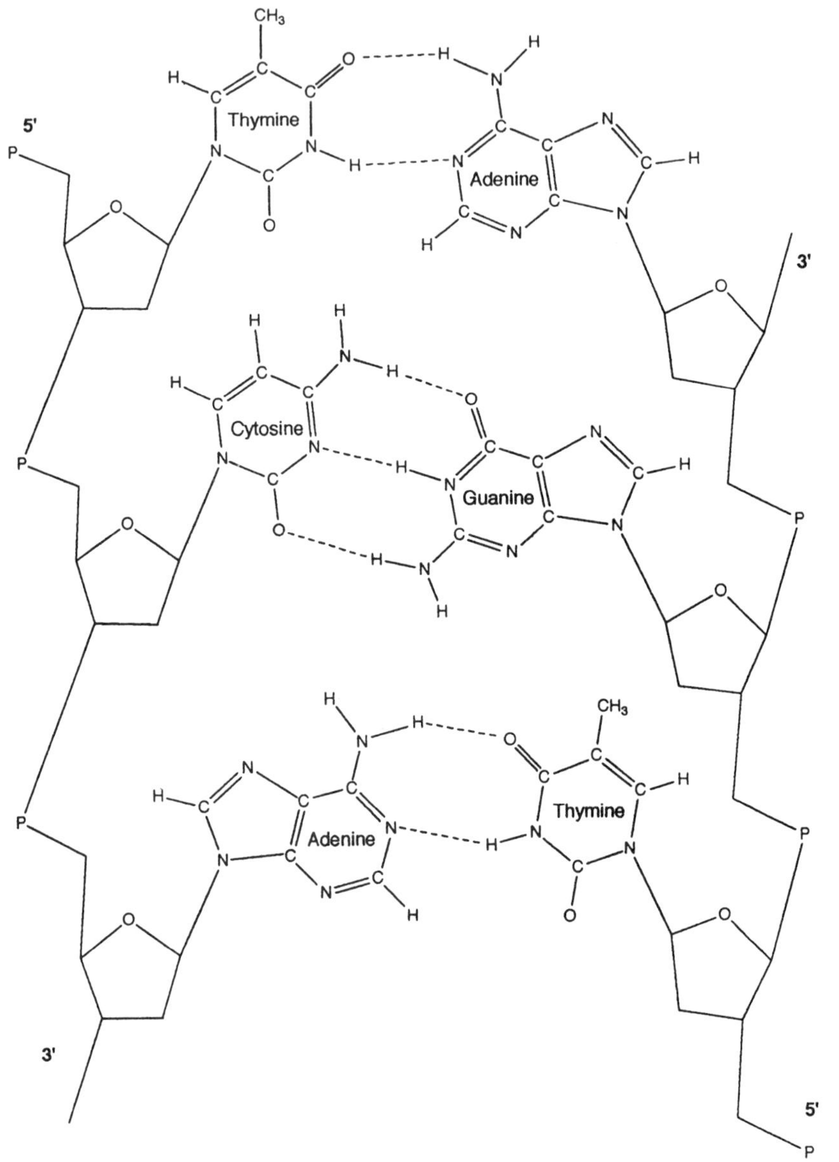

Figure 1.2 Two-dimensional structure of a 3 bp double-stranded DNA molecule showing antiparallel pairing of the two complementary strands held together by specific hydrogen bonds.

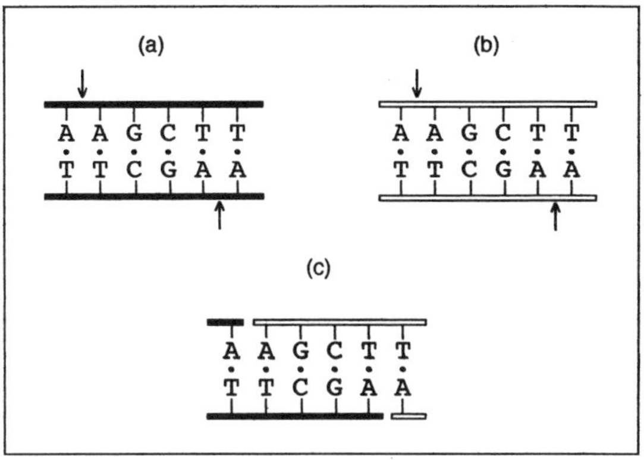

Figure 1.3 (a) The restriction enzyme *Hin*dIII creates a staggered cut in DNA at the points shown. If a second molecule is cut with the same enzyme (b), the two compatible sticky ends provide the initial structure for joining the two molecules (c). A DNA ligase stabilizes the structure by completing the sugar backbone of the molecule.

1.2.3 Cutting and joining DNA

DNA can be cut or 'restricted' using purified bacterial enzymes called restriction endonucleases. These identify specific base 'recognition' sequences and cleave the DNA within this sequence. Some 200 or more restriction enzymes are commercially available, each capable of cleaving DNA at a particular site, although some are isoschizomers (recognize the sequence) as described in Chapter 4. An example is *Hin*dIII, derived from the bacterium *Haemophilus influenzae* Rd which cleaves DNA as shown in Figure 1.3. Note that *Hin*dIII generates a staggered cut in the double-stranded DNA molecule giving rise to short single-stranded tails known as cohesive or sticky ends. These are particularly useful for recombinant DNA work since they provide the initial structure for annealing two pieces of DNA that have been restricted with the same enzyme; the bond is completed by the action of a DNA ligase. Methods for joining blunt-ended fragments, which usually involves enzymic modification of the fragment ends, are also available.

1.2.4 Basic analysis of DNA–agarose gel electrophoresis

Agarose gel electrophoresis is widely used as a basic analytical tool for the characterization of DNA molecules. This technique effectively separates pieces of DNA in a heterogeneous mixture on the basis of molecular size; small pieces of DNA migrate through the gel matrix faster than larger pieces under the influence of an electric current (DNA molecules have a net negative charge and migrate towards the anode). Migration rates also depend on the conformation of the DNA since linear molecules migrate more slowly than circular molecules of the same nucleotide length. Moreover, circular molecules can adopt a very compact, twisted 'supercoiled' configuration and can migrate even more rapidly through the gel matrix.

Gel electrophoresis involves the loading of a small sample of a heterogeneous DNA mixture (the result of a restriction endonuclease digestion for example) into a preformed well in the gel (Figure 1.4). It is then subjected to an electric current in a horizontal or vertical tank, via an ionic buffer solution, for between 10 min and several hours depending on agarose concentration, gel size, voltage applied and DNA conformation. Electrophoresis of restriction digests of relatively simple DNA molecules gives rise to discrete 'bands' of DNA which are readily visualized by staining with a UV-fluorescent stain, such as ethidium bromide, which is incorporated in the gel and the running buffer at 0.5 μg/ml (alternatively the DNA can be stained after the electrophoresis process by incubating the gel in a tank of fresh buffer containing the stain). After staining, the gel is exposed to

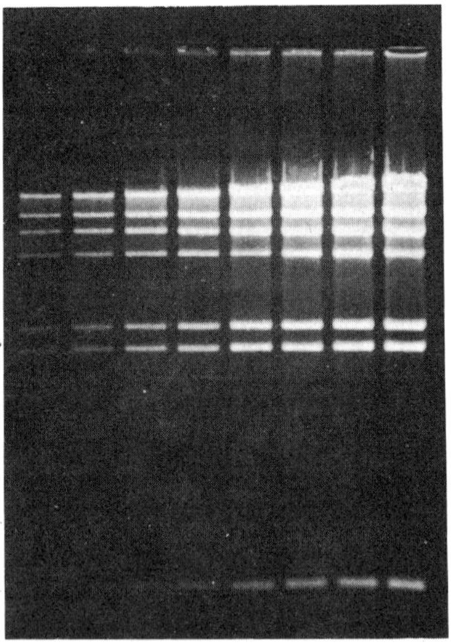

Figure 1.4 A Polaroid® photograph of an ethidium bromide-stained agarose gel following electrophoresis of a DNA restriction digest [sample concentrations loaded into the gel range from 10 to 500ng; fragment sizes range from approximately 23 000 (top) to 100 bp (bottom)].

mid-range UV light (300 nm) using a transilluminator. The DNA bands fluoresce brightly and can be photographed through a suitable filter using Polaroid® film. The molecular size of the DNA fragments can be estimated by comparing the migration of the bands with that of size standards separated on the same gel.

1.2.5 Types of DNA found in organisms

(a) Nuclear DNA

The vast majority of DNA in eukaryotic organisms, the genomic DNA, is found in the nucleus where it exists largely in the form of monotonous linear sequences. Surprisingly, less than 10% of this DNA is actually translated leaving at least 90% of the genome apparently redundant. In multicellular organisms, much of this non-coding redundant DNA comprises 'satellite' repeat regions (regions where characteristic nucleotide sequences are serially repeated) which may be present at one or more sites throughout the genome. These regions have no apparent function. Other repetitive sequences called 'interspersed repeats', which are widely distributed as individual copies throughout the genome, are transcribed and have regulatory and/or structural roles. Most transcribed genes, however, are unique single-copy sequences, although not all of the single-copy DNA is transcribed.

Isolation of nuclear DNA

Total nucleic acid extracts from fresh, frozen or lyophilized tissue contain a large proportion of nuclear DNA. Following protein precipitation by phenol and/or chloroform, proteinase treatment may be necessary to remove the final traces of protein. Following precipitation and resuspension, RNase treatment may be used to remove any co-precipitated RNA.

(b) Organellar DNA

Most organellar proteins are encoded by nuclear genes, synthesized in the cytoplasm and imported into the organelle. However, some organellar proteins and all organellar RNAs are encoded by the organelle DNA and synthesized within the organelle. Mitochondrial genomes are exceptionally diverse, particularly in terms of size. Indeed large size differences can exist within a species. Plant mitochondrial genomes are generally composed of several double-stranded circular and/or linear DNA molecules (Figure 1.5) which can recombine with each other. In contrast, chloroplast genomes usually comprise a single type of double-stranded circular molecule which is the same size within any one species and similar in size between species.

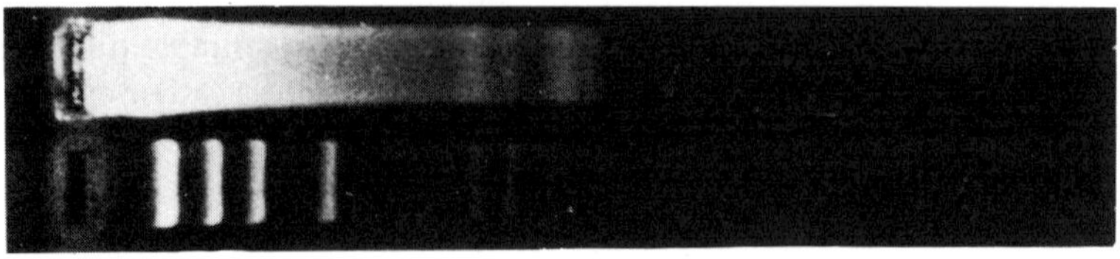

Figure 1.5 Agarose gel electrophoresis of native mtDNAs isolated from the rice variety Wu10A showing a number of plasmid-like molecules. Lane marked M = DNA size markers, from left to right: 2367, 946, 666, 426, 230, 196 bp (Finch, unpublished observations).

Organellar genomes are always present in multiple copies within the organelle.

Isolation of mitochondrial and chloroplast DNA
This is most easily accomplished by preparing intact organelles by partial tissue homogenation and selectively pelleting the organelles by centrifugation in a simple sucrose gradient. The organelles are then treated with DNase to remove nuclear DNA, washed and lysed by proteinase treatment. The released DNA is precipitated in the normal way.

(c) Plasmid DNA

Plasmids are small DNA molecules that are usually autonomous, double-stranded and circular. In bacteria these molecules, although separate from the main genome, often contain genes which confer important bacterial traits such as antibiotic resistance. Due to their small size, plasmids are relatively easy to characterize and manipulate. Because of their special properties, bacterial plasmids have become important tools of recombinant technology since they comprise convenient genetic transformation vectors (see later section).

Isolation of plasmid DNA
This involves selective propagation of the bacterial cells in liquid medium followed by lysis, deproteinization and precipitation. The total DNA preparation is stained with ethidium bromide and subjected to ultracentrifugation in a caesium chloride density gradient. The discrete plasmid fraction, which can be visualized in UV light, is selectively pipetted from the centrifuge tubes.

1.2.6 RNA structure

The basic structure of ribonucleic acid (RNA) is very similar to that of DNA, but there are two main differences: the sugar backbone of an RNA molecule is comprised of ribose rather than deoxyribose and the organic base thymine is substituted by a different pyrimidine, uracil. RNA can exist as single-stranded and double-stranded molecules and it is important to note that the former can hybridize with complementary single-stranded DNA to form a double-stranded DNA:RNA hybrid helix. RNA is highly versatile and diverse in terms of structure and function, but its three main forms cooperate in the process of protein synthesis. Messenger RNA (mRNA) is transcribed as a single-stranded copy of one of the genomic DNA strands and, following the complex post-transcriptional modification that occurs in eukaryotes (see below), carries the genetic message into the cytoplasm to the ribosomes which possess ribosomal RNA (rRNA) as a structural component. Here, transfer RNA (tRNA) molecules with specialized secondary structure translate the nucleotide code, incorporating amino acids into the growing polypeptide.

(a) Post-transcriptional modification of primary RNA transcripts

It should be noted that the initial or primary RNA transcript is a faithful complementary copy of the DNA template but can be up to 10 times longer than the mature mRNA molecule. Post-transcriptional modification mainly involves the removal of the intervening sequences, the 'introns', from the primary transcript in a process called splicing. This process does not occur in prokaryotes which means that non-spliced eukaryotic genes cannot be correctly expressed in recombinant bacteria.

(b) Isolation of RNA

One must be particularly careful when working with naked RNA, which is notoriously unstable. The extraction buffer may be modified by the addition of nuclease inhibitors and/or substances that form stable complexes with RNA. Reagents are often prepared using water containing diethyl pyrocarbonate (DEPC), a potent nuclease inhibitor.

RNA can be precipitated from total nucleic acid preparations using ethanol in the presence of a salt (e.g. lithium chloride); co-precipitated DNA is digested from the resupended pellet with RNase-free DNase. The use of phenol solutions is sometimes avoided by using columns/membranes to purify the crude extract.

1.2.7 Further analyses of nucleic acids

(a) Blotting methods

Southern blot

This is a technique whereby DNA fragments that have been separated by gel electrophoresis are transferred from the gel to a nylon or nitrocellulose membrane to obtain a 'print' of the gel (Southern, 1975). In this process, the membrane is usually placed on the gel and a flow of buffer is drawn through the system by the blotting effect of a wad of paper towels stacked on the top. The DNA fragments are drawn up with the buffer but remain deposited on the filter where they are denatured by treatment with NaOH and fixed. The membrane is treated with a small radioactively-labelled oligonucleotide (short piece of single-stranded DNA or RNA) referred to as a 'probe', which under appropriate conditions hybridizes specifically to fragments on the membrane which contain complementary base sequences. This specificity of binding depends upon the 'stringency' of the hybridization conditions (altered by varying the salt concentration and temperature of the incubation buffer). Hybridized fragments can subsequently be visualized as bands by exposure of the membrane to X-ray film in a process known as autoradiography.

The utility of this technique largely depends upon the nature of the probe. Probes that hybridize to satellite repeat regions (see earlier section on nuclear DNA) reveal the size differences or 'polymorphisms' that occur between individuals in these highly variable loci. The subsequent banding pattern observed on the autoradiograph provides a characteristic and highly specific 'fingerprint' of the analysed individual. Probes which hybridize to single-copy DNA, which may show variations in primary DNA sequence and thus restriction fragment length polymorphisms (RFLPs), are particularly useful for genetic mapping studies (see later section). In a similar fashion, specific probes can be used to detect the presence of a particular gene sequence in an individual, e.g. a foreign gene in a transgenic plant (see later section).

Northern blot

Heterogeneous mixtures of RNA molecules can also be analysed by electrophoresis and, in addition, transferred to a membrane [usually a chemically reactive (diazobenzyloxymethyl) paper since RNA does not normally bind to nitrocellulose] using a technique analogous to Southern blotting, called Northern blotting (Alwine *et al.*, 1979). After blot transfer, hybridization with a radioactively-labelled RNA or DNA probe can be induced in the normal manner.

Note that an analogous method, called 'Western blotting', is available for proteins (Burnette, 1981). Protein mixtures can be separated by gel electrophoresis and blotted onto a nitrocellulose membrane. Specific molecules can be detected, if a specific or heterologous antibody is available, using a 'double antibody' technique. The specific antibody, which can bind to the protein of interest, is, in turn, bound to a second antibody which is either radioactively-labelled or conjugated to an enzyme that is detectable by a simple colour or fluorescence reaction with its substrate.

Electroblotting, which involves electrophoretic transfer of molecules from a gel onto a nitrocellulose filter or nylon membrane, or vacuum blotting are rapid alternatives to passive blotting, and can be applied to proteins and nucleic acids.

(b) DNA sequencing

The sequence of nucleotides in long pieces of DNA is commonly determined using the 'Sanger method' (also known as the 'dideoxy' or 'chain termination' method; Sanger *et al.*, 1977). This method determines the sequence of single-stranded DNA because the sequencing process involves the enzymatic synthesis of a second, complementary strand using the original DNA strand as a template. For this process it is necessary to ligate the strand to be sequenced into a characterized plasmid, which can either exist naturally as a single-stranded molecule or can be denatured, prior to sequencing. Insertion into a plasmid allows propagation in bacterial cells and provides known flanking regions at either end of the target sequence.

The DNA polymerase enzyme which catalyses the synthesis of the complementary strand on the template requires a double-stranded region from which to commence further strand synthesis. Consequently, the first step in sequencing the piece of single-stranded DNA is to allow a small, complementary length of DNA, a primer, to hybridize to a known sequence at one end of the template. This provides a short double-stranded region from which the DNA polymerase can start to work.

The complementary strand synthesis reaction is carried out in four separate reaction tubes. Each tube contains the following:

1. The template with primer attached.
2. All four nucleotides. These are supplied in the 'deoxynucleoside triphosphate' form (see below) in which they can be readily incorporated into the growing DNA strand and are radioactively labelled with ^{35}S which will later aid the sequencing process.
   deoxyadenosinetriphosphate (dATP) → adenine
   deoxyguanosinetriphosphate (dGTP) → guanine
   deoxycytidinetriphosphate (dCTP) → cytosine
   deoxythymidinetriphosphate (dTTP) → thymine
3. The DNA polymerase enzyme which catalyses the reaction.

In addition, the four vessels contain one of four different, modified nucleotides. These are 'dideoxynucleoside triphosphates' (ddATP, ddGTP, ddCTP, ddTTP).

Synthesis of the complementary DNA strand

As the reaction progresses, the deoxynucleosides (dATP, dGTP, etc.) are sequentially incorporated into the growing DNA strand which is complementary to the template. However, the modified dideoxynucleoside present in each vessel (ddATP, ddGTP, etc.) has an equal chance of being incorporated into the growing DNA strand as its related deoxynucleoside

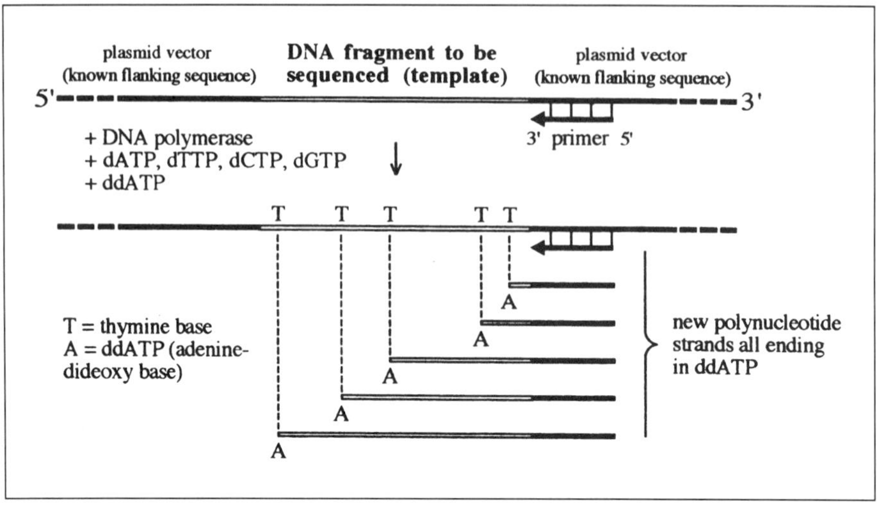

Figure 1.6 Synthesis of DNA strands, complementary to the template, in the ddATP reaction vessel. Similar syntheses occur in the other three reaction vessels.

form. Wherever this occurs, further strand synthesis is terminated (due to the chemical nature of the dideoxynucleoside molecule). Thus, in the reaction vessel which contains ddATP as well as dATP (and the other deoxynucleosides), a whole group of complementary strands is produced of different lengths depending on where the ddATP has become incorporated instead of the dATP. This occurs in all four reaction vessels, and thus four groups of terminated strands are produced. One group contains different length strands all ending in ddATP, one with strands ending in ddGTP, etc. (Figure 1.6).

Comparing the lengths of the terminated DNA strands by gel electrophoresis

The newly synthesized DNA strands in each group are denatured, to separate them from the template molecules, and their lengths determined by gel electrophoresis. For this particular process the electrophoresis must be very carefully controlled because it is necessary to separate strands which differ in length by a single nucleotide. The thin gel used (about 0.5 mm) is formed using polyacrylamide, which is capable of giving much better resolution than agarose, and contains urea which retains the DNA in a denatured state. Samples from each of the four reaction vessels are electrophoresed in four adjacent lanes of the gel, marked A, T, G or C depending on which dideoxybase was added. The bands produced contain very small amounts of DNA but can be visualized by autoradiography. This sensitive process detects the ^{35}S previously incorporated into one or more of the bases.

Reading the DNA sequence from the autoradiograph

The band that has moved the furthest on the gel is located. This represents the smallest detectable strand of DNA produced, i.e. the strand which was terminated by incorporation of the dideoxybase at one of the first few positions on the template. The track in which this band occurs is noted. If it is in the track marked A, then the first detectable base in the sequence is A. The next smallest band is then located. This corresponds to the DNA strand which is one base longer than the previous one. The track in which it appears is noted. If it is in track T, for example, then the next base in the sequence is T. This process is continued up the autoradiograph (Figure 1.7). Note that the first few nucleotides sequenced will represent the known flanking region of the target sequence. The sequence obtained from the autoradiograph is *complementary* to that of the template but both strands of the double stranded target molecule are sequenced.

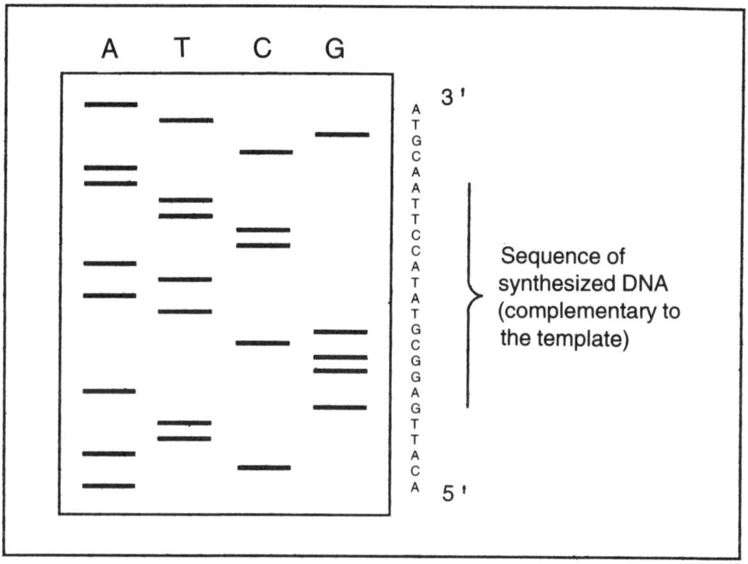

Figure 1.7 Interpretation of the autoradiograph. Each track contains fragments produced by complementary strand synthesis in the presence of one of the four dideoxynucleosides. The sequence is read by identifying the track in which each fragment lies, starting with the one that has migrated the greatest distance (and is clearly resolved), and then progressing up the autoradiograph, reading the sequence in a 5′ to 3′ direction.

(c) Polymerase chain reaction (PCR)

PCR allows rapid and specific amplification, and thus ready detection, of selected DNA sequences present in very small amounts in complex mixtures of nucleic acids. This powerful tool of modern molecular biology, developed in the 1980s (Saiki *et al.*, 1985, 1988; Mullis and Faloona, 1987), is carried out using the following basic principles.

1. A solution of double-stranded 'template' DNA containing the sequence to be copied (the target sequence) is denatured by incubation at approximately 95°C.
2. A pair of defined oligonucleotide primers, designed to bracket the target sequence, are annealed through complementary hybridization to the template by incubation at 40–60°C.
3. The primers are extended across the target sequence by the action of a DNA polymerase which is induced to incorporate free nucleotides into the growing strand by incubation at 72°C (Figure 1.8).

With a sufficient supply of the primers, nucleotides and a special thermostable DNA polymerase that retains significant activity despite repeated heating to 95°C (*Taq* polymerase), the above processes of denaturation, primer annealing and primer extension can be carried out in repeated cycles or rounds. Since the copied strands of one round, i.e. the products, serve as templates for the next, each successive round essentially doubles the amount of target sequence produced. The use of a thermostable DNA polymerase, which can be added in excess to the reaction vessel with the other components at the start of the reaction, has allowed the whole process to be automated through the use of a programmable heating block or 'thermal cycler'.

The reaction mechanism of PCR

When genomic DNA is used as the PCR template, the initial rounds are a screening phase where the target sequence is selected for further amplification (Figure 1.8). The primers direct DNA polymerization across the target sequence and beyond, the length of the synthesized product being limited only by time. In subsequent rounds, however, the newly synthesized products are the preferred template, being present in excess of the genomic DNA. Unlike the genomic DNA, these templates are demarcated segments amplified in the previous cycles and defined by the specificity of the two primers hybridizing to sites on either complementary DNA strand. The specificity of amplification depends on the stability of this hybridization process. Thus, hybridization conditions must be highly stringent (this depends on incubation temperature and magnesium ion concentration, both of which influence hybridization), particularly in the early cycles where the genomic DNA, which is likely to possess many sites which are homologous to the intended hybridization loci, forms the template.

Analysis of PCR products

The desired product of a PCR reaction is often a single type of amplified fragment, the presence of which in the reaction mixture can be recognized by agarose gel electrophoresis and ethidium bromide staining. The latter

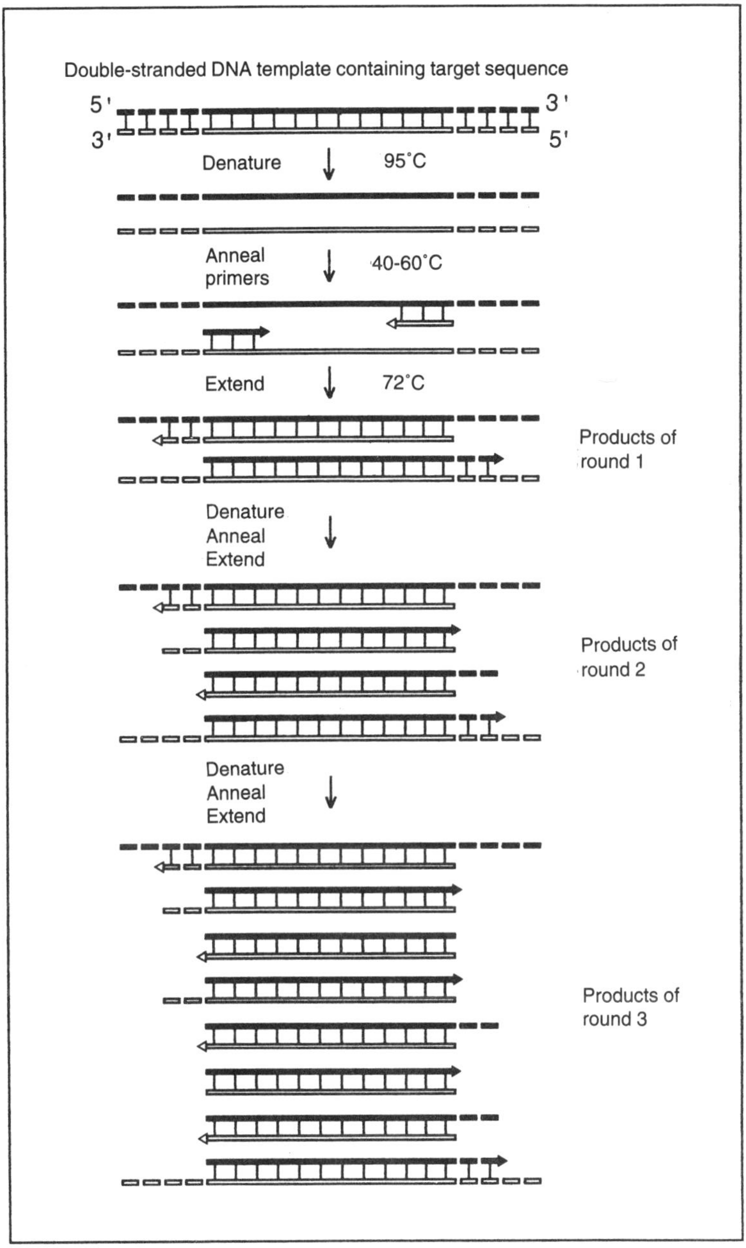

Figure 1.8 The first three rounds of a PCR. Solid lines within the DNA template represent the target sequence; regions beyond the target sequence are represented by broken lines. The black and white DNA strands are complementary to each other and arrowheads indicate the direction of polymerization. Note the properly demarcated fragments are only produced after the second round of amplification.

will allow the actual product size to be compared with the expected size based on sequence data. Southern blot or sequence analysis may be used to confirm the identity of the product.

Non-specific amplification

A multiplicity of fragment types may be erroneously produced under some circumstances. Poorly designed primers, which are not sufficiently specific to the target sequence, may lead to cross reactivity. Moreover, non-specific reaction conditions, such as suboptimal annealing temperature, may give rise to multiple-product amplification. 'Primer–dimer' may also be produced as a result of complementarity, and thus hybridization, between primers. The resulting dimer molecules compete for amplification with the target sequence and interfere with the detection of small products. Specific amplification can sometimes only be achieved by re-amplifying an internal region of a non-specific product. This approach can overcome difficulties of non-specific amplification since it provides the opportunity of sequencing the first product, previously amplified using a given pair of primers, and then using this sequence information to design a highly specific second set of 'nested' primers (see Chapter 4).

The merits of PCR

PCR has become a very popular tool in molecular biology for a number of reasons. It is relatively rapid and easy to perform and can be very specific. However, its biggest advantage *and* disadvantage is its extreme sensitivity. In practice, the technique can amplify the target sequence approximately 10^6-fold in 30 cycles, so that even a single copy of the target sequence might be easily detectable after the amplification process. Consequently, all reagents, pipettes and consumables must be free from any contaminating DNA either from previous experiments, or from positive control reagents, to avoid false positive results.

Some applications of PCR

PCR amplification has found specific application in mutation analysis. 'Allele specific oligonucleotide' (ASO) primers that have sufficient sequence similarity such that they can bind to the same target locus but are sufficiently different (by one or more bases) such that under highly stringent hybridization conditions they will only bind to an allele that is perfectly matched in sequence, can be used to reveal very small sequence differences between alleles, such as point mutations (see Chapter 4). Other PCR applications include detection of nucleic acid sequences of plant pathogens in putatively infected individuals prior to symptom development, analysis of putatively transformed plants to detect the presence of foreign genes (see section on genetic transformation) and detection of genetically modified microorganisms (GMMs) in the environment (see Chapter 11). Moreover, the generation of multiple copies of specific DNA sequences using PCR is now important for a wide range of recombinant DNA techniques, including the production of DNA probes, production of large amounts of DNA for sequencing, construction of complementary DNA (cDNA) libraries from small amounts of mRNA and gene cloning (see section on gene isolation/cloning). Some of these applications are facilitated by recent developments which allow PCR products to be cloned directly into plasmid vectors.

Reverse transcription (RT)-PCR

PCR can be applied to RNA if the latter is reverse transcribed into cDNA prior to amplification. This can be achieved using a separate, purified reverse transcriptase, although some DNA polymerases also exhibit reverse transcriptase activity which allows the whole procedure to be carried out in one reaction vessel. RT-PCR is being used increasingly in plants to study aspects of gene expression and to detect the presence of pathogenic RNA viruses.

Random amplification of polymorphic DNAs (RAPDs)

A developing application of PCR is in the generation of complex DNA fingerprints and novel molecular markers. PCR, carried out with genomic templates and a single oligonucleotide primer of arbitrary sequence, conveniently gives rise to the production of a range of amplified fragments some of which exhibit length polymorphisms between individuals (Figure 1.9). The mechanism of this process is not well understood but may be due to the fortuitous occurrence, in (probably) all genomes, of inverted repeat sequences of varying inter-repeat distance which effectively form the target sequence, each repeat unit defining one end of a given PCR product. The production of such fingerprints constitutes the technique of RAPD analysis and has been useful for intraspecific biotype identification. In addition, RAPD fragments are increasingly used as molecular markers which are linked to, and thus may show similar inheritance characteristics to, given genetic traits. An ability to determine the presence or absence, in particular individuals, of RAPD markers known to be linked to phenotypic traits of interest can serve to accelerate plant breeding programmes (see below).

(d) Molecular linkage mapping using restriction fragment length polymorphisms (RFLPs) as markers

Production of genetic maps using molecular markers is becoming extremely important to plant science. Such maps, and the techniques used for producing them, are invaluable for studying the introgression of novel genes into crop species and ultimately provide inroads to the isolation and cloning of previously uncharacterized genes which will, in turn, lead to the production of crops that are resistant to biotic and abiotic stresses. A concise description of the theory and techniques involved is warranted here.

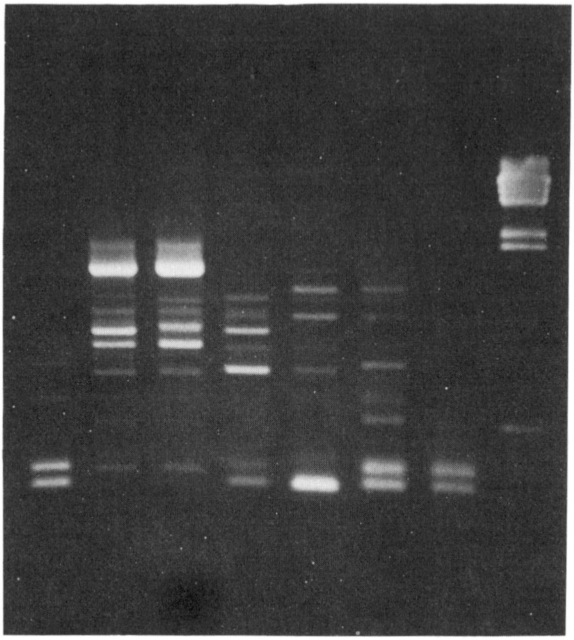

Figure 1.9 RAPD fingerprints of seven *Sinapis arvensis* ecotypes amplified using a 10mer random primer. The far right lane contains DNA size markers, from top to bottom: 23 130, 9416, 6557, 4361, 2322, 2027, 564 and 125bp (Finch and Marshall, unpublished observations).

Genetic linkage

Genes which co-exist on the same chromosome in an organism are said to be 'linked' and genomes consist of a number of 'linkage groups' which is equal to the number of homologous chromosome pairs present. Genes in different linkage groups (on different chromosomes) reassort independently of each other (randomly) at meiosis. Genes within the same linkage group (on the same chromosome) may become separated during meiosis due to crossing over but this depends on how far apart they are on the chromosome. Genes which are physically very close together tend to co-segregate and are said to be tightly linked.

Genetic linkage maps

Linkage relationships between genes are conventionally represented in the form of genetic linkage maps in which each linkage group (chromosome) is represented as a line on which a series of loci (positions of genes or gene groups) are marked. Interlocus distances, which are proportional to the frequency of recombination between the loci, are usually given in 'map units' or 'centimorgans', where 1 centimorgan corresponds to 1% recombination per generation. Traditionally such linkage relationships were derived by studying the recombination of recognizable phenotypic (morphological) characters or 'markers' in the progeny of selfed F_1 plants, themselves produced from homozygous parents. The homologous chromosomes of such F_2 individuals comprise different mixtures of parental chromosome segments. Linkage studies of these plants were based on the fact that chromosome segments which are not on the same chromosome will undergo random recombination and those on the same chromosome will undergo recombination at a frequency proportional to their physical distance apart. Closely linked segments will tend to recombine less frequently than distant segments and so map distance is a function of the recombination (due to crossing over) which occurs during gamete formation (meiosis) in the F_1 plants. Traditionally, genes controlling easily recognizable morphological markers, such as novel leaf shape, could be used to construct useful linkage maps due to their tight linkage to genes of interest that may not themselves give rise to traits that can be easily recognized (e.g. disease resistance).

In summary, linkage maps are based on indirect analysis of the recombination of chromosomal segments, and comprise positional information on genes encoding recognizable morphological markers and useful genes which are linked to them. They thus provide important insights into genetics able for use in breeding programmes where the breeder would want to select indirectly for genes of interest by selecting for linked markers which encode easily recognizable traits. Protein markers (isozymes) have been widely adopted in breeding programmes, but most recently phenotypically neutral molecular markers have been developed (Tanksley *et al.*, 1990). Such markers, based on numerous naturally occurring differences or 'polymorphisms' in primary DNA sequence between individuals, are called restriction fragment length polymorphisms (RFLPs).

RFLPs are identified using restriction endonucleases which, as aforementioned, recognize specific DNA base sequences and cleave the DNA at these sites. The distribution of these sites in a given DNA molecule is reflected in the number and size of restriction fragments produced following digestion which, in turn, reflects the primary sequence of the DNA molecule being analysed. The restriction fragments are usually separated by agarose gel electrophoresis. Restriction of small DNA molecules, e.g. mitochondrial DNA, usually results in the production of relatively few fragments which can be readily visualized and resolved by simple ethidium bromide staining of the gel itself. In such studies, polymorphic fragments may be easily identifiable (Figure 1.10). However, restriction of nuclear DNA from complex organisms yields so many fragments that individual bands cannot be easily identified on the electrophoresis gel. The resolution of polymorphic fragments in this situation requires Southern blotting (see earlier section), followed by hybridization to a radiolabelled probe which only hybridizes to some of the bands.

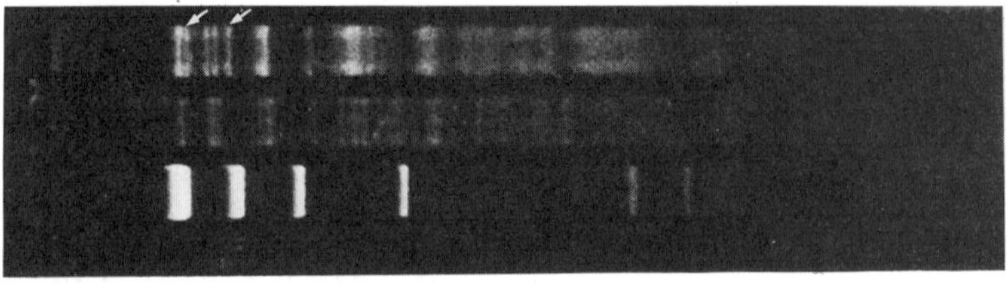

Figure 1.10 Restriction analysis of mtDNA from the two rice varieties Wu10A and Taipei 309 showing a number of polymorphisms, most obviously the presence of two T309-specific restriction fragments. Lane M = DNA size markers, from left to right: 2367, 946, 666, 426, 230 and 196 bp (Finch, unpublished observations).

Producing RFLP probes

Whether polymorphisms occur or not depends on the restriction enzyme used and the nucleotide sequence of the probe. Thus a range of probes and restriction enzymes must be tested to establish polymorphic enzyme/probe combinations. A collection or 'library' of probes is usually prepared by restricting DNA isolated from the species of interest. For mapping, it is important that these probes represent regions of single copy DNA. Fortuitously, repetitive DNA often exhibits significant levels of methylation and methylation-sensitive restriction endonucleases, such as *Pst*I, cannot digest it. Libraries of *Pst*I fragments, which are single copy enriched, are thus a particularly useful source of RFLP probes. Individual fragments are multiplied, using established gene cloning technology (see section on genomic DNA library), and some of them are tested as probes to see if they hybridize to, and thus identify, polymorphic fragments, i.e. fragments which appear in one individual but not another.

Producing an RFLP linkage map

RFLP maps are produced by following the segregation of RFLP markers, which are inherited in a codominant Mendelian manner, in an F_2 'mapping population'. This population comprises approximately 50 individuals produced by selfing the F_1 population from a cross between two homozygous parents which are genetically divergent enough to exhibit restriction fragment length polymorphisms, although obviously still related enough to produce fertile progeny. Restriction enzyme/probe combinations which are found to reveal polymorphism are then scored in the mapping population. This involves DNA isolation, restriction, Southern blotting and DNA:DNA hybridization for each individual, for each polymorphic probe/enzyme combination. Scoring an RFLP marker in the F_2 population is relatively straightforward since only three different types of plant can be present: two homozygotes (each with one of the two parental alleles) and the heterozygote (both alleles present; see Figure 1.11). The degree to which

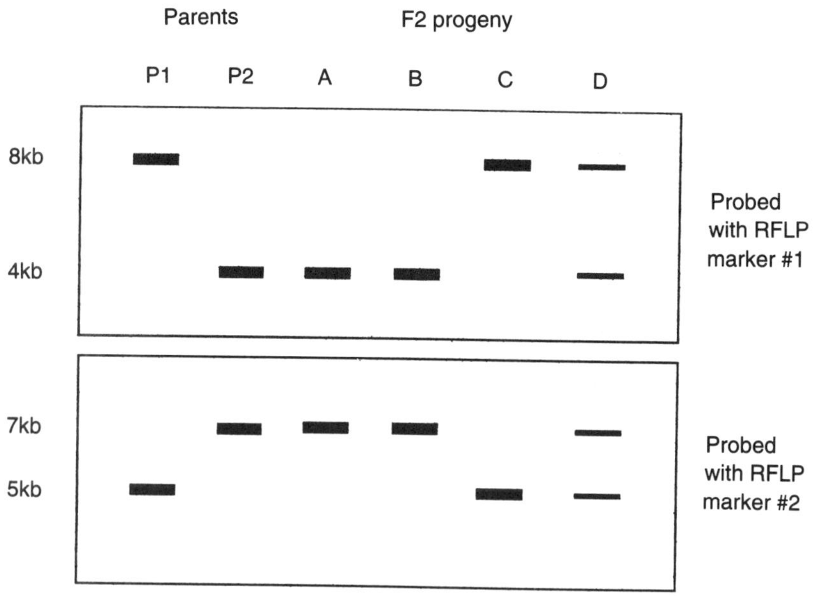

Figure 1.11 Diagrammatic autoradiographs of two F2 mapping membranes. The restricted genomic DNA has been probed with RFLP marker 1 (top) and 2 (bottom). Note that the two probes co-segregate in the four individuals shown, and so may linked.

the different probes tend to co-segregate is determined (e.g. an individual which is homozygous for a given probe will also tend to be homozygous for other probes which are linked to the first) and this information can be used to construct a linkage map. Statistical analyses are used to test whether recombination is random or not. Computer-based algorithms are used to determine the most likely recombination frequencies and thus map distances. Several computer software packages are now available for these purposes.

Assigning linkage groups to chromosomes

After surveying the mapping population with a range of enzyme/probe combinations, a map of a few hundred markers may be obtained. The number of linkage groups should equal the number of homologous chromosome pairs in the species under study. These linkage groups can be assigned to particular chromosomes, which are usually recognizable by microscopy, using a process called *in situ* hybridization, whereby a chosen radioactively labelled RFLP marker from each linkage group is used to individually probe intact chromosomes on a cell squash preparation. Alternatively, similar chosen markers can be used to probe DNA isolated from monosomic plants which may be available for the species under study. Monosomic plants lack a single chromosome and a full set of plants may be available, each of which is 'monosomic' for a different chromosome. Nitrocellulose membranes are prepared from gels on which DNA from each of the monosomic plants has been electrophoresed in adjacent lanes. When these membranes are probed, the hybridization signal, as detected by autoradiography, is recognizably fainter for the plant which is monosomic for the chromosome (linkage group) from which the probe is derived. This 'dosage effect' can be used to assign all the linkage groups to all the chromosomes if a full set of plants is available.

Utility of RFLP linkage maps

It is desirable for the mapping population used for the construction of the RFLP linkage map to be segregating for conventional gene markers as well as RFLP markers. This allows conventional genes to be placed on the RFLP map by observing co-segregation with the RFLP markers. This information can be particularly useful for breeding projects since the breeder can select indirectly for traits of interest by selecting for RFLP markers that are closely linked to them. The RFLP markers have the added advantage of being detectable at an early developmental stage.

RFLP linkage maps which are almost saturated with respect to RFLP markers, such that any conventional gene is closely linked to one or more of them, are now becoming established for a number of crop plants. The availability of such 'near saturated' maps means that many previously uncharacterized genes are now becoming amenable to established gene cloning techniques such as 'chromosome walking', which allows DNA sequences close to a point of known sequence (an RFLP marker) to be isolated using progressive overlapping DNA segments as hybridization probes (see later).

RAPD markers

Unlike RFLP markers which are inherited in a co-dominant manner, RAPD markers tend to be dominant and therefore do not distinguish heterozygotes from homozygotes. Consequently, these markers are not so useful for mapping purposes, but may be very useful for placing additional loci on existing RFLP maps. Current investigations in this field include using PCR to amplify length-polymorphic short tandem repeat (STR) loci, now known to exist in a number of plants, for purposes of mapping and fingerprinting.

1.3 ISOLATION/CLONING OF GENES FOR GENETIC TRANSFORMATION

In an increasing number of cases, transgenic plants expressing functional heterologous genes, not only from other plant species but also from animals or microorganisms (e.g. *Bacillis thuringiensis, B.t.* genes), have been produced. In some cases, e.g. where an antisense approach is used (see later) or for purposes of satisfactory gene expression, it is necessary to have a homologous gene. The isolation of a particular gene from the genome of a given organism would at first appear to be a daunting prospect. However, important clues may already exist to the identity of the gene sequence or its locus and these clues are often exploitable for the isolation of the gene itself. A number of methods are currently being utilized for the isolation of useful genes, as outlined below. Specific examples of the application of these methods are given in Chapters 2 and 4.

1.3.1 Genomic DNA library

A specific gene would normally be isolated from a genomic DNA library of the species of interest. Such a library consists of a collection of overlapping restriction fragments of the genome, randomly cloned into a suitable vector (e.g. individually ligated into plasmids and propagated in *Escherichia coli* on culture plates) such that collectively the resulting recombinant clones represent virtually the entire genome. One or more clones should contain all or part of the desired sequence. The clones are then screened for the presence of this sequence. There are several approaches to this depending on the information available, but most methods adopt a DNA:DNA hybridization approach.

(a) Hybridization approaches for gene isolation

The culture plates are replica plated onto membrane discs (this involves laying a disc on the top of each plate to obtain prints of the cell colonies). The plates themselves are retained as master copies. The cells on the disc are lysed to release the DNA which is denatured and fixed to the nitrocellulose in a similar fashion to that described for Southern blotting. The membranes are then treated with a suitable radioactively-labelled DNA or RNA probe which contains complementary sequences to the gene of interest. Autoradiography reveals the position of the required clone/s on the membrane which are then identified and removed from the corresponding position on the master plate and further propagated. The vital prerequisite is a suitable probe which may come from several sources.

Heterologous probes

Isolation of a gene may be possible if a heterologous probe is available. This consists of a corresponding whole or partial gene sequence previously isolated from another plant species or organism. The success of this approach relies on there being sufficient sequence similarity for specific hybridization to take place.

Artificial oligonucleotide probes

If the protein product of the required gene can be isolated and purified, partial amino acid sequences can be determined and corresponding artificial oligonucleotide probes synthesized. However, due to the degeneracy of the genetic code, each probe must consist of a mixture of nucleotide constructions and this can cause problems of specificity, although knowledge of codon preference in the species can minimize this problem.

A PCR approach

If two well spaced artificial oligonucleotides are synthesized, these can be used as PCR primers to amplify a significant portion of the gene of interest. The amplified product can then be utilized as a more specific DNA probe. A similar approach is also applicable if sequence information of flanking regions of the gene is available although similar information about the gene itself is not. In this case the PCR reaction will amplify the entire gene sequence.

Chromosome walking

A highly saturated molecular linkage map provides sequence information, in the form of closely linked RFLP or RAPD markers, for genomic regions which are adjacent to mapped genes of interest. It may be possible to utilize such information in a PCR approach (see above) to isolate and clone a given gene. A more applicable approach to isolating a gene which is adjacent to a molecular marker however, is to use the marker as a probe to identify a homologous clone and use this, in turn, as a probe to identify clones containing overlapping fragments which are closer to the desired locus. One of these fragments is itself used as a probe to identify clones containing fragments which again overlap and are closer to the desired locus, and so on. In this fashion it is possible to 'walk' along a chromosome until the desired gene is reached. Note that at each stage, partial nucleotide sequences at the end of each probe must be determined to ensure the walk occurs in the correct direction. Moreover, for the technique to work satisfactorily, each probe must correspond to a single-copy sequence.

Differential screening of cDNA libraries

A cDNA library comprises a collection of clones containing reverse transcripts of the mRNAs isolated from a given tissue (c.f. genomic DNA library). Each clone acquires only one cDNA but different clones may have acquired the same type of cDNA molecule. Such libraries can be used to identify plant genes that are specifically expressed in particular tissues, or are inducible in some way, by sequence comparison with cDNAs produced from different, or non-induced, tissue from the same plant. This is achieved by screening the cDNA library of the target tissue with a range of cDNA hybridization probes generated from the mRNAs of both tissues. Clones that hybridize to cDNA probes from only one tissue type are known to be specifically expressed in that tissue. The tissue-specific or induced cDNAs identified in this way can be used to probe a genomic DNA library to obtain the corresponding full-length genes.

Screening cDNA expression libraries

It is worth noting here that cDNA 'versions' of eukaryotic genes are useful in recombinant work when it is necessary to express a eukaryotic gene in a prokaryote, since they already lack the introns that would otherwise prevent correct expression of the full length eukaryotic gene due to the inability of prokaryotes to splice genes. This can be useful for gene identification and cloning purposes since if a library of cDNA clones is expressed in a normally susceptible bacterium, the gene of interest may be identifyable by applying appropriate chemical selection pressure to the bacterial cells. Such an approach is applicable, for example, for the isolation of pesticide-resistance genes. Cells that survive the chemical selection are cloned further and provide cDNA probes for a genomic library.

Such cDNA expression libraries can also be screened using an immunological approach. The recombinant clones are replica plated onto nitrocellulose discs and lysed to release protein which binds to the membrane. The presence of foreign protein can then be detected using a homologous or heterologous antibody, if available, using a double-antibody detection system (see section on blotting methods).

(b) Utilizing mutations for gene isolation

Functional complementation

This usually involves transformation of mutants, which are auxotrophic for the substance conferred by the gene of interest, with clones from a genomic library of the species under study. Subsequent prototrophy in the transgenic mutant confirms that the transforming clone contains the required gene. This approach relies on the availability of suitable mutants although heterologous mutants, e.g. fungal auxotrophs, have been utilized for cloning plant genes.

Transposon tagging

Transposable elements or 'transposons' are naturally occurring DNA sequences with an ability to jump or 'transpose' from one genomic locus to another. This seemingly random process can result in the insertion of a transposon into a gene or its regulatory sequences such that a heritable change in the expression of that gene, a mutation, occurs. Thus, transposons have been widely used for creating mutant phenotypes which have proved useful for studying gene function. Transposon tagging is a process whereby transposon insertion in a gene creates a corresponding and identifiable alteration of phenotype, such that the transposon itself can be used as a hybridization probe to isolate that gene. This approach has been adopted in maize where transposable elements have been well characterized. Maize (*Zea mays*) transposable elements are now being used as heterologous tags for gene isolation in other plant species.

Genomic subtraction

This identifies DNA sequences that are present in a wild-type organism (target sequences) but are absent in a homozygous deletion mutant. The procedure involves multiple hybridizations between denatured wild-type DNA restriction fragments and an excess of sonication fragments of biotinylated DNA from the deletion mutant. In each successive hybridization, avidin-coated beads are added to remove biotinylated hybrid molecules (avidin and biotin have a strong affinity for each other) which represent sequences common to both plants. After several cycles the remaining DNA is enriched for sequences that are absent in the deletion mutant.

1.4 PLANT GENETIC TRANSFORMATION

1.4.1 An introduction to the molecular strategies for introducing useful traits into crops

Considerable progress has been made in gene cloning technology in recent years and this has allowed the development of strategies for introducing useful traits into crop plants. Major advances have been realized in the areas of induced virus resistance, insect resistance and herbicide tolerance as described in detail later in this book. However, a number of the strategies employed to achieve these goals are outlined in this section.

(a) Virus resistance

Two main strategies have been adopted for introducing virus resistance into plants (Beachy *et al.*, 1990).

Coat protein mediated resistance (CPMR)

This has been achieved in a number of crops by transforming plants with the viral genes which encode coat protein (see Chapter 5). The accumulation, in uninfected cells, of this protein, which is normally produced in infected plant cells to encapsidate the replicated viral RNA, results in effective resistance. This may be due to a swamping of 'receptor sites' within the plant cells such that invading viral particles cannot bind and become infective. There is also evidence to suggest that entry of viral particles into the vascular tissue is impeded in transgenic plants, thus limiting the spread of infection.

Antisense genes

Virus resistance has also been achieved using an 'antisense' approach which requires a more detailed explanation. In eukaryotes, genes comprise double-stranded DNA sequences. For a given locus, the two strands are designated 'sense' and 'antisense' where the antisense strand provides a template for the synthesis of a complementary mRNA molecule. The transcribed mRNA molecule thus has the same sequence as the sense DNA strand (note that at different loci along the same chromosome, the sense/antisense designation may be different since the specific regulatory 'promoter' sequence upstream of each gene instructs RNA polymerase which strand to use as a template).

If a given gene (e.g. for viral coat protein) is isolated, cloned and inserted into a transformation vector (e.g. a tumour inducing 'Ti' plasmid) in a reverse or 'antisense' orientation and then used to transform a plant, during

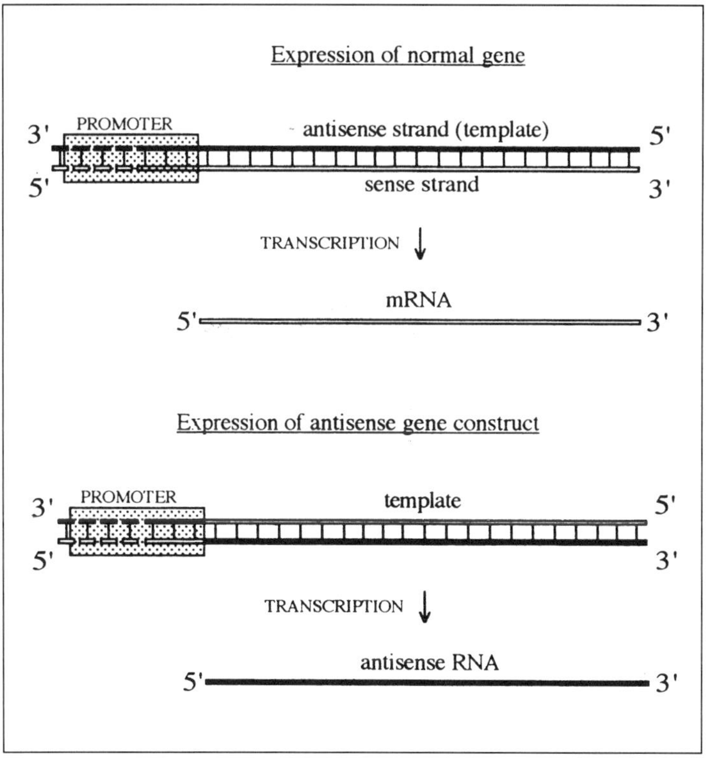

Figure 1.12 Production of an 'antisense RNA' which has a complementary sequence to the normal mRNA for a given gene results in the production of two different but complementary RNAs. Within the cell, these can hybridize due to their sequence complementarity, inhibiting translation and effecting a post-transcriptional downregulation of gene activity.

transcription in the transformant the wrong DNA strand will be used as a template. This will result in the production of an 'antisense RNA' which has a complementary sequence to the normal mRNA for that gene. A transformant would therefore be producing two different but complementary RNAs. Within the cell, these can hybridize due to their sequence complementarity, a phenomenon which inhibits translation, thus effecting a post-transcriptional downregulation of the gene's activity (Figure 1.12).

The introduction of antisense viral genes into plants will therefore have the effect of inhibiting coat protein production and thus will limit the production of new viral particles in infected cells. Note that this antisense approach has enormous potential in many other areas of crop genetic manipulation where downregulation of gene expression may be useful and was largely developed during studies to alter the physiology of fruit ripening by inhibiting ripening-related genes (other applications are discussed in Chapter 2).

(b) Insect resistance

Bacterial '*B.t.*' (*Bacillus thuringiensis*) genes, for example, confer insect control proteins which have long been used in insecticidal sprays to control mainly Lepidopteran pests due to their specific endotoxic properties (see Chapter 7). The advent of genetic transformation technology has led to the introduction of *B.t.* genes into several crop species with the result that larval damage can be effectively controlled. A second strategy has involved the use of plant-derived genes conferring antinutritional agents, such as protease inhibitors, which significantly reduce the insect's ability to digest plant material.

(c) Herbicide resistance

This has been achieved using at least three general approaches depending on the herbicide in question and the genes available. Heterologous genes are often employed, as are mutated genes which have been isolated from tissue cultured cells exposed to the herbicide *in vitro*. Such mutations can occur spontaneously in plant cells grown *in vitro* due to a phenomenon known as somaclonal variation.

Overproduction of the herbicide's biochemical target

This can be achieved by transferring additional copies of the gene that confers the biochemical target for the herbicide in question or transferring a mutant gene conferring an enzyme with enhanced activity. This requires a detailed knowledge of the target enzyme and the availability of the gene that confers it. In general, heterologous genes have been used for this approach but homologous mutant target loci with amplified expression levels isolated from *in vitro* selected cell lines (tissue cultured cells that have acquired resistance *in vitro*) have also been utilized.

Reduction of the herbicide's affinity for the target
There are a number of examples where 'target' genes in a given species have acquired mutations, through *in vitro* selection, such that they are no longer affected to the same extent by the herbicide in question. If such genes are isolated and cloned, they can be used to transform other species to confer herbicide resistance.

Transfer of herbicide detoxification/degradation gene
Bacterial genes capable of degrading/detoxifying certain herbicides have been identified, cloned and transferred into plants to produce herbicide resistant transformants. Further details of herbicide resistant plants follow in Chapter 6.

1.4.2 Use of a natural vector system for genetic transformation

Once useful genes have been identified and suitable strategies established, reproducible methods must be available for the introduction of the foreign genetic material into a wide range of important crops. The most widely used method for the introduction of foreign genes into plants exploits the natural transformation system of soil-borne *Agrobacterium* spp. and the popular technique of plant tissue culture. Tissue culture, a method of maintaining and manipulating the growth of plant organs, tissues or cells under nutritionally and environmentally optimized axenic conditions *in vitro*, ensures that new plants can be regenerated from explant cells that have been genetically altered following interaction with the agrobacteria.

In its natural environment, wild-type *Agrobacterium tumefaciens* causes crown gall disease. The bacterium characteristically possesses a tumour-inducing or 'Ti' plasmid, part of which, the T-DNA, is transferred to the plant cells upon infection. The T-DNA, which contains opine biosynthesis and oncogene loci, becomes integrated in the plant genome where expression of the foreign genes characteristically gives rise to tumourous outgrowths comprising dedifferentiated opine-synthesizing plant cells. The portion of the Ti plasmid retained by the bacterium contains virulence genes, which have a number of vital roles concerned with the infection process, and opine metabolism genes that allow the microorganism to utilize opine (of which there are a number of types including octopine and mannopine) as a unique carbon/nitrogen source.

Plant biotechnologists are able to exploit the natural transformation system of *A. tumefaciens* by constructing modified Ti plasmids that lack the undesirable tumour-inducing oncogenes but contain a foreign gene of interest, e.g. for virus resistance or herbicide tolerance, and a closely linked selectable marker gene, e.g. for antibiotic resistance, inserted within the

T-DNA region. The virulence genes must be retained within the plasmid to ensure effective gene transfer, although in many cases workers utilize a 'binary' vector system whereby the virulence genes are present on a second plasmid acting in trans with the T-DNA-containing plasmid.

A.tumefaciens cells containing such recombinant plasmids are subsequently used to transform plant tissues *in vitro* by briefly exposing plant tissue, usually leaf disc explants, to a bacterial suspension. Transgenic plant material, regenerated from the transformed cells, can be identified at an early developmental stage by using selective (antibiotic-containing) culture medium on which only cells that have acquired the antibiotic-resistance genes from the modified T-DNA will be able to regenerate into plants. The presence of the foreign gene is then confirmed through molecular analyses (see below) and the plants can be screened for conferred tolerance to the biotic or abiotic stress. This transformation method, and a similar approach which utilizes *A. rhizogenes* (a closely related bacterium which normally causes hairy root disease through the action of a root inducing or 'Ri' plasmid, but which can also be suitably disarmed and exploited), has been the most succesful approach for plant genetic transformation. However, the host range of agrobacteria is limited mainly to dicotyledonous plant species and alternative methods have necessarily been developed for transforming the most important group of crop plants, the cereals.

1.4.3 Alternative transformation methods

(a) Direct DNA uptake into protoplasts

The most important method for transforming cereals also involves an interaction between plant cells and a recombinant plasmid *in vitro*. The plasmid must be small and contain the gene of interest, a selectable marker and suitable regulatory sequences, and must be capable of replication in the plant and in *E. coli* for propagation purposes. Single plant cells called protoplasts (cells which, following enzymic treatment, lack a cell wall and are bound only by a plasma membrane) are suspended in a solution containing plasmid and subjected to chemical or electrical treatments which induce transient membrane pore formation. The plasmid DNA is then taken up and integrated into the plant genome at a low but significant frequency such that when transformed protoplasts are regenerated back into plants, using advanced tissue culture techniques, transgenic individuals are produced.

(b) Particle bombardment

Many ingenious alternative methods for introducing foreign genes into plants have been tested (Potrykus, 1990) but only one other method has proved to be reproducible. This 'particle bombardment' or 'biolistic' approach (see Chapter 6) involves the acceleration of microscopic DNA-coated metal pellets into 'target' plant cells. Particle acceleration is achieved by a variety of means including electrical discharge, compressed air (air-gun) and gunpowder discharge (modified starting pistol!). The target cells are usually callus cultures (dedifferentiated tumour-like plant cell cultures) which, following bombardment, can be readily regenerated into plants following growth on selective medium (see below). More recently, imbibed seed embryos have been used as targets. The use of seed embryos is particularly attractive since no tissue culture procedures are required but the extremely low transformation frequencies obtained to date mean that this approach, although straightforward and potentially very widely applicable, is particularly labour intensive and expensive since many plants have to be grown up under containment and screened using molecular methods (see below).

1.4.4 Identifying transformants

Putative transformants regenerated on selective media must be analysed to confirm the presence of the foreign gene construct. Positive confirmation requires a multidisciplinary approach involving one or more biochemical tests and molecular analyses. The biochemical tests are usually based on established assays for enzymes conferred by marker genes closely linked to the gene of interest on the transformation plasmid. They thus provide indirect confirmation of the presence of the gene of interest. Molecular analyses can also be used indirectly to detect the marker gene sequence or directly to detect the sequence of the foreign gene itself.

(a) Biochemical methods

Marker genes

An increasing number of marker genes are routinely used as selectable tags incorporated into plant transformation vectors. Closely linked to the gene of interest, their expression provides good evidence of successful gene transfer at an early stage of development of the putatively transformed plant. Transient and stable gene expression markers are commonly utilized. The former are used to obtain initial indications of successful gene uptake since they provide a gene expression signal, without relying on stable gene introgression, within very short periods of the transformation event. The latter are detected by their stable expression over longer periods of post-transformation development. Three of the most commonly used selectable markers for plant transformations are briefly described below.

Chloramphenicol acetyl transferase gene The bacterial chloramphenicol acetyl transferase (*cat*) gene confers resistance to chloramphenicol and is often used as a transient expression marker due to the availability of an established chloramphenicol acetyl transferase (CAT) assay. This involves a straightforward chromatographic analysis to detect the incorporation of a radioactive label from [^{14}C]chloramphenicol into its acetylation products following the action of any CAT present in cell extracts from the putative transformed tissue. If necessary the level of expression can be quantified by measuring the level of radioactivity in the acetylation products using a scintillation counter.

β-glucuronidase gene The cloned *E. coli* β-glucuronidase (*gus*) gene has been widely used as a transient and stable gene expression marker for plant transformation studies, fused to the gene of interest in a chimaeric gene construct and under the control of suitable regulatory sequences. The β-glucuronidase enzyme (GUS) can be readily assayed using a range of methods since spectrophotometric, fluorometric and histochemical substrates are commercially available. Enzyme extracts from putative transformants are required for the photometric assays, but histochemical assays, which provide information on localized gene expression, are carried out on fixed tissues using a substrate which gives an easily recognizable insoluble indigo product following GUS action and oxidation.

There have been reports of intrinsic GUS-like activity in a diversity of plant species, a finding that demands careful use of controls in experiments utilizing the *gus* gene. In addition, in earlier transformation studies, contamination of transformed tissue with the recombinant *Agrobacterium* vector led to false positive results due to *gus* expression in contaminating bacterial cells rather than the plant cells. This problem has been overcome by the addition of an intron to the gene such that bacterial expression is not possible (prokaryotes cannot splice the primary RNA transcript).

Neomycin phosphotransferase II gene Used as a stable gene expression marker, the bacterial (*nptII*) gene indicates its presence in transformed plant tissues by conferring antibiotic resistance. This is due to the phosphorylating action of the neomycin phosphotransferase II (NPTII) enzyme on a range of aminoglycoside antibiotics including kanamycin and G418. The presence of this marker gene, which is fused to the gene of interest in chimaeric gene constructs, is often confirmed using a radioactive assay for NPTII whereby transfer of radioactive phosphorus (^{32}P) label from AT^{32}P to the antibiotic can be detected in fractionated tissue extracts from transformants. It has been found in a number of cases that the use of kanamycin for selection can lead to plant regeneration difficulties unless the exposure period is very carefully optimized.

(b) Immunological method

The presence of foreign protein in transgenic plants (conferred by the gene of interest or the marker gene/s) can be determined by a simple immunological 'dot blot immunoassay' if antibodies are available for the protein in question. A solution of the total protein is applied to a nitrocellulose filter via a removable plastic well, the liquid is drawn through under vacuum and the bound protein is detected using a double-antibody technique.

1.4.5 Molecular methods

Unlike biochemical or immunological analyses, these techniques do not provide an assessment of gene expression but provide direct evidence for the presence of the foreign gene sequence.

(a) Nucleic acid hybridization

Southern blot analysis of each putative transformant can be carried out using the gene of interest, or the marker gene sequence, as a specific radiolabelled probe. The presence of complementary sequences, as detected by autoradiography, provides good evidence for the presence of the foreign gene construct in the putative transformant. A more rapid approach involves a nucleic acid dot blot (c.f. dot blot immunoassay) where a solution of total DNA from the putative transformant is denatured, applied to a filter, fixed and hybridized to the radioactively-labelled probe.

(b) DNA amplification

The PCR is now increasingly used as a method for detecting foreign gene sequences in putative transformants. Using total DNA from each putative transformant as template and primers that specify a portion of the foreign gene construct, the amplification of a DNA segment of the expected size, as determined by gel electrophoresis, provides good evidence for the presence of the foreign gene. Further confirmation can be obtained by blotting and hybridizing the PCR product to a specific probe, by restriction analysis or ultimately by sequencing part or all of the amplified segment.

1.4.6 Limitations to current plant genetic transformation technology

Despite exciting developments in the particle bombardment approach and reproducible success in producing transgenic cereals by direct DNA uptake into protoplasts, there is still a need for a generally applicable, facile transformation method for the cereals. There is still much art and labour involved in producing the type of tissue cultures required for the current transformation methodologies. Moreover, there is a general requirement for the development of improved gene promoter systems for transformation purposes. A promoter, situated upstream of the gene, is a DNA sequence that regulates gene expression by directing transcription.

Currently, the most widely used promoter is the CaMV 35S promoter derived from the plant pathogen, cauliflower mosaic virus. However, promoters of greater efficiency are required to improve the levels of expression of foreign genes in transgenic plants to make full use of the range of cloned genes now available. The development of tissue-specific and inducible promoter systems is also a priority so that foreign-gene expression can be limited to appropriate tissues or can be 'switched on' by specific biochemical or physical treatments.

FURTHER READING

Draper, J., Scott, R., Armitage, P. and Walden, R. (1988) *Plant Genetic Transformation and Gene Expression: A Laboratory Manual*, Blackwell Scientific Publications, Oxford.

Innes, M.A., Gelfand, D.H., Sninsky, J.J. and White, T.J. (1990) *PCR Protocols – A Guide to Methods and Applications*, Academic Press, San Diego, CA.

Old, R.W. and Primrose, S.B. (1985) *Principles of Gene Manipulation*, Blackwell Scientific Publications, Oxford.

Sambrook, J., Fritsch, E.F. and Maniatis, T. (1989) *Molecular Cloning: A Laboratory Manual*, Cold Spring Harbor Laboratory Press, Cold Spring Harbor, NY.

Watson, J.D., Tooze, J. and Kurtz, D.T. (1983) *Recombinant DNA: A Short Course*, Scientific American Books, New York.

Weinberg, R.A. (1985) The molecules of life. *Scientific American*, **253**, 34–43.

Weintraub, H.M. (1990) Antisense RNA and DNA. *Scientific American*, **262,** 34–40.

White, B.A. (1993) *PCR Protocols – Current Methods and Applications*, *Methods in Molecular Biology*, *Vol. 15*, Humana Press, New Jersey.

REFERENCES

Alwine, J.C., Kemp, D.J., Parker, B.A., Reiser, J., Renar, J., Stark, G.R. and Wahl, G.M. (1979) Detection of specific RNAs or specific fragments of DNA by fractionation in gels and transfer to diazobenzyloxymethyl paper. *Methods in Enzymology*, **68,** 220–42.

Beachy, R.N., Loesch-Fries, L.S. and Tumer, N.E. (1990) Coat protein-mediated resistance against virus infection. *Annual Review of Phytopathology*, **28**, 451–74.

Burnette, W.N. (1981) 'Western blotting': electrophoretic transfer of proteins from sodium dodecyl sulphate–polyacrylamide gels to unmodified nitrocellulose and radiographic detection with antibody and radioiodinated protein *A. Analytical Biochemistry*, **112**, 195–9.

Mullis, K.B. and Faloona, F. (1987) Specific synthesis of DNA in vitro via a polymerase catalyzed chain reaction. *Methods in Enzymology*, **155**, 335–50.

Potrykus, I. (1990) Gene transfer to plants: assessment and perspectives. *Physiologia Plantarum*, **79**, 125–34.

Saiki, R.K., Gelfand, D.H., Stoffel, S., Scharf, S.J., Higuchi, R., Horn, G.T., Mullis, K.B. and Erlich, H.A. (1988) Primer-directed enzymatic amplification of DNA with a thermostable DNA polymerase. *Science*, **239**, 487–91.

Saiki, R.K., Scharf, S.J., Faloona, F., Mullis, K.B., Horn, G.T., Erlich, H.A. and Arnheim, N. (1985) Enzymatic amplification of β-globin genomic sequences and restriction site analysis for diagnosis of sickle cell anemia. *Science*, **230**, 1350–54.

Sanger, F., Nicklen, S. and Coulson, A.R. (1977) DNA sequencing with chain terminating inhibitors. *Proceedings of the National Academy of Science USA*, **74**, 5463–67.

Southern, E.M. (1975) Detection of specific sequences among DNA fragments separated by gel electrophoresis. *Journal of Molecular Biology*, **98**, 503–17.

Tanksley, S.D., Young, N.D., Paterson, A.H. and Bonierbale, M.W. (1989) RFLP mapping in plant breeding: new tools for an old science. *Bio/Technology*, **7**, 257–64.

PART TWO

Chemical and Biological Control

Molecular approaches to the design of chemical crop protection agents

J.B. Pillmoor and S.G. Foster

2.1 INTRODUCTION

Chemical crop protection agents play a vital role in the efficient production of our food and there is a continuing need for new compounds. Resistance problems and ever more stringent registration requirements have led to some older compounds being withdrawn and, thus, the need for replacements. Agronomic practices are continually changing and this leads to new problems that often cannot be solved by existing compounds, despite the fact that there are some 700 active ingredients listed in the *Pesticide Manual* (Worthing and Hance, 1991).

The discovery of new agrochemicals is, however, a long and costly business. The development of a new product is likely to take over 8 years and cost 20–60 million pounds, and the rate of success appears to be ever decreasing. It is generally accepted that 20–40 000 compounds need to be synthesized in order to find just one commercial product. These costs and statistical pressures mean that the agrochemical industry must employ all available techniques in order to make the discovery process as efficient as possible.

The discovery of new agrochemicals relies on four main sources of new lead compounds:

1. Novel synthetic chemistry.
2. Analogue chemistry.
3. Natural products.
4. Biochemical design.

Molecular Biology in Crop Protection. Edited by George Marshall and Dale Walters. Published in 1994 by Chapman & Hall, London. ISBN 0 412 54390 7(HB) 0 412 54400 8(PB).

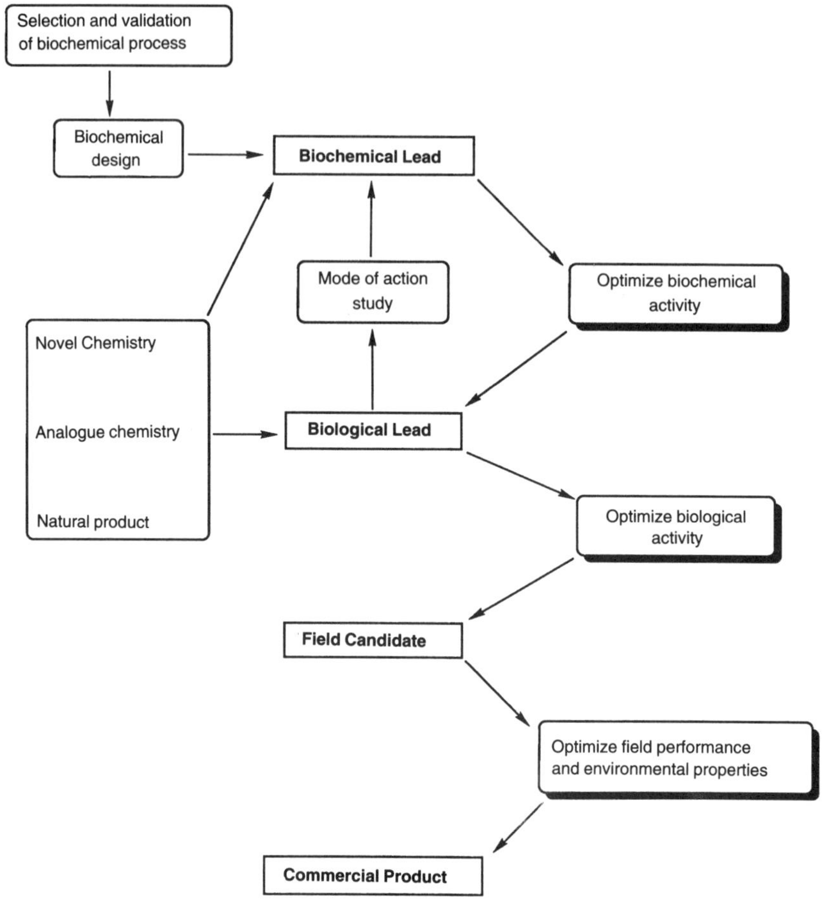

Figure 2.1 Outline of the discovery process.

The inspiration for the first approach involves only the chemical aspects of the proposed synthesis. Once a few compounds have been made and screened, further synthesis will depend solely on whether biological activity has been obtained. In contrast, the second and third approaches will have started from a compound that is already known to have biological activity. The fourth approach is unique in that the initial work will concentrate on discovering biochemical inhibitors with the biological aspects being addressed at a later date.

The sequence of events involved in the overall discovery process is outlined in Figure 2.1. Where any particular project feeds into the process will obviously depend on the source of the lead compound but it is clear that there are common elements, irrespective of the starting point. The purpose of this chapter is to provide a basic review of how the recent advances in molecular biology, biochemistry and computer technology are helping, or might be expected to help, in the discovery of new agrochemicals. We will limit the discussion to three areas:

1. The selection and validation of new biochemical targets.
2. The biochemical design of new inhibitors.
3. The optimization of lead compounds.

Although the discussion will tend to focus individually on the potential contributions from each of the specialist disciplines of molecular biology, biochemistry and molecular modelling, it is important to emphasize that success will only be achieved if the inputs from these disciplines are integrated and linked to the central disciplines of chemical synthesis and biological evaluation throughout the course of a project.

2.2 SELECTION AND EVALUATION OF NEW BIOCHEMICAL TARGETS

The process of biochemical design starts with the selection of a suitable biochemical process. Established sites of pesticide action are, of course, a valuable starting point (for reviews of the sites of actions of pesticides see Corbett *et al.*, 1984; Kirkwood, 1991; Köller, 1992). However, there is probably a greater chance of obtaining wholly novel compounds if a novel site of action is worked on. The key challenge is then to choose a suitable enzyme or receptor from among the many thousands that are present in every organism, as not all biochemical processes will be suitable targets for agrochemicals.

Until recently, the choice of which new biochemical targets to work on depended greatly on theory and the intuition of the biochemist. For targets chosen by these means, however, the doubt always remains that they might not be as essential as first thought due, for example, to the presence of a by-pass route. This lack of certainty can undermine the confidence required to undertake a complex inhibitor design project. A good example of how the theory can be wrong is the enzyme ketol-acid reductoisomerase, which is involved in the biosynthesis of the branched chain amino acids. As will be described in more detail later, the theory that this enzyme should be a good target was very convincing, but the practice has proved this not to be the case. Fortunately, it is now becoming feasible to determine experimentally whether any particular enzyme is a good target for an agrochemical prior to initiating a synthesis programme, at least in the herbicide and fungicide areas. The two main approaches that can be employed are the study of lethal mutants and the newer techniques of molecular genetics, as discussed below. To date, similar approaches do not appear to have been applied to the discovery of new insecticide targets, but this will hopefully be addressed in the future.

2.2.1 Lethal mutants

In classical genetics many plant and fungal mutants have been generated and these provide a potential means of evaluating novel biochemical targets as new sites of action for herbicides and fungicides.

In plant research, one of the most accessible classes of mutant for this type of work are lethal auxotrophs, which are deficient in one particular aspect of their biosynthetic metabolism. For example, amino acid auxotrophs will die if the required amino acid is not supplied to them. There are many examples in the literature of plant auxotrophic mutants (Blonstein, 1986) and one could imagine that withdrawing the nutrient supplement would indicate the symptomology and speed of kill that could be expected if the missing enzyme was inhibited with a herbicide. Tobacco (*Nicotiana tabaccum* L.) thiamin auxotrophic mutants, for example (McHale *et al.*, 1990), grow slowly without added thiamin and, once the maternal supply is exhausted, become chlorotic and die. Bleaching of new leaf growth is also evident when the thiamin supplement is withdrawn from established plants. These data suggest that thiamin biosynthesis could be a good herbicide target, as might have been expected. Unfortunately, the exact point at which the biosynthesis is disrupted has not been reported to date, so it is not possible at present to identify a specific enzyme for further study. This problem is relatively common with many of the plant mutants reported in the literature and, in addition, it is often quite difficult to determine from the data available whether death does occur rapidly upon removal of the supplement. Equally, it can be quite difficult to determine whether the internal levels of the supplied nutrient are similar to the normal endogenous levels. If they are much higher, of course, the auxotroph could survive for a much longer period of time following withdrawal of the nutrient than would be the case for a normal plant in which biosynthesis had been inhibited by a herbicide. This could lead to good targets being discarded. Nevertheless, further effort to evaluate these mutants is definitely worthwhile although proof that the approach can work really requires analysis of an auxotroph that is deficient in an established herbicide target, something that does not appear to have been reported to date. Another class of plant mutants have, however, been shown unequivocally to be good herbicide targets. These are the photorespiratory mutants that lack the ability to photorespire. They can, however, be grown under conditions in which photorespiration is not required (e.g. high carbon dioxide and low oxygen) but they die rapidly when transferred to normal atmospheric conditions (e.g. Sommerville and Ogren, 1982).

In the fungicide area, the potential for using mutants to identify new biochemical targets has been proven, albeit retrospectively. Melanin biosynthesis is known to be important in the early stages of *Pyricularia oryzae* (rice blast) infection since its biosynthesis is inhibited by the commercial fungicide tricyclazole. As melanin-deficient mutants of *P. oryzae* (albino mutants) are available, it would have been possible to have used these mutants to validate melanin biosynthesis as a new fungicide target. This would, however, have required pathogenicity testing to have been undertaken, as these mutants are only unable to penetrate and infect their plant host and can grow perfectly well *in vitro*. Such testing can sometimes be a problem as the pathogenicity of many fungi can be lost during repeated *in vitro* culturing. Nevertheless, this example does demonstrate the potential power of this approach, albeit in hindsight and with the obvious limitation that it is only applicable to fungi that can be cultured *in vitro*.

Using mutants to identify new agrochemical targets does, of course, have some additional problems that need to be borne in mind. Generation of the mutants is essentially a random process using UV irradiation or chemical mutagens and it may not be possible to generate mutants for all enzymes. For example, in situations where more than one functional copy of the gene is expressed it would be unlikely that mutation of all the functional genes could be achieved. Only enzymes for which the deficiency can be overcome by appropriate manipulation of the growth conditions can be investigated by this approach. Once a mutant has been generated its phenotype has to be analysed under realistic growth conditions to determine whether it is of agrochemical interest and this can bring its own problems as discussed earlier. Even then, a considerable amount of biochemical work can be required to determine the exact site of the mutation. Finally, mutants tend to be 'all-or-nothing', so no information is gained about the level to which a particular enzyme activity needs to be reduced in order to achieve agrochemically useful effects. In practice, 100% inhibition of an enzyme is unlikely to be achieved with an agrochemical and results with mutants that totally lack an enzyme could therefore be misleading.

Despite the above challenges, the potential for using lethal mutants as a means of validating some biochemical pathways in certain circumstances is clear and certainly deserves more attention.

2.2.2 Molecular genetics

The ideal approach to the identification and evaluation of new targets is to alter specifically the levels of the selected enzyme, and recent progress in molecular biology is now opening-up this possibility. There are two ways to achieve this, the first is by 'antisense RNA', the method of choice in plants and the second is gene disruption, the method of choice in fungi. As both methods require that the gene for the target enzyme is available, the possible approaches to obtaining the gene are outlined before discussing the two methods.

(a) Gene cloning

While many genes have been cloned and the numbers are growing rapidly, only in a limited number of cases will the specific gene of interest be available for the relevant organism. It is worth emphasizing that, for both antisense and gene disruption, it is essential that the gene from the organism to be studied is available. Consequently, many projects will start by cloning the gene of interest and there are four basic techniques that can be employed, examples of which will be described below:

1. Heterologous probing.
2. Polymerase chain reaction (PCR).
3. Functional complementation.
4. Protein purification.

The first two depend on the gene sequence for the protein of interest being already available from other organisms. Where this is not the case, one of the final two techniques will be required. Some of these techniques have already been reviewed in Chapter 1 and we will therefore limit ourselves to giving an example of where each has been used to clone a gene of potential interest to the agrochemical industry.

Heterologous probing

This technique uses a gene cloned previously from one organism as a probe for the corresponding gene in the target organism. The gene for the herbicide target acetolactate synthase, for example, was cloned from tobacco and *Arabidopsis thaliana* by using the *Saccharomyces cerevisiae ILV2* gene known to code for the enzyme (Mazur *et al.*, 1987).

The polymerase chain reaction (PCR)

In the last 5 years this very powerful technique has been added to the molecular biologist's armoury. Many genes have been cloned by PCR, including chitin synthase, the target of the fungicidally active polyoxins and nikkomycins (Bowen *et al.*, 1992). The amino acid sequence for chitin synthases from *S. cerevisiae* and *Candida albicans* were compared, revealing highly conserved regions. Oligonucleotides representing these sequences were synthesized and were used to amplify a part of the chitin synthase gene from 15 fungi. It is clear that these PCR primers could potentially be used to clone chitin synthase from any plant pathogenic fungus.

Functional complementation

In this technique, gene libraries of the organism of interest are transferred into a suitable mutant (normally auxotrophic) that lacks the endogenous target gene function. Where the mutant regains the ability to grow on minimal media (i.e. is no longer auxotrophic), it has acquired the gene coding for the enzyme of interest. A good example is given by the work of Minet *et al.* (1992), in which some genes of interest as potential herbicide targets from *A. thaliana* were identified by functional complementation of eight different *S. cerevisiae* auxotrophs.

Protein purification

Purification of the enzyme of interest to homogeneity can allow part of the sequence of the protein to be determined. Using this information, oligonucleotide probes can be designed and used to clone the gene of interest. For example, cabbage histidinol dehydrogenase, a potential herbicide target, was cloned using this method by Nagai *et al.* (1991).

(b) *Antisense RNA*

The principle is summarized in Figure 2.2 (see also Chapter 1) and reviewed by Mol *et al.* (1990). The gene for the enzyme is isolated and then re-introduced into the organism in the reverse orientation. Providing that a suitable promoter is included, this new gene will produce antisense mRNA (i.e. 'back-to-front'). This is believed to titrate out the natural mRNA by binding

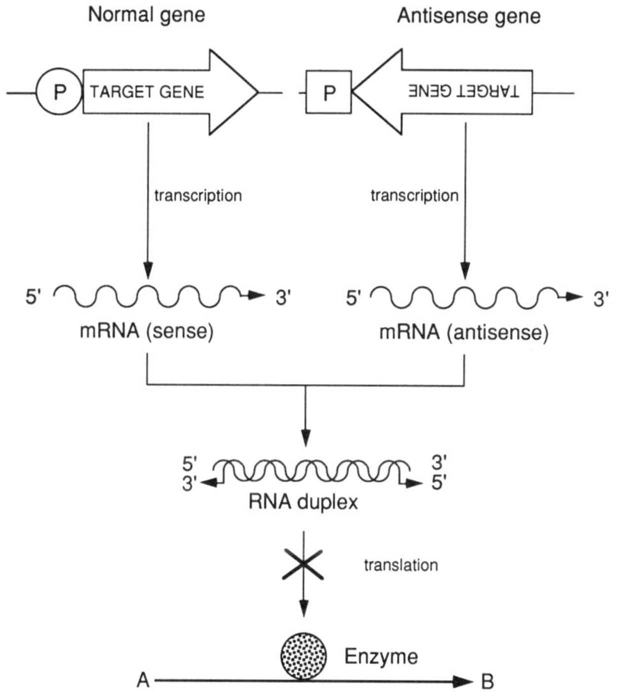

Figure 2.2 Antisense validation of target genes in plants. In the plant, a promoter Ⓟ drives the expression of the target gene to produce mRNA (transcription). The mRNA is normally used as the template for protein synthesis by the ribosome to produce the target enzyme (translation). The target enzyme then catalyses the conversion of A to B. To create an antisense plant, a copy of the target gene is inserted in the reverse direction with respect to a second promoter ([P] , usually derived from the cauliflower mosaic virus). This promoter drives the expression of the antisense gene to give antisense mRNA. It is believed that the sense and antisense mRNA form a duplex, thus preventing translation of the sense mRNA.

to it although the actual mechanisms involved have not, so far, been fully elucidated. The net result is reduced levels of the natural mRNA and, hence, reduced levels of the protein.

One of the best known examples of antisense technology, although not a herbicidal one, is the suppression of polygalacturonase expression during tomato ripening. This enzyme naturally causes softening of the fruit and two companies, ICI and Calgene, have developed plants which contain an antisense gene for polygalacturonase. These plants produce fruits that have very little or no polygalacturonase activity and soften only slowly on the shelf (Smith *et al.*, 1988).

There is a rapidly increasing number of examples of the use of antisense RNA technology to specifically regulate the levels of a range of enzymes. We have validated the approach, in our collaborative work with The Institute for Gene Biology in Berlin, by creating plants that contain an antisense gene for the known herbicide target acetolactate synthase (involved in the biosynthesis of the branched chain amino acids). The antisense plants created were initially rescued by supplying the branched chain amino acids, but upon transfer to normal soil, the plants rapidly developed symptoms typical of those associated with the herbicides that inhibit this enzyme and subsequently died (Foster and Höfgen, 1993). A large amount of effort is ongoing worldwide to create a range of different antisense plants and the initial indications are that some of these will result in the identification of new herbicide targets. However, at present, the literature does not appear to contain any clear examples that can be used to demonstrate that a particular enzyme would be a good herbicide target. Although dramatic effects on photosynthesis and plant growth were achieved by reducing the key enzyme in photoassimilation, ribulosebisphosphate carboxylase, below 50% (Quick *et al.*, 1991), this enzyme would probably not be an attractive herbicide target due to the extremely high levels that are present in plants. Indeed, the examples in the literature to date are actually more help in identifying enzymes that would not be good herbicide targets! For example, almost complete elimination of cytosolic pyruvate kinase (previously thought to be a key enzyme in glycolysis) did not adversely affect growth, indicating that the plants, unexpectedly, have alternative pathways available (Gottlob-McHugh *et al.*, 1992). A similar situation was also evident for the enzyme phenylalanine ammonia lyase. This enzyme has a role in lignification and could be considered as a potential herbicide target. However, work by Elkind *et al.* (1990) has shown that, although transgenic tobacco plants expressing an antisense phenylalanine ammonia lyase gene had reduced levels of enzyme activity and a number of visual symptoms, no lethality was observed. These, and other examples, show clearly the potential of this approach. A further refinement that we can hopefully look forward to in the future will be the use of inducible promoters with the antisense gene; promoters, for example, like the tetracycline inducible promoter (Gatz *et al.*, 1992). Such a system would open up the prospects of being able to apply a small molecule to an established plant to induce the expression of the antisense gene and hence reduce the levels of the target enzyme at a chosen time. This would not only mimic the application of a post-emergence herbicide more closely, but would also obviate the need to protect the transgenic plant against the consequences of the constitutive expression of the antisense gene during its initial growth, and allow a much greater range of different enzymes to be investigated.

An interesting consequence of the antisense approach is that the level of expression of the antisense gene will determine the extent to which the levels of the natural mRNA, and therefore the enzyme, are reduced. This can lead to a range of phenotypes being produced and provide a valuable means of determining the degree of enzyme inhibition that would be required to achieve a herbicidal effect. Clearly, an enzyme for which only a small perturbation in level is lethal would be a more attractive target than an enzyme for which a large decrease in activity is required. However, it is worth noting that there has been no report where an antisense approach has achieved complete abolition of the activity of an enzyme; the maximum reduction appears to have been approximately 90–95%.

The antisense approach does, of course, have potential limitations. In cases where several different copies of one gene are expressed, antisense RNA may be required to each individual gene. Should an enzyme be particularly stable *in vivo* and only be degraded and resynthesized at very low rates, the reduction in enzyme mRNA may not actually reduce the levels of enzyme activity over a reasonable period of time. This could be a particular limitation when using inducible promoters. Careful biochemical analysis of the plants produced by this approach is undoubtedly required if any firm conclusions are to be made. Finally, it must borne in mind that this whole approach is still in its infancy and requires a large investment of time and effort for every enzyme that is investigated.

(c) Gene disruption

This technique is, at present, limited to fungal species. It was first developed for *Saccharomyces cerevisiae* and non-pathogenic filamentous fungi, but now the technique is being extended to plant pathogenic fungi and is summarized in Figure 2.3. In the first step, the gene of interest is isolated and then interrupted (in the test tube) by replacing a segment of it with another gene, usually a gene conferring antibiotic resistance. This construct is then returned to the fungus where homologous recombination will exchange the natural gene with the interrupted one. Such an exchange will result in the fungus losing its functional copy of the target gene and gaining the interupted target gene (and hence antibiotic resistance). These transformed fungi can be isolated by growing in the presence of the antibiotic. In reality, this gene replacement does not occur at a very high frequency as the interrupted gene can also integrate randomly into the genome.

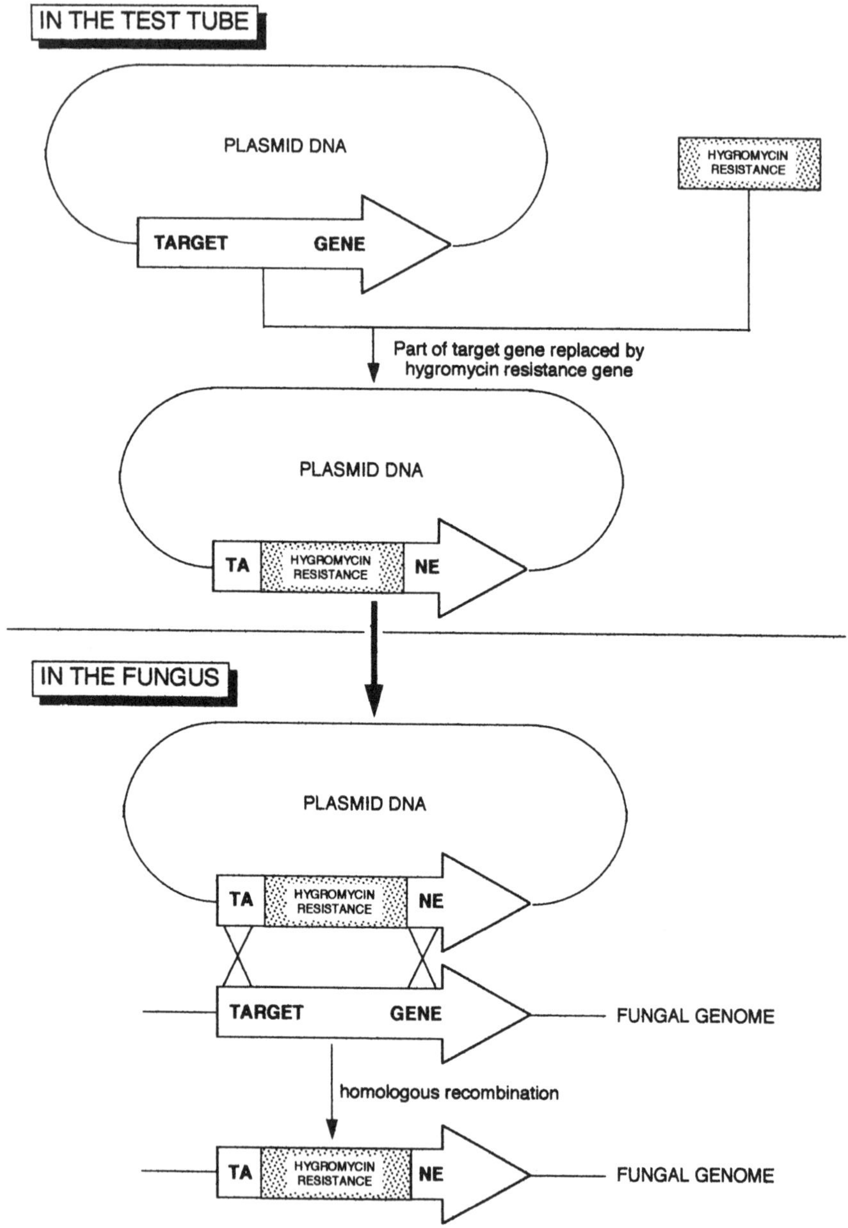

Figure 2.3 Gene disruption of target genes in fungi. To disrupt the target gene, it is isolated and part of it replaced with a 'marker' gene, such as a hygromycin resistance gene, in a plasmid vector capable of transforming the fungus. This construct is returned to the fungus where homologous recombination will replace the wild-type gene for the disrupted version. Transcription and translation will produce only the enzyme for hygromycin resistance and the target gene function will be lost.

Consequently, one has to analyse many antibiotic resistant fungi to find the few that have also lost the target gene function.

There are three examples in the literature of gene disruption in plant pathogenic fungi. Two are for the enzyme cutinase, of potential importance in fungal penetration of plant surfaces. Stahl and Schäfer (1992) disrupted the cutinase gene of *Nectria haematococca* (the causal agent of foot rot in pea) to create a cutinase deficient mutant. Although this mutant had no cutinase activity, it was as pathogenic as the wild-type organism suggesting that cutinase is not important for the pathogenicity of this particular fungus. In another example, Sweigard *et al.* (1992) disrupted a cutinase gene in *P. oryzae* (rice blast). These mutants were still as pathogenic as the wild-type fungus. However, biochemical analysis showed that although this cutinase gene was disrupted, there was a second gene giving rise to a second enzyme activity which had been unperturbed by the disruption. No conclusion could therefore be drawn as to whether cutinase is essential for the pathogenicity of *P. oryzae*. In the third example, Scott-Craig *et al.* (1990) disrupted an extracellular endopolygalacturonase in the maize (*Zea mays* L.) pathogen *Cochliobolus carbonum*. Endopolygalacturonases could be expected to be involved in fungal penetration through plant cell walls due to their role in pectin degradation. However, while the disruption in *C. carbonum* resulted in complete elimination of all the endopolygalacturonase activity, there was no effect on pathogenicity. The endopolygalacturonase would therefore appear not to be a good fungicide target in this pathogen.

A problem with this approach is that where several copies of the same gene are expressed, they will all need to be disrupted. So, for example, with the cutinases of *P. oryzae* described above, although an inhibitor of the enzyme would be expected to inhibit all or most forms of the enzyme, gene disruption would be required of every gene individually. A further problem is that it is only feasible to work on enzymes where manipulation of growth conditions will allow the fungus to survive in the absence of the target enzyme and this certainly limits the number of enzymes that can be investigated.

Finally, the technique of gene disruption is currently limited to only a few plant pathogens of agronomic importance which can be genetically transformed with exogenous DNA.

2.3 BIOCHEMICAL DESIGN OF NEW INHIBITORS

The concept of designing biologically active compounds from biochemical first principles is well established in the pharmaceutical industry, where degrees of success have been achieved. Although the approach has also been adopted in the agrochemical industry, albeit to a lesser extent, it is unfortunately true that no agrochemical product has yet been discovered by this approach.

Nevertheless, there have been scientific successes which provide optimism for the future of this approach. In this section we will outline the principles involved and give some selected examples to illustrate the potential power of the biochemical design approach.

2.3.1 The principles of enzyme inhibition

In their role as biological catalysts, enzymes facilitate the progress of chemical reactions and they can be inhibited in a number of ways. It is beyond the scope of this review to consider enzyme reactions and their inhibition in detail and the reader is referred to specialist texts on the subject (e.g. Walsh, 1979). However, some key principles are outlined below.

For a chemical reaction to take place between two compounds in free solution, the two compounds need to meet in the correct orientation and with sufficient energy to overcome the high activation barrier normally required for reaction to occur. These conditions will only be satisfied very infrequently and, consequently, the uncatalysed rate of reaction will be very slow. What an enzyme essentially does is to immobilize the two compounds together in the ideal orientation for reaction. In addition, the enzyme also stabilizes the high energy intermediates and the transition state complexes that are formed during the course of the reaction so as to decrease the activation energy barrier. The influence of an enzyme on the activation energy for a chemical reaction is shown in Figure 2.4, the difference in activation energy between the uncatalysed and catalysed reactions being the amount of energy involved in binding the unstable reaction species. Through these processes enzymes can achieve truly remarkable increases in the rate of reaction of up to 10^{14}.

Interference with any of these processes will prevent the enzyme carrying out its normal reaction. An inhibitor simply binds to an enzyme by forming a number of interactions with the amino acids that make up the enzyme in such a way as to prevent the normal chemical reaction taking place. The potency of the inhibitor depends on the strength of the interactions formed and these can be either covalent or non-covalent. Non-covalent inhibitors are often classified in text books into two main types, competitive and non-competitive. Purely competitive inhibitors interact with the enzyme at the normal substrate binding site and therefore give an apparent reduction in the affinity of the enzyme for the substrate, an effect that can be overcome by increasing the substrate concentration. Fully non-competitive inhibitors interact at a site distinct from the substrate binding site and only reduce the overall reaction rate and not the affinity of the enzyme for the substrate. Representatives of most of the classical types of inhibitors can be found among commercial agrochemicals and a few key examples are given below.

Although relatively few agrochemicals bind covalently to their target enzyme, the insecticidal carbamates and organophosphates that act

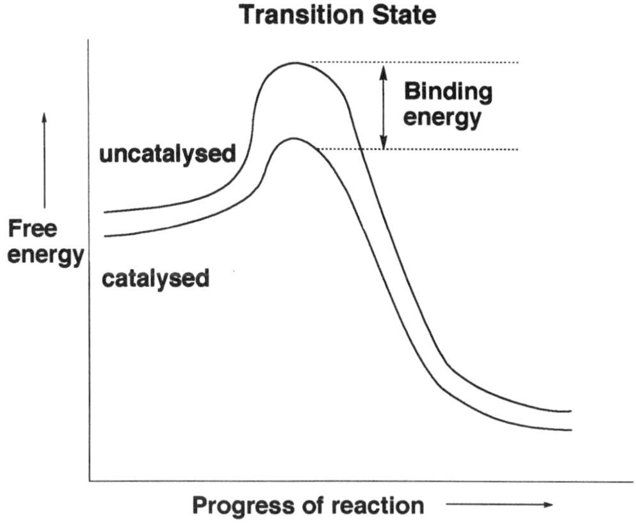

Figure 2.4 Schematic representation of the differences in activation energy between uncatalysed and catalysed reactions.

through inhibition of the enzyme acetylcholine esterase (a key enzyme in nerve transmission), are examples of this type of interaction.

Purely competitive inhibitors are exemplified by the fungicidally active polyoxins that inhibit the enzyme chitin synthase (involved in cell wall biosynthesis) by mimicking the normal substrate of the enzyme, UDP *N*-acetyl glucosamine. In order to effectively out-compete the normal substrate, additional interactions of the inhibitor with the enzyme, quite apart from those involved in binding the substrate, are often required. For example, prochloraz is a member of a large group of fungicides that inhibit the 14 α-demethylation step in ergosterol biosynthesis by interacting with the enzyme both at the normal sterol substrate binding site and with the iron that is also present at the active site. Very tight binding to the enzyme can also potentially be achieved by mimicking the transition state complex or the other high energy intermediates that are very tightly bound by the enzyme during the course of the normal chemical reaction. This type of inhibitor is described in more detail when considering the enzymes ketol-acid reductoisomerase and glutamine synthetase in the next section.

Fully non-competitive inhibitors are exemplified by the herbicides that inhibit the enzyme phytoene desaturase in the carotenoid biosynthesis pathway (e.g. norflurazon).

Finally, it is worth mentioning two other compounds as examples of some of the other types of inhibition that can occur. The herbicide glyphosate, that inhibits the enzyme enol-pyruvyl shikimate phosphate synthase (aromatic amino acid biosynthesis), is an example of an inhibitor that specifically interacts only with the enzyme–substrate complex (known as uncompetitive inhibition). Glyphosate binds to the complex between the enzyme and its first substrate, shikimate-5′-phosphate, and then actually forms a pseudo-type transition state complex. Finally, the somewhat special case of the herbicides that inhibit the enzyme acetolactate synthase (branched chain amino acid biosynthesis) is also worth noting. These compounds (e.g. chlorsulfuron) bind to the enzyme after it has interacted with the first substrate molecule, pyruvate, although it is now known that they bind at a site that is distinct from, but very close to, the active site of the enzyme. The term 'extraneous site inhibitors' has been coined to describe these compounds (Schloss and Aulabaugh, 1990).

2.3.2 Enzyme mechanisms as an inspiration for inhibitor design

Currently the most practical approach to inhibitor design is through the study and understanding of enzyme mechanisms. A principal route is to try to design stable mimics of the transition state complex or the other high energy intermediates that are formed during the normal chemical reaction. Probably the best example of this type of design approach is the work undertaken by Schloss and co-workers on the enzyme ketol-acid reductoisomerase, involved in branched chain amino acid biosynthesis (Aulabaugh and Schloss, 1990). This enzyme catalyses the step immediately after the established herbicide target, acetolactate synthase, in the biosynthetic pathway. It therefore represents an attractive target in a pathway whose inhibition is already known to give herbicidal effects. By studying the reaction mechanism of the enzyme (Figure 2.5a), the oxalyl hydroxamate (compound I in Figure 2.5b) was proposed as a potential reaction intermediate analogue. Synthesis and testing confirmed the theory as the compound was found to be a powerful inhibitor with an overall affinity of 22 pM. Unfortunately the compound has only limited herbicidal activity. Although poor delivery and stability of the compound may have initially been expected to be the cause of the poor biological activity, this was subsequently shown not to be the case and it is now believed that the enzyme itself is actually a poor herbicide target (Wittenbach *et al.*, 1991). These authors have shown that application of compound I to intact plants resulted in almost complete inhibition of the enzyme, clearly demonstrating that the compound can access the enzyme *in vivo*. Despite this high level of inhibition, plant death does not rapidly ensue, indicating that the plants can either tolerate starvation of the branched chain amino acids or that the enzyme is present in such excess that sufficient flux through the pathway can be maintained.

This finding certainly raises questions about why the inhibition of the previous enzyme, acetolactate synthase, should give so much more lethal

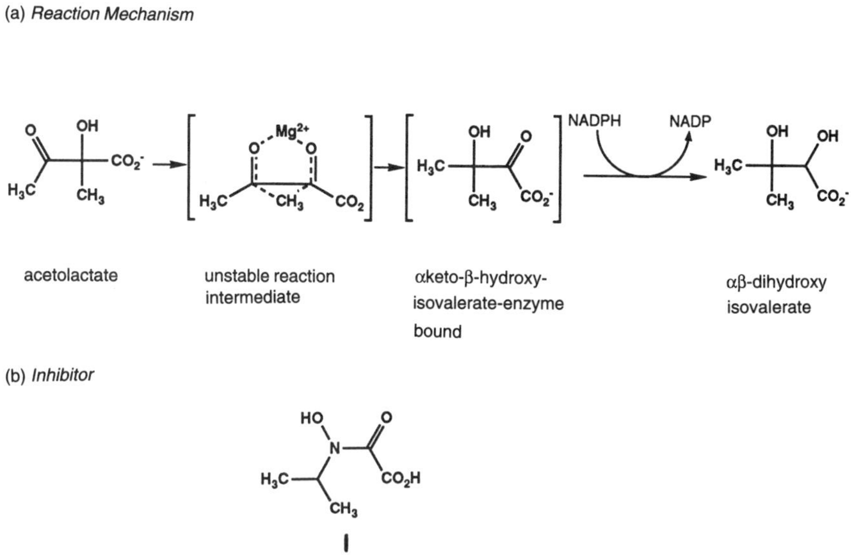

Figure 2.5 Ketol-acid reductoisomerase – mechanism and inhibitor. Redrawn from Pillmoor *et al.* (1991).

effects. At present, the explanation is not clear. However, it has been suggested that a contributory factor to the death of the plant following the inhibition of acetolactate synthase is the accumulation of a toxic intermediate, α-keto butyrate, rather than just simple starvation for the branched chain amino acids (La Rossa and Van Dyk, 1987). This example clearly emphasizes the importance of only working on fully validated targets, as discussed previously.

A further example of this approach is the work that was undertaken in our laboratories on glutamine synthetase, a key enzyme in the assimilation of inorganic nitrogen, that has been described in more detail elsewhere (Wright *et al.*, 1991). The starting point for the work was the principle that a tetrahedral phosphorus atom can act as a stable mimic of the tetrahedral carbon formed as part of enzyme reactions involving nucleophilic attack on a carbonyl carbon. Glutamine synthetase uses such a reaction, as shown in Figure 2.6, and compound II was proposed as a potential inhibitor. Searching the literature showed, however, that compound II had already been synthesized elsewhere (Rupp *et al.*, 1977) and it is now a commercially successful compound with the common name phosphinothricin. This example clearly demonstrates that biochemical design of agrochemicals is possible, even though no group took this project from conception to commercial reality.

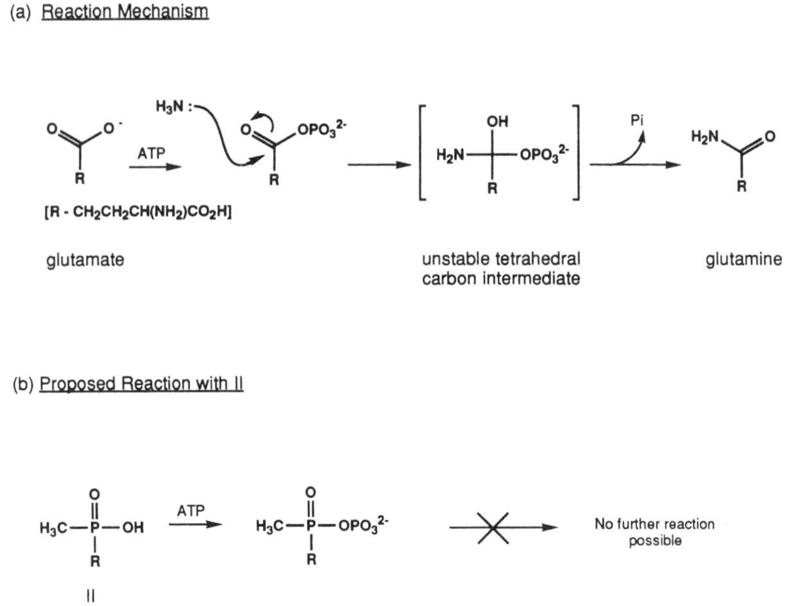

Figure 2.6 Glutamine synthetase – mechanism and inhibition. Redrawn from Wright *et al.* (1991).

2.3.3 Known inhibitors as an inspiration for design of novel compounds

Having chosen an enzyme to work on, a study of any known inhibitors of the enzyme can provide inspiration for the design of new compounds. For example, the work undertaken in our laboratories on pyruvate dehydrogenase, a central enzyme in glycolysis, was stimulated by a report that compound III (Figure 2.7) was a known inhibitor of the enzyme. Compound III was subsequently reprepared and found to be an effective plant growth retardant. This led to a project to design and synthesize more effective inhibitors as described in detail elsewhere (Baillie *et al.*, 1988). Compound IV subsequently emerged as the best material for field testing. It was unfortunately too damaging to crops at rates that gave reliable weed control to be progressed, but this example again clearly demonstrates that good levels of biological activity can be achieved by this approach.

Another example of this approach is the work by Huxley-Tencer *et al.* (1992) to design novel fungicides that inhibit $\Delta^{8,7}$-sterol isomerase (ergosterol biosynthesis). Using the known fungicide fenpropimorph as the starting point, consideration of the enzyme mechanism, particularly the desire to better mimic the sterol ring system, led to the synthesis of compound V

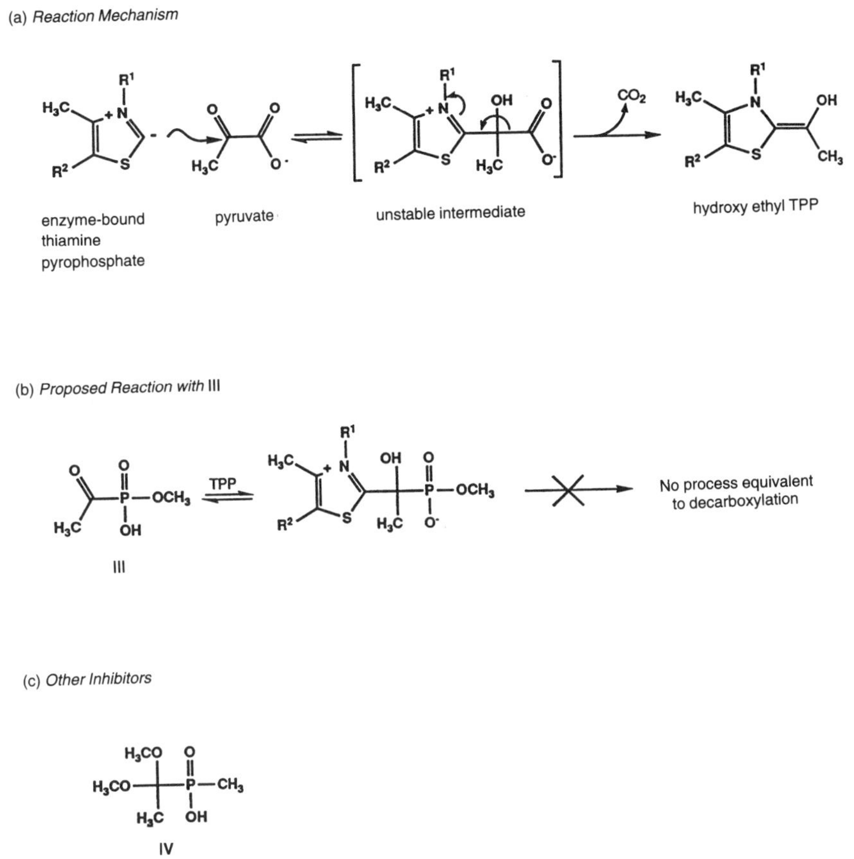

Figure 2.7 Pyruvate dehydrogenase – mechanism and inhibition. Redrawn from Wright *et al.* (1991).

(Figure 2.8). This compound was some 20 times more active than fenpropimorph at both the biochemical and biological levels.

2.3.4 Protein structure as an inspiration for inhibitor design

The ability to visualize enzymes and receptors in three dimensions is slowly becoming more common and this opens up a whole range of new possibilities for the design of novel inhibitors. Currently, the main means of obtaining structural data on proteins is through crystallization and X-ray diffraction analysis, although other techniques such as NMR analysis are also showing some promise. All these techniques are totally dependent on a plentiful supply of pure enzyme, in some cases gram amounts of enzyme being required. These levels are almost unobtainable without the use of molecular biological techniques to clone and overexpress the enzyme of

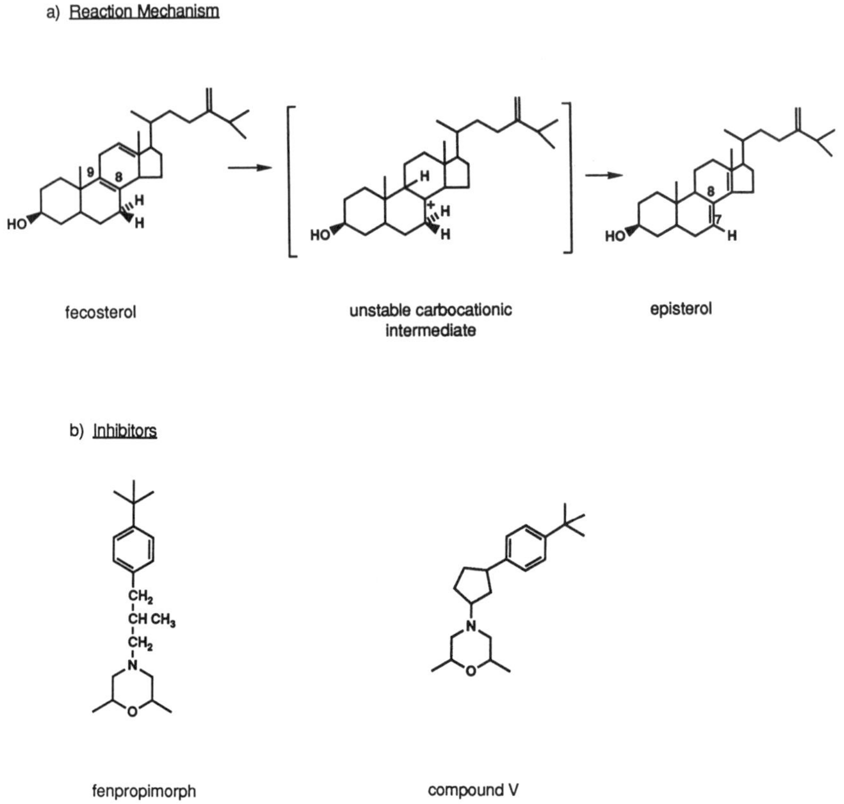

Figure 2.8 $\Delta^{8,7}$-Sterol isomerase – mechanism and inhibitors.

interest. Using high expression vectors it is now routinely possible to achieve yields of the protein of interest in genetically engineered *Escherichia coli* of up to 20% of the total protein. As an example of the power of this approach, the results in terms of enzyme yield for one of our targets are shown in Figure 2.9. In the past it would take a biochemist 7 days and 5 kg of plant material to isolate 0.8 units of enzyme activity, only enough enzyme to test 10 new chemicals. With the cloned enzyme, 6 g of cells and 1 day will yield 170 units of enzyme, enough to test 2000 new compounds. Fortunately, some of the work described previously to validate a new target can provide a clone of the gene and give a head start to the over-expression work.

It is worth emphasizing that the crystal structure of an enzyme is of maximum use in designing new inhibitors, if the crystals are produced with a ligand bound, in order to provide a starting point for the work. To date, the crystal structures of only a few proteins of interest to the agrochemical industry have been reported and there is no successful example in the literature of the use of this approach to design new agrochemicals. Perhaps

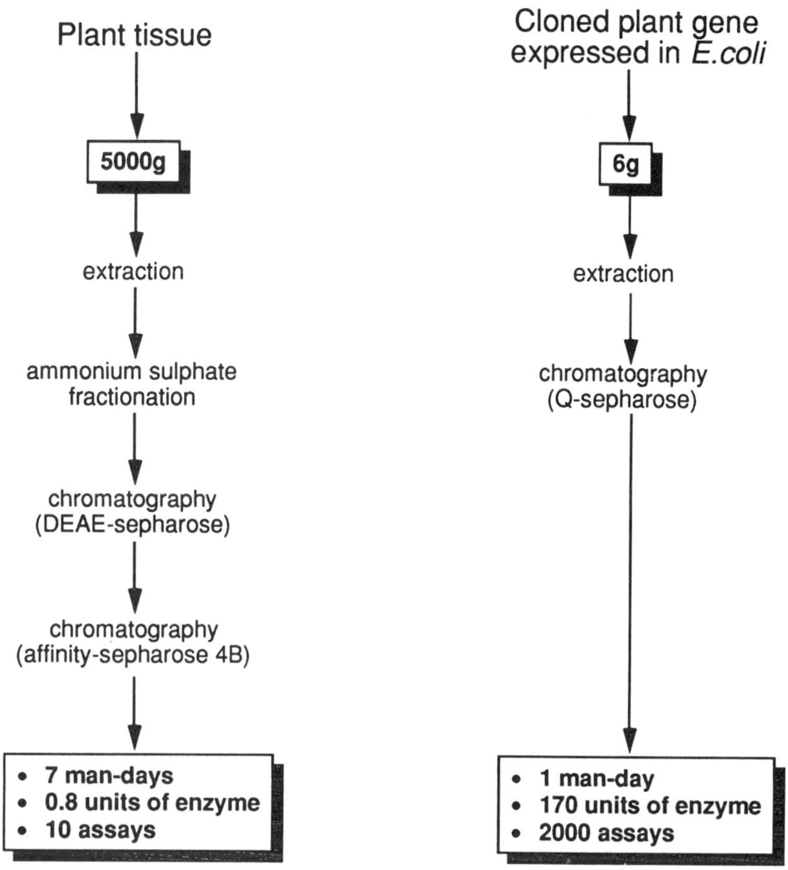

Figure 2.9 A comparison between the isolation of an enzyme from a cloned and uncloned source.

the area that has received the most attention in the agrochemical field is photosystem II. Several groups of herbicides are known to exert their activity by displacing the secondary quinone from the Q_B binding site within photosystem II, thereby blocking electron transfer. Although crystallization of photosystem II from plants has not so far been achieved, the photosynthetic reaction centre from purple bacteria has been crystallized. Using the coordinates for the bacterial system (Deisenhofer and Michel, 1989) obtained from entry 1 PRC of the October 1992 version of the Protein Data Bank (Bernstein *et al.*, 1977; Abola *et al.*, 1987), the location of the ubiquinone in the Q_B binding site can be visualized (Plate 1). The power of using such crystallographic data to identify the key interactions between a protein and a small ligand is clear from Plate 1, in which four of the main residues involved, i.e. (216)Phe, (223)Ser, (225)Gly and (190)His, are shown.

An understanding of these interactions opens up the prospects of being able to design new compounds to maximize the interactions or even to create new interactions. It is worth emphasizing that improving the potency of a compound in terms of its inhibition constant by a factor of 10, requires an additional 5.8 kJ/mol of binding energy. Under optimum conditions, the establishment of just one new hydrogen bond can give up to 25 kJ/mol; this equates to up to four orders of magnitude in binding potency which would change a micromolar inhibitor into one with a potency of sub-nanomolar! Unfortunately, despite the detailed knowledge of the bacterial reaction centre, no new inhibitor for the plant photosystem II has so far been designed using this information. This probably reflects the fact that the bacterial system can only provide a crude model for the plant system, although the recent reports of new mutants may help to refine the model further (for review, see Sinning, 1992). Recently, initial crystallographic data for the target enzyme for glyphosate, enol-pyruvoyl shikimate phosphate synthase have been published (Stallings *et al.*, 1991), which may be useful in designing new inhibitors of this enzyme.

Despite the lack of success to date, the power of this approach is becoming clear through work in the pharmaceutical area. This can best be demonstrated by referring to the work on the enzyme dihydrofolate reductase, which is a well established target for both antibacterial agents (e.g. trimethoprim) and treatment of certain forms of cancer (e.g. methotrexate). Work on the design of inhibitors of this enzyme has been thoroughly reviewed elsewhere (e.g. Kuyper, 1990), but one of the classical examples is worth repeating. Study of the binding of trimethoprim to the enzyme from *E. coli* [reproduced in Plate 2 using the coordinates for the enzyme (Bystroff *et al.*, 1990) obtained from entry 7 DFR of the October 1992 version of the Protein Data Bank (Bernstein *et al.*, 1977; Abola *et al.*, 1987)] indicated that an extra interaction with (57)Arg could be obtained by adding a 3′-carboxyalkyl substituent to trimethoprim. Synthesis of the carboxypentoxy analogue increased the potency by a factor of 54 (K_i of 24 pM for the carboxypentoxy analogue as compared with 1.3 nM for trimethoprim). Location of this analogue in the crystal structure of the enzyme shows that the carboxyl group of the side chain is able to form hydrogen bonds with (57)Arg, as predicted (Plate 3).

While the above, and other examples, provide clear encouragement for this type of approach, it is also true that our understanding of protein interactions and binding energies are not yet sufficiently advanced to make the design process anything like facile. A particular problem is that of conformational flexibility of both the inhibitor and the enzyme which makes prediction of how a compound will actually bind to the protein very difficult (Jorgensen, 1991). This means that the process is still largely dependent on empirical data and most workers in the field clearly accept that inhibitor design through the crystallization approach is a time consuming and iterative process that requires many rounds of theory, synthesis,

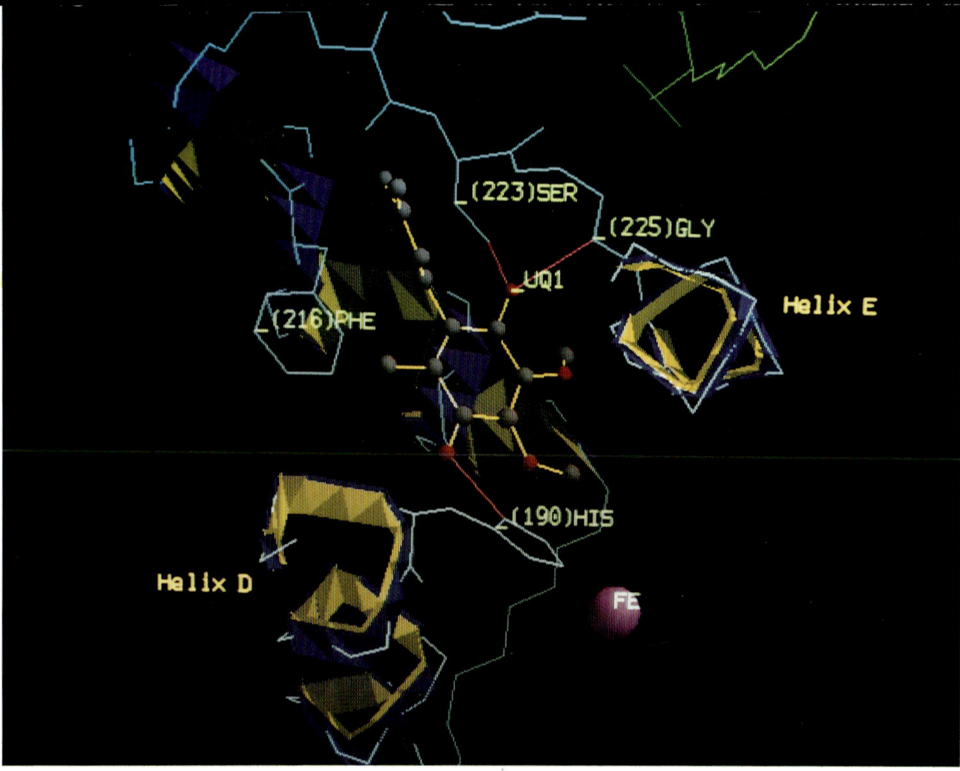

Plate 1 Computer representation of the binding of ubiquinone at the Q_B-binding site in the reaction centre of *R. viridis*. Ubiquinone (UQ1, shown in yellow with atoms given) is shown bound in the reaction centre. Four of the key amino acids involved in binding the ubiquinone, (216)Phe, (223)Ser, (225)Gly and (190)His, are labelled. Hydrogen bonds to the latter three residues are shown in red.

Plate 2 Computer representation of the binding of trimethoprim (TMP) at the active site of dihydrofolate reductase. Trimethoprim is shown bound at the active site of the enzyme with the arginine residue at position 57 [(57)Arg] being uninvolved in the interaction.

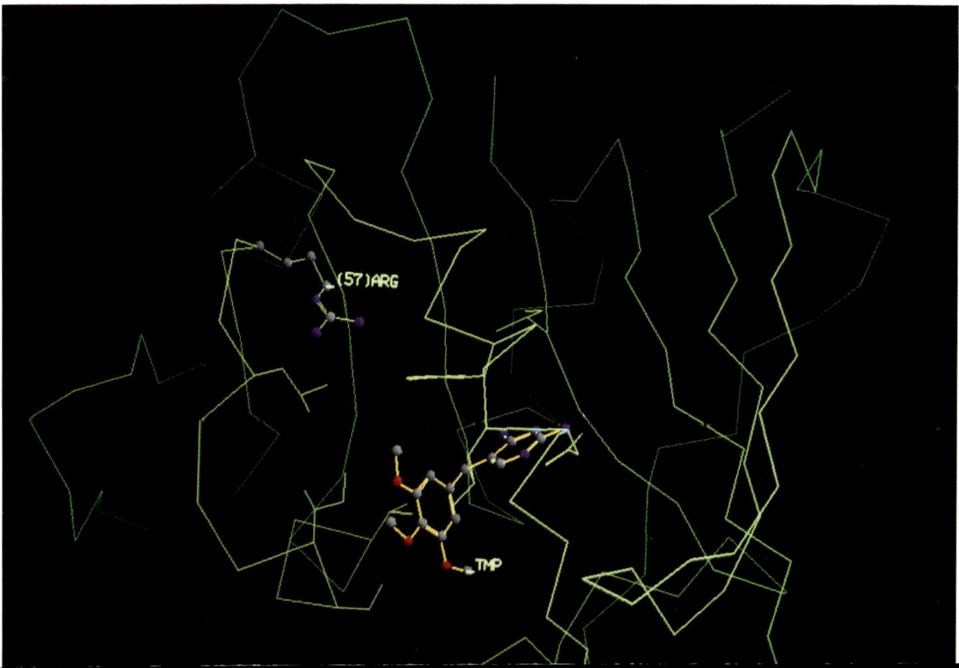

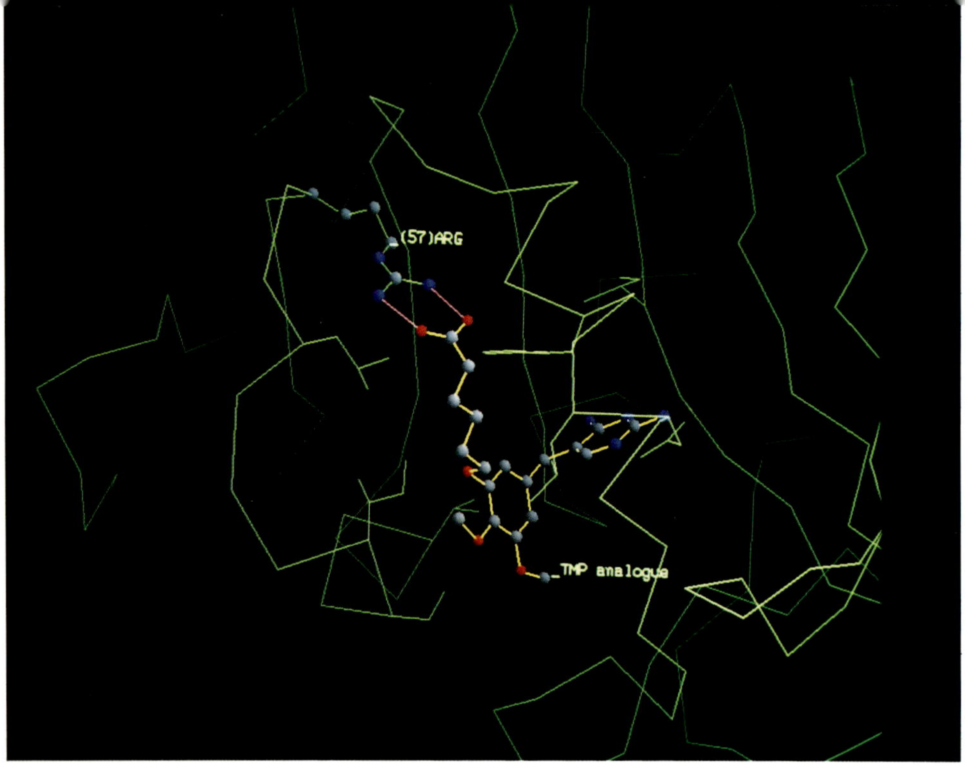

Plate 3 Computer representation of the binding of a trimethoprim (TMP) analogue at the active site of dihydrofolate reductase. The carboxypentoxy analogue of trimethoprim is shown bound at the active site and can clearly interact with (57)Arg to form the two hydrogen bonds shown in red.

Plate 4 Computer representation of prochloraz modelled into the lanosterol binding pocket within the sterol 14 α-demethylase. The Van der Waals' radii of prochloraz are shown bound in the Van der Waals' cage of the normal sterol substrate in such a way that the unsubstituted nitrogen of the triazole ring of prochloraz is bound to the haem at the active site of the enzyme.

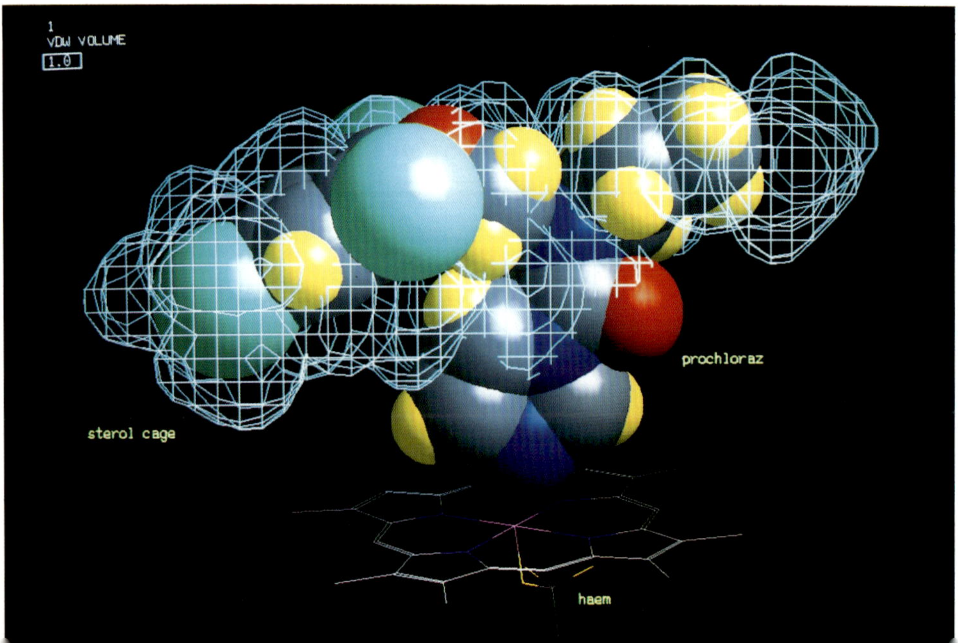

testing and revision of the theory. Nevertheless, the approach can be used in a predictive manner by building up a model of what is required for binding to the enzyme, from structural data about the inhibitors and the protein. This model can then be used to try to identify new inhibitors by screening suitable data bases for compounds that can fit in the binding site (e.g. DesJarlais *et al.*, 1988). In addition, a process of placing key molecular fragments from the lead inhibitor in the model of the binding site can also be used to try to generate novel compounds, by using the computer to suggest ways of connecting the fragments in the correct positions (e.g. Lewis, 1991). While these approaches offer much potential for the future, there appears, as yet, to be no clear example where a novel, commercially useful compound has been designed using these techniques.

2.4 OPTIMIZATION OF LEAD COMPOUNDS

Referring back to Figure 2.1, it is clear that optimization of lead compounds needs to be undertaken throughout the life of a project. In this section we will outline how the optimization work can be approached by moving through each of the major steps. However, it is important to emphasize that all optimization work is of an iterative nature and has to take into account all the information that is available at each stage. Optimization of one property of the compound in isolation is not a viable way of progressing, since it is true that it will always be a compromise between the different characteristics required for commercial success.

Starting with the optimization of biochemical leads, it is clear that this will involve refinement of the initial theory used to design the inhibitor. Where three-dimensional information is available, the processes involved have already been described in the previous section. Where knowledge about the enzyme mechanism was the starting point, the results from the first few inhibitors can be used to confirm or alter the starting theory. It is worth noting at this point the role that molecular biology can play at this stage through site directed mutagenesis. Selective modification of the amino acids that are known, or believed, to be present at the active site through specific mutagenesis experiments, can provide another powerful means of obtaining more information about both the general topography of the active site, as well as the actual chemical mechanism employed (see review by Wagner and Benkovic, 1990). Indeed, much of the early work on photosystem II was heavily dependent on the analysis of naturally occurring mutants (e.g. Böger, 1985).

As compounds are synthesized and biochemical inhibition data are obtained, a computer model of the binding site can be built up even in the absence of any direct information about the structure of the protein. This can then be used to screen compound data bases to identify new candidates for synthesis, in a similar manner as described previously in the earlier section on the design of inhibitors using the protein structure. An example of this approach is the work on the auxin receptor by Bures *et al.* (1991), where the authors were able to identify novel, physiologically active ligands by searching a chemical database.

Models of how a group of inhibitors may interact with their biochemical target can also be used to rationalize the results of the biochemical structure–activity relationships and sometimes suggest ways of identifying the most promising areas for further research. This approach has been used to create a model of the binding site for a series of herbicidally active compounds that act by inhibition of the enzyme acetyl-CoA carboxylase that is involved in fatty acid biosynthesis (Winkler *et al.*, 1989). The model can discriminate between active and inactive compounds and may prove to be useful in designing new herbicidal compounds. Several groups have used molecular modelling techniques to study ways in which fungicides that inhibit the 14 α-demethylation step in ergosterol synthesis might bind to the enzyme. From the work undertaken in our laboratories, it can be seen that prochloraz fits very well within the Van der Waals' cage of the lanosterol substrate in such a way that the unsubstituted nitrogen of the triazole ring could bind to the iron in the active site (Plate 4). This model can clearly show which compounds would be expected to be biochemically active and which would be expected to be inactive. While the model could have indicated the most promising areas of research had it been available in time, it was of only limited use in fine tuning the work in any particular area (Baillie, 1987). This is likely to be the position with any crude model that is created solely from inhibition data but, as long as the limitations are borne in mind, such models can be useful in identifying new general areas for research.

As well as the above modelling approach, biochemical data can be used in a more classical, quantitative structure analysis to define the general properties required to inhibit the enzyme (e.g. Hansch and Klein, 1991). This approach has been used widely for inhibitors of photosystem II (e.g. Trebst and Draber, 1979) and acetylcholine esterase (e.g. Weiden, 1968). However, as noted wryly by the latter author, it was much easier to synthesize inhibitors of acetylcholine esterase than it was to synthesize insecticides! Nevertheless, once an understanding is obtained of what type of compound is required to achieve good levels of biochemical activity, it is then possible to approach, more rationally, the problems of delivering the compound to its biochemical target within the organism. Efficient uptake and transport of the compound are dependent on the overall physical properties of the compound, e.g. its lipophilicity and whether an acidic group is present (Bromilow *et al.*, 1986). Inevitably, the characteristics required for efficient transport will require compromises to be made with respect to the characteristics required for inhibition of the biochemical target. Even when efficient delivery can be achieved, a further problem of potential detoxification of the compound by the target organism has to be borne in mind. At present, is it not possible to accurately predict how a compound might be metabolized by the target organism. Examples of insecticide resistance due to increased metabolism, however, show clearly that this is a potential barrier to achieving biological activity (e.g. Devonshire, 1991). In the herbicide area, the most common reason why one plant is susceptible to a particular herbicide and another is tolerant, is that the latter species can detoxify the compound. This area has been studied intensively because of the attractive prospect of being able to design-in the required crop selectivity, through an understanding of the different metabolic capabilities of weeds and crops. Unfortunately, our knowledge is still not far enough advanced for this to be a primary means of achieving the desired crop selectivity, and we still have to rely on a wide-ranging synthesis and testing programme (Brown *et al.*, 1991).

Where no biochemical information is available, the rational optimization of the biological activity becomes even more of a problem. However, classical quantitative structure–activity relationships can be developed and used to provide suggestions for further synthesis. A good example of this approach is the programme undertaken that led to the discovery of the herbicide diflufenican (Cramp *et al.*, 1987).

During the development of a project it is becoming increasingly important to take into account environmental considerations at an early stage. Computer models can provide early warnings of potential mammalian toxicity (e.g. The DEREK system; Sanderson and Earnshaw, 1991), and an increasing number of models are becoming available to predict whether a compound may be likely to be a problem with respect to ground water contamination (e.g. The Pesticide Root Zone Model; Carsel *et al.*, 1985). As would be expected, these models are by no means infallible, but they can provide some useful, early warnings.

2.5 SUMMARY AND FUTURE PROSPECTS

The rapidly developing techniques of molecular biology, protein crystallization and molecular modelling are undoubtedly providing new approaches for both biochemists and chemists to design and discover novel agrochemicals. The ability to prove that novel biochemical processes are good targets for agrochemicals, before chemical synthesis begins, will provide the necessary confidence to undertake the complex programmes of work that are required to design new inhibitors of these processes. A plentiful supply of the target enzymes or receptors will allow, in favourable cases, a three-dimensional view of the target to be obtained more readily and thereby provide an additional approach to the more classical means of designing inhibitors. In this, and in all aspects of the agrochemical discovery process, the use of various computer modelling techniques will undoubtedly increase.

However, these new techniques have yet to prove themselves commercially. By themselves they do not guarantee success, they merely help in the process and provide a means by which the odds can potentially be reduced. Indeed, they will only succeed in achieving this if their inputs are carefully integrated into the traditional, and proven, approaches of chemical synthesis and biological evaluation. In the final analysis, commercial success is totally dependent on the inventive abilities of the people involved in the work along, of course, with a little luck!

ACKNOWLEDGEMENTS

The authors thank Dr C.G. Earnshaw for providing the computer-generated diagrams used in Plates 1, 2 and 4, along with Mr M.B. Payne and their other colleagues at Chesterford Park for their helpful discussions and inputs.

REFERENCES

Abola, E., Bernstein, F.C., Bryant, S.H., Koetzle, T.F. and Weng, J. (1987) Protein Data Bank, in *Crystallographic Databases – Information Content, Software Systems, Scientific Applications*, (eds F.H. Allen, G. Bergerhoff and R. Sievers), Data Commission of the International Union of Crystallography, Cambridge, pp. 107–32.

Aulabaugh, A. and Schloss, J.V. (1990) Oxalyl hydroxamates as reaction-intermediate analogues for ketol-acid reductoisomerase. *Biochemistry*, **29**, 2824–30.

Baillie, A.C. (1987) Prochloraz and its analogues. Chemistry, mode of action, and biological efficacy, in *Synthesis and Chemistry of Agrochemicals*, (eds D.R. Baker, J.G. Fenyes, W.K. Moberg and B. Cross), ACS Symposium Series **355**, American Chemical Society, Washington, pp. 328–39.

Baillie, A.C., Wright, K., Wright, B.J. and Earnshaw, C.G. (1988) Inhibitors of pyruvate dehydrogenase as herbicides. *Pesticide Biochemistry and Physiology*, **30**, 103–12.

Bernstein, F.C., Koetzle, T.F., Williams, G.J.B., Meyer, Jr, E.F., Brice, M.D., Rogers, J.R., Kennard, O., Shimanouchi, T. and Tasumi, M. (1977) The Protein Data Bank: a computer-based archival file for macromolecular structure. *Journal of Molecular Biology*, **112**, 535–42.

Blonstein, A.D. (1986) Auxotroph isolation *in vitro*, in *A Genetic Approach to Plant Biochemistry*, (eds A.D. Blonstein and P.J. King), Springer-Verlag, Wien, pp. 259–79.

Böger, P. (1985) The photosynthesis membrane as the target of herbicidal action. *Plant Research and Development*, **21**, 69–84.

Bowen, A.R., Chen-Wu, J.L., Momany, M., Young, R., Szaniszlo, P.J. and Robbins, P.W. (1992) Classification of fungal chitin synthases. *Proceedings of the National Academy of Sciences* USA, **89**, 519–23.

Bromilow, R.H., Chamberlain, K. and Briggs, G.G. (1986) Techniques for studying the uptake and translocation of pesticides in plants. *Aspects in Applied Biology*, **11**, 29–44.

Brown, H.M., Deitrich, R.F., Kenyon, W.H. and Lichtner, F.T. (1991) Prospects for the biorational design of crop selective herbicides, in *The British Protection Conference – Weeds-1991*, Vol. 2, The British Crop Protection Council, Farnham, pp. 847–56.

Bures, M.G., Black-Schaefer, C. and Gardner, G. (1991) The discovery of novel auxin transport inhibitors by molecular modelling and three-dimensional pattern analysis. *Jounal of Computer-Aided Molecular Design*, **5**, 323–34.

Bystroff, C., Oatley, S.J. and Kraut, J. (1990) Crystal structures of *Escherichia coli* dihydrofolate reductase: the NAP^+ holoenzyme and the folate $NADP^+$ ternary complex. Substrate binding and a model for the transition state. *Biochemistry*, **29**, 3263–77.

Carsel, R.F., Mulkey, L.A., Lorber, M.N. and Baskin, L.B. (1985) The pesticide root zone model (PRZM): a procedure for evaluating pesticide leaching threats to groundwater. *Ecological Modelling*, **30**, 49–69.

Corbett, J.R., Wright, K. and Baillie, A.C. (1984) *The Biochemical Mode of Action of Pesticides*, 2nd edn, Academic Press, London.

Cramp, M.C., Gilmour, J., Hatton, L.R., Hewett, R.H., Nolan, C.J. and Parnell, E.W. (1987) Design and synthesis of *N*-(2,4-difluorophenyl)-2-(3-trifluoromethylphenoxy)-3-pyridinecarboxamide (diflufenican), a novel pre- and early post-emergence herbicide for use in winter cereals. *Pesticide Science*, **18**, 15–28.

Deisenhofer, J. and Michel, H. (1989) The photosynthetic reaction centre from the purple bacterium *Rhodopseudomonas viridis*. *Science*, **245**, 1463–73.

DesJarlais, R.L., Sheridan, R.P., Seibel, G.L., Dixon, J.S., Kuntz, I.D. and Venkataraghavan, R. (1988) Using shape complementarity as an initial screen in designing ligands for a receptor binding site of known three-dimensional structure. *Journal of Medicinal Chemistry*, **31**, 722–9.

Devonshire, A.L. (1991) Role of esterases in resistance of insects to insecticides. *Biochemical Society Transactions*, **19**, 755–9.

Elkind, Y., Edwards, R., Mavandad, M., Hendrick, S.A., Ribak, O., Dixon, R. A. and Lamb, C.J. (1990) Abnormal plant development and down-regulation of phenylpropanoid biosynthesis in transgenic tobacco containing a heterologous phenylalanine ammonia lyase gene. *Proceedings of the National Academy of Sciences USA*, **87**, 9057–61.

Foster, S.G. and Höfgen, R. (1993) Towards a rational design of pesticides, in *Opportunities for Molecular Biology in Crop Production* (eds D.J. Beadle, D.H.L. Bishop, L.G. Copping, G.K. Dixon and D.W. Holloman), BCPC Monograph No. 55, The British Crop Protection Council, Farnham, UK, pp. 75–81.

Gatz, C., Frohberg, C. and Wendenburg, R. (1992) Stringent repression and homogeneous de-repression by tetracycline of a modified CaMV35S promoter in intact transgenic tobacco plants. *The Plant Journal*, **2**, 397–404.

Gottlob-McHugh, S.G., Sangwan, R.S., Blakeley, S.D., Vanlerberghe, G.C., Ko, K., Turpin, D.H., Plaxton, W.C., Miki, B.L. and Dennis, D.T. (1992) Normal growth of transgenic tobacco plants in the absence of cytosolic pyruvate kinase. *Plant Physiology*, **100**, 820–5.

Hansch, C. and Klein, T.E. (1991) Quantitative structure–activity relationships and molecular graphics in evaluation of enzyme–ligand interactions. *Methods in Enzymology*, **202**, 512–43.

Huxley-Tencer, A., Francotte, E. and Bladocha-Moreau, M. (1992) 1(*R*)-(2,6-*cis*-dimethylmorpholino)-3(S)-(p-tert-butylphenyl) cyclopentane: a representative of a novel, potent class of bio-rationally designed fungicides. *Pesticide Science*, **34**, 65–74.

Jorgensen, W.L. (1991) Rusting of the lock and key model for protein–ligand binding. *Science*, **254**, 954–5.

Kirkwood, R.C. (ed.) (1991) *Target Sites for Herbicide Action*, Plenum Press, London.

Köller, W. (1992) Target Sites of Fungicide Action, CRC Press, Boca Raton, FL.

Kuyper, L.F. (1990) Receptor-based design of dihydrofolate reductase inhibitors, in *Protein Design and the Development of New Therapeutics and Vaccines*, (eds J.B. Hook and G. Poste), Plenum, New York, pp. 297–327.

LaRossa, R.L. and Van Dyk, T.K. (1987) Metabolic mayhem caused by 2-ketoacid imbalances. *BioEssays*, **7**, 125–30.

Lewis, R.A. (1991) Rational methods for site-directed drug design: novel approaches for the discovery of potential ligands. *Biochemical Society Transactions*, **19**, 883–7.

Mazur, B.J., Chui, C.F. and Smith, J.K. (1987) Isolation and characterisation of plant genes coding for acetolactate synthase, the target enzyme for two classes of herbicides. *Plant Physiology*, **85**, 1110–7.

McHale, N.A., Kawata, E.E. and Cheung, A.Y. (1990) Plastid distribution in a thiamine-requiring mutant of *Nicotiana sylvestris* blocks accumulation of specific nuclear and plastid mRNAs. *Molecular and General Genetics*, **221**, 203–9.

Minet, M., Dufour, M.-E. and Lacroute, F. (1992) Complementation of *Saccharomyces cerevisiae* auxotrophic mutants by *Arabidopsis thaliana* cDNAs. *The Plant Journal*, **2**, 417–22.

Mol, J.N.M., van der Krol, A.R., van Tunen, A.J., van Blokland, R., de Lange, P. and Stuitje, A.R. (1990) Regulation of plant gene expression by antisense RNA. *FEBS Letters*, **268**, 427–30.

Nagai, A., Ward, E., Beck, J., Tada, S., Chang, J.-Y., Scheidegger, A. and Ryals, J. (1991) Structural and functional conservation of histidinol dehydrogenase between plants and microbes. *Proceedings of the National Academy of Sciences USA*, **88**, 4133–7.

Pillmoor, J.B., Wright, K. and Lindell, S.D. (1991) Herbicide discovery through rational design: Some experiences, in *The British Crop Protection Conference – Weeds-1991*, Vol. 2, The British Crop Protection Council, Farnham, pp. 857–66.

Quick, W.P., Schurr, U., Scheibe, R., Schulze, R., Rodermel, E.-D., Bogorad, S.R. and Stitt. M. (1991) Decreased Rubisco in tobacco transformed with 'antisense' rbcS.I. Impact on photosynthesis in ambient growth conditions. *Planta*, **183,** 542–4.

Rupp, W., Finke, M., Bieringer, H. and Langeleuddeke, P. (1977) Patent: Ger. Offen. DE 2717440. (*Chemical Abstracts*, **88**, 70494e).

Sanderson, D.M. and Earnshaw, C.G. (1991) Computer prediction of possible toxic action from chemical structure; the DEREK system. *Human and Experimental Toxicology*, **10**, 261–73.

Schloss, J.V. and Aulabaugh, A. (1990) Acetolactate synthase and ketol-acid reductoisomerase: targets for herbicides obtained by screening and *de novo* design. *Zeitschrift für Naturforschung*, **45c**, 544–51.

Scott-Craig, J.S., Panaccione, D.G., Cervone, F. and Walton, J.D. (1990) Endopolygalacturonase is not required for pathogenicity of *Cochliobolus carbonum* on maize. *The Plant Cell*, **2**, 1191–200.

Sinning, I. (1992) Herbicide binding in the bacterial photosynthesis reaction centre. *Trends in Biochemical Sciences*, **17**, 150–4.

Smith, C., Watson, C., Ray, J., Bird, C., Movviss, P., Schuch, W. and Grierson, D. (1988) Antisense RNA inhibition of polygalacturonase gene expression in transgenic tomatoes. *Nature*, **334**, 724–6.

Sommerville, C.R. and Ogren, W.L. (1982) Genetic modification of photorespiration. *Trends in Biochemical Sciences*, **7**, 171–4.

Stahl, D.J. and Schäfer, W. (1992) Cutinase is not required for fungal pathogenicity on pea. *The Plant Cell*, **4**, 621–9.

Stallings, W.C., Abdel-Meguid, S.S., Lim, L.W. *et al.* (1991) Structure and topological symmetry of the glyphosphate target 5-enol-pyruvylshikimate-3-phosphate synthase: a distinctive protein fold. *Proceedings of the National Academy of Sciences USA*, **88**, 5046–50.

Sweigard, J.A., Chumley, F.G. and Valent, B. (1992) Disruption of a *Magnaporthe grisea* cutinase gene. *Molecular and General Genetics*, **232**, 183–90.

Trebst, A. and Draber, W. (1979) Structure–activity correlations of recent herbicides in photosynthetic reactions, in *Advances in Pesticide Science, Part 2*, (ed. H. Geissbühler), Pergamon Press, Oxford, pp. 223–34.

Wagner, C.R. and Benkovic, S.J. (1990) Site directed mutagenesis: a tool for enzyme mechanism dissection. *Trends in Biotechnology*, **8**, 263–70.

Walsh, C. (1979) *Enzymatic Reaction Mechanisms*, Freeman & Co., San Francisco.

Weiden, M.H.J. (1968) Insecticidal carbamoyloximes. *Journal of the Science of Food and Agriculture*, 1968 Supplement, 19–31.

Winkler, D.A., Liepa, A.J., Anderson-McKay, J.E. and Hart, N.K. (1989) A molecular graphics study of factors influencing herbicidal activity of oximes of 3-acyl-tetrahydro-2H-pyran-2,4-diones. *Pesticide Science*, **27**, 45–63.

Wittenbach, V.A., Aulabaugh, A. and Schloss, J.V. (1991) Examples of extraneous site inhibitors and reaction intermediate analogues: acetolactate synthase and ketol-acid reductoisomerase, in *Pesticide Chemistry* (ed. H. Frehse), VCH, Weinheim, pp. 151–60.

Worthing, C.R. and Hance, R.J. (1991) *The Pesticide Manual*, 9th edn, The British Crop Protection Council, Farnham.

Wright, K., Pillmoor, J.B. and Briggs, G.G. (1991) Rationality in herbicide design, in *Herbicides, Topics in Photosynthesis*, Vol. 10, (eds N.R. Baker and M.P. Percival), Elsevier, Amsterdan, pp. 337–68.

Molecular approaches to the design of biotic crop protection agents

R.D. Possee and L.A. King

3.1 INTRODUCTION

The use of microbial biopesticides to control insect pests has, for many years, been eclipsed by the application of chemical insecticides to achieve that goal. Bacterial, viral and fungal species all provide candidates for use as biological insecticides. They are perceived as 'environmentally friendly' products, with few of the undesirable side effects of the chemical pesticides. Frequently, they also exhibit a degree of host specificity denied to chemicals. This permits targeting of certain insect pests, without harming non-target organisms.

The principal disadvantage of biocontrol agents is their slow speed of action in comparison with chemicals. Crop growers are used to spraying an infested area with chemical insecticides and achieving almost instantaneous results. Biological agents frequently require many days or even weeks to achieve similar levels of control. In this lag period the insect pests continue to feed and decimate the crop. Where the appearance of the crop is economically important (e.g. apples), even minor damage can reduce the value of the product to negligible levels.

Application of the biocontrol agents may also be more problematic than the use of chemicals. Biopesticides must remain viable prior to spraying. Incorrect storage and application in unsuitable conditions can seriously reduce the effectiveness of these products.

In spite of the problems associated with biocontrol agents, interest in their use has increased in recent years for a variety of reasons. The repeated

Molecular Biology in Crop Protection. Edited by George Marshall and Dale Walters. Published in 1994 by Chapman & Hall, London. ISBN 0 412 54390 7(HB) 0 412 54400 8(PB).

application of chemicals has resulted in the development of resistance in the target insect populations. Higher doses of chemicals, repeated more often, are required to maintain effective control. Ultimately, this strategy is self defeating as levels of resistance increase and the chemical is no longer useful. The cost of chemical agents continues to rise and many countries can no longer afford to pay for these products in scarce foreign currency. Finally, particulary in Western countries, an environmental movement is beginning to turn the tide of public opinion against the use of chemicals. There are demands for organically-grown food, free of artificial chemicals. The use of biological control agents is considered acceptable and highly desirable.

In this chapter, we will outline the properties and use of the principal biocontrol agents which are currently attracting interest. We will also describe the research underway to improve the effectiveness of these products. These efforts involve primarily the use of genetic manipulation techniques. Perversely, the improvement of a natural biopesticide requires the use of another technology which also has its opponents. Critics of gene manipulation argue that it is a dangerous practice and has no place in future agricultural developments. It is hoped that this chapter will dispel some of the myths concerning the technology. At the very least, it should provide valuable information for those who wish to learn more about its use in agriculture.

3.2 ENTOMOPATHOGENIC BACTERIA

There are few pathogenic bacteria of insects which have potential for use as biological pesticides. They are limited to three species of spore formers, i.e. *Bacillus thuringiensis*, *B. sphaericus* and *B. popilliae*. Other bacteria which are pathogenic to insects are known, but they have not been exploited significantly as control agents because of inconsistent virulence, safety or complications in *in vitro* propagation. *B. thuringiensis* has captured the imagination of most workers interested in this field and will constitute the major part of this section. However, the characterization of the insect-specific toxin genes harboured by these organisms has suggested novel ways in which their mode of action might be extended to other bacteria which could be used as biocontrol agents. The transfer of genes from *B. thuringiensis* to crop plants to improve host resistance is dealt with in Chapter 7.

3.2.1 Historical perspective

B. thuringiensis, more commonly known as *B.t.*, was first isolated from diseased larvae of the silkworm, *Bombyx mori*, and flour moth, *Anagasta kuehniella* in the early part of this century (Ishawata, 1901; Berliner, 1911, 1915). The formation of parasporal bodies by the Gram-positive bacteria was also recognized at this time, but their proteinaceous crystal nature was not characterized until much later (Hannay, 1953; Hannay and Fitz-James, 1955). The role of the protein crystal in pathogenicity was identified by Angus (1954).

3.2.2 Mode of action

The *B.t.* bacterium contains a large DNA genome which undoubtedly specifies a multitude of different gene products. The life cycle of the bacterium can be regarded in two stages. There is vegetative cell division, in the manner of most bacterial species, when growth conditions are optimal. Sporulation occurs when environmental conditions change, resulting in low nutrient levels or dessication. The spore is a resting phase, when there is minimal metabolic activity and the organism can survive almost indefinitely. It is during the process of sporulation that the δ-endotoxin is produced, accounting for up to 35% of the cell dry weight as a proteinaceous crystal. The spores germinate in response to improved conditions in the environment and initiate the process of vegetative cell division once more.

The spores enable *B.t.* to persist for long periods in soil. There appears to be little vegetative growth in the soil (Petras and Casida, 1985; Akiba, 1986). There have been isolations of *B.t.* from insects, soil, stored product dusts, deciduous and conifer leaves and aquatic environments. Despite the isolations of *B.t.* from insects, there are very few epizootics in the field (Burges, 1973).

The crystal comprises a protoxin protein with a molecular weight of about 130 kDa. The protoxin is solubilized in the alkaline pH of the larval midgut and subsequently cleaved enzymatically to the active toxin of 60–70 kDa. The toxin diffuses through the peritrophic membrane lining the gut and binds to receptors present in the midgut epithelium. The gut is paralysed and the insect stops feeding (Dulmage *et al.*, 1978; Salama and Sharaby, 1985). The toxin is thought to insert into the membrane with the resultant formation of a pore (Knowles and Ellar, 1986; Ellar, 1989; Hendrick *et al.*, 1990). The pore interferes with the inward potassium ion gradient (Sacchi *et al.*, 1986) and results in swelling of the microvilli (Luthy *et al.*, 1982). These cells eventually lyse. The leakage of ions from the gut to the haemolymph results in ionic imbalance in the haemolymph. The disruption to the gut may also permit invasion of the haemolymph by the bacteria, resulting in septicaemia and death.

The anti-feeding effect of the active toxin is an important feature of *B.t.* pathogenesis. Without this phenomenon, the bacteria would be flushed from the gut by the rapid passage of food.

3.2.3 Examples of *B.t.* used as control agents

The majority of *B.t.* products are based on the subspecies *kurstaki* strain

Table 3.1 Commercial formulations of *B.t.*

Commercial name	*Producer*	B.t. *ssp./strain*	*Target insects*
Dipel	Abbot Kurstaki	HD-1	>100 species
Thuricide	Sandoz Kurstaki	HD-1	>100 species
Biobit	Novo Kurstaki	HD-1	>100 species
Javelin	Sandoz Kurstaki	HD-1	>100 species + *Spodoptera* spp. *Galleria mellonella*

HD-1. This is currently produced by companies in the USA and Western Europe (Thompson, 1989 and Table 3.1). This product is active against 100 lepidopteran species. The advantage of a broad host range product is obvious; more insect pest species can be targeted with a single application. There are other *B.t.* products, such as HRD12 strain of *B.t. kurstaki*, which can be used to complement the HD-1 strains where these are not so effective.

Until the 1980s, most *B.t.* products were used to control forest pests. In North America, these sales accounted for approximately 60% of the world-wide market. In more recent years, however, *B.t.* has been used to control cotton pests.

3.2.4 Problems with *B.t.* as a control agent

Despite the recent sucesses in the use of *B.t.* as an alternative to chemical insecticides, it has not been widely adopted as a pesticide because of a number of disadvantages. These include: host range specificity, ineffectiveness in controlling pests which feed internally or on roots, rapid inactivation on foliage by sunlight and other factors, degradation in the soil by soil microorganisms and low residual acivity in water due to settling of spores and crystals and adsorption to organic particles (Gelerntner and Schwab, 1993). Crops with more than one pest species to control present a particular problem. Several species of insect pests are not susceptible to the HD-1 strain (e.g. *Spodoptera* spp.), or are difficult to target because of their feeding habits (e.g. *Heliothis* spp. on cotton).

These problems contribute to the reluctance of growers of the major crops (cotton, grain, oilseed) to adopt *B.t.* as a weapon in their armoury to combat pests. Despite the impressive predictions of a $300 million market for *B.t.* products by the year 2000; this represents a tiny fraction of the total market for chemical insecticides.

3.2.5 Strategies to improve *B.t.* products

(a) *Direct modification of the* B.t. *bacterium*

Non-recombinant DNA methods may be used to modify the toxin genes harboured by *B.t.* Plasmid curing has been used to eliminate certain toxin genes from the parental organism, followed by their replacement with other toxin genes via conjugal plasmid transfer (Gonzalez and Carlton, 1982; Klier *et al.*, 1983; Carlton *et al.*, 1990). These products have improved activity against certain target insects or a broader host range. The advantage of this approach is that it does not utilize recombinant DNA technology and is not liable to the special regulations which these techniques require.

Recombinant DNA methods do, however, offer more flexibility in the design of improved *B.t.* products. For example, Honee *et al.* (1990) constructed a chaemeric toxin gene containing the N-terminal regions of two different protoxins, *CryIA(b)* with activity against tobacco budworm (*Heliothis virescens*) and the large cabbage white butterfly (*Pieris brassicae*) and *CryIC* with activity against *Spodoptera exigua*. The fusion protein produced in *Escherichia coli* had activity against all three insect species. Electroporation, a method which uses electric currents to introduce DNA into cells, was utilized to insert heterologous toxin genes into *B.t.* strains (Crickmore *et al.*, 1990).

To date, gene manipulation has not resulted in spectacular increases in levels of toxicity of the various strains. The exception is the construction of a hybrid toxin gene between *CryIA(a)* and *CryIA(c)* to produce a protein with 30-fold higher toxicity to *Heliothis virescens* (Ge *et al.*, 1991).

(b) *Transgenic microorganisms containing* B.t. *genes*

The problem of targeting insect pests which feed internally has been addressed by inserting a *B.t.* gene into a bacterium which colonizes the plant vascular system. These microorganisms, otherwise known as endophytes, serve to deliver the toxin to the required site of action. The Crop Genetics International Corporation (USA) inserted a single copy of the *B.t.* kurstaki δ-endotoxin gene *CryIA(c)* into the chromosome of *Clavibacter xyli* ssp. *cynodontis* (InCide®) (Beach, 1990). Corn seed inoculated with *C. xyli* recombinants containing the functional *B.t.* gene produce plants colonized with the bacteria. These plants have been used in small-scale field trials in the US to monitor the control of the European corn borer, *Ostrinia nubilalis*. Moderately effective control of this pest has been attained.

In an attempt to increase the foliar persistence of *B.t.* endotoxins, plasmids containing the *B.t.* endotoxin genes have been inserted into the non-pathogenic, leaf-colonizing *Pseudomonas fluorescens*. The recombinant *P. fluorescens* cells are grown in fermenters. These cells synthesize the endotoxin, which accumulates as a crystal and attains levels of 10–20% of the total cell protein. At the end of the fermentation, the intact bacterial cells are harvested (Barnes and Cummings, 1987). The endotoxin remains protected within the host, unlike natural *B.t.*, where the cells lyse at the end of the replication cycle to release spores and unprotected endotoxin.

This technology has been developed by the Mycogen Corporation (USA) under the trade name, CellCap®. Given the uncertainty of the consequences of releasing live recombinant microorganisms into the environment, the transgenic *P. fluorescens* were inactivated by fixation. This served to cross-link cell wall proteins.

3.2.6 The control of soil-dwelling pest insects

The extension of the use of *B.t.* to the control of soil-dwelling insect pests has been hampered by two major factors. Firstly, δ-endotoxins are rapidly degraded in the soil. Secondly, as the target insects are hidden within the soil, it is difficult to deliver the *B.t.* preparation in an effective manner.

The answer to these problems is analagous to the approach used to deliver *B.t.* endotoxin to leaf borers (see above). Microorganisms have been identified which are closely associated with the roots of crop plants. These bacteria are modified to express the endotoxin genes of *B.t.* and thus deliver the toxin to the sites where it can affect the insect pests. For example, a corn root colonizing recombinant *P. fluorescens* was produced containing the *CryIA(b)* endotoxin gene (Obukowicz *et al.*, 1987). The gene was targeted to the chromosome of the bacterium, to prevent mobilization and transfer to non-target microorganisms. Toxin production was confirmed and demonstrated to have an effect on the larvae of the tobacco hornworm, *Manduca sexta*. The recombinant bacterium was never tested in field trials.

In a similar study, Waalwijk *et al.* (1991) modified a *P. fluorescens* strain by inserting a dipteran-active endotoxin *CryIVB* gene into the chromosome. This recombinant bacterium was designed to affect the soil dwelling larvae of the crane fly, *Tipula oleracea*. This insect pest can cause significant damage to grass. The expression of active toxin was confirmed by performing bioassays using crane fly larvae. Only 10–30% of the test insects were killed by the toxin, indicating a relatively low level of synthesis.

Pigeon pea (*Cajanus cajan* L.) plants have also been partially protected using transgenic bacteria expressing the *B.t.* δ-endotoxin genes. This crop is an important source of protein in the tropics. Nodulation by *Bradyrhizobium* spp. enables the plants to grow in relatively poor soils. Unfortunately, larvae of the the dipteran species, *Rivella angulata*, can reduce yields by feeding on the contents of root nodules of pigeon peas. To counter this behaviour by the insect larvae, recombinant *Bradyrhizobium* were produced containing the *CryIVD* gene, which is active against dipteran species (Nambiar *et al.*, 1990). The use of these modified bacteria reduced *Rivella* infestations on pigeon peas by 40%.

3.2.7 Mechanisms of insect resistance to *B.t.* endotoxins

The application of chemical pesticides to control insect populations has invariably resulted in the development of resistance to those agents. Georghiou and Lagunes (1988) have recorded over 500 species of insects that are resistant to all known classes of chemical insecticides. While the use of *B.t.*-based products offer viable alternatives, it would be irresponsible to assume that the biological nature of the agents render them less likely to induce resistance in the target insects. Users of *B.t.* in agriculture and forestry cannot afford to squander this valuable product by over-use. Although there have been very few examples in the past 30 years of insects becoming resistant to *B.t.* after its use in the field, it must be remembered that it only accounts for about 1% of the world insecticide market.

Kinsinger and McGaughey (1979) discovered a 42-fold increase in resistance of laboratory-reared *Plodia interpunctella* (Indian meal moth) to the *B.t.* product Dipel®, when compared with field strains of the same insect. Later studies, in which populations of insects were selected for resistance to *B.t.*, demonstrated a 27-fold reduction in sensitivity after two generations and a 97-fold reduction after 15 generations (McGaughey, 1985). These studies are often considered to represent an extreme case, because of the continuous exposure of the Indian meal moth to the *B.t.* insecticide. However, the development of resistance to *B.t.* in a pest in the field has been recorded by Tabashnik *et al.* (1990). The larvae of *P. xylostella* (the diamondback moth), feeding on watercress, were treated with *B.t.* Dipel® over a 4 year period. The treatment with *B.t.* was then suspended for 7 years before a resumption of the programme with *B.t.* Javelin®. Those areas which had been heavily treated with the *B.t.* products yielded insects which were 2 to 33 times less susceptible. Furthermore, the mortality recorded in the field was only about 30%, whereas in laboratory populations of insects 90–100% mortality was achieved. Interestingly, however, the resistance of field populations of insects to *B.t.* declined when treatment with the insecticide was stopped.

3.3 ENTOMOPATHOGENIC FUNGI

3.3.1 Historical perspective

The role of entomopathogenic fungi in causing disease in insects was first recognized in the early part of the 19th century from studies on silkworms. Metschnikoff (1879) and Krassilstchik (1888) produced the fungus, *Metarhizium anisopliae*, to control the wheat cockchafer, *Anisoplia austriacea*, and the sugarbeet curculio, *Cleonus punctiventris*. Despite the early enthusiasm for the technology, the introduction of cheap and effective chemical insecticides reduced interest in the use of pathogenic fungi in much the same way as for bacterial and viral biocontrol agents. In recent years there has been a resurgence of exploration of their potential, prompted by the realization that chemical insecticides pose problems for the environment and are expensive to use. Entomopathogenic fungi are unlikely to capture a major part of the pesticide market, but they do have a future in specialized applications and in integrated approaches to insect pest control. The rapid developments in the field of molecular biology have also improved the prospects for deriving more effective fungal insecticides. Although it is more difficult to manipulate filamentous entomopathogenic fungi, recent advances suggest that it is feasible and should permit the construction of genetically altered strains. In the short term, these may not be economically viable fungal insecticides, but they will enable important analyses of pathogenicity and virulence factors to be conducted.

3.3.2 Mode of action

In contrast to the oral route of infection by which entomopathogenic bacteria and viruses enter the host insect, entomopathogenic fungi are able to penetrate directly the outer integument. The insect cuticle contains chitin fibrils within a protein matrix together with lipids and waxes and small quantities of phenols, inorganics and pigments. The ability to effect penetration of the cuticle is crucial; fungi lacking the necessary mechanisms are unable to cause disease in insects. The process involves both enzymatic and physical mechanisms.

Deuteromycete fungi produce conidia (spores) which are distributed randomly by wind and water. These structures, if they contact the insect integument, must adhere and remain in contact with the surface to permit germination and subsequent invasion (Charnley, 1984). Non-specific hydrophobic forces probably have a major role in the adhesion of conidia. Other mechanisms involving specific glycoproteins and enzymes may also be significant. This is suggested by a report that some strains of fungal species produce conidia which only bind to certain insect species (Fargues, 1984). After binding to the insect cuticle, entry is effected by a variety of processes. The conidium has to germinate and produce hyphae which are able to penetrate the outer layer of the insect. This may involve the use of cuticle degrading enzymes or simply be accomplished by mechanical force. A variety of enzymes may be produced to assist this process, including; proteases, chitinases and lipases (Charnley and St Leger, 1991).

Once the fungal pathogen has penetrated the insect cuticle, it proceeds to spread throughout the insect host, developing a network of hyphal structures. The insect is eventually overcome by these structures and dies. In the terminal stages of infection, the fungus procuces more conidia, which are released to spread infection to other hosts.

Other factors which may have a role in the demise of the insect hosts are a range of low molecular weight insecticidal toxins produced by entomopathogenic fungi. These include beauvericin and the destruxins. The synthesis of these products has been demonstrated *in vitro*. Beauvericin is produced by *Beauveria bassiana* and is a hexadepsipeptide comprising three repeat units of D-2-hydroxyisovaleric acid linked to N-methyl phenylalanine. The destruxins have been implicated in fungal pathogenesis by injection into Lepidopteran larvae and adult Diptera and demonstrating paralysis and death (Samuels *et al.*, 1988). Furthermore, lethal amounts of destruxins have been isolated from *Metarhizium anisopliae*-infected insects (Suzuki *et al.*, 1971).

3.3.3 Examples of entomopathogenic fungi used as control agents

Like entomopathogenic viruses and bacteria, entomopathogenic fungi have a worldwide distribution. There are several hundred species currently identified, belonging to approximately 100 genera. However, only about 20 species of fungi have been used for the control of insect pests. The Deuteromycetes contain the largest number of entomopathogenic species. *Beauveria bassiana, B. brongniartii* and *Metarhizium anisopliae* are the most well known examples from this group. They have considerable potential as biological pesticides.

The entomopathogenic fungi *B. bassiana* is used in the countries that used to comprise the USSR and in China. The species is produced on a large scale for the control of a variety of pest insects, including *Leptinotarsa decemlineata* and *Cydia pomonella* (Ferron, 1981), *Ostrinia nubilalis*, pine caterpillars and leaf hoppers (Franz and Krieg, 1980; Hussey and Tinsley, 1981). In China, the fungus is frequently used in conjunction with low doses of chemical insecticides. In Europe, *B. brongniartii* has been used to control the cockchafer, *Melolontha melolontha* (Ferron, 1978; Keller, 1983). Soil was treated with conidia or adults were sprayed with blastospores. These treatments resulted in the disease becoming established in the population with subsequent fatalities in succeeding generations. These results emphasize the long-term nature of the control and the lack of a quick 'knockdown' as is seen in the use of chemical insecticides.

Metarhizium anisopliae is an extremely promising agent for the control of a number of insect pest species. In parallel with the success of baculoviruses against the velvet bean caterpillar, it has been used in Brazil to control insects such as *Mahanarva posticata* (Ferron, 1981), a pest of pasture and sugar cane. The fungus has been used over several thousand hectares and the treatment of crops was so successful that the use of chemical agents declined to 18% of the area that needed protection from the insect pests (Risco, 1980).

In the Pacific Islands and South-East Asia, the fungus has also been used to control the rhinoceros beetle, *Oryctes rhinoceros*, a major pest of palms. The product has been used in conjunction with the baculovirus, *O. rhinoceros* (Bedford, 1980).

In addition to surface insect pests, *M. anisopliae* has been used to control insects within the soil itself. Examples include, the pasteur cockchafer, *Aphodius tasmaniae* in Australia (Coles and Pinnock, 1982) and the black vine weevil, *Otiorhynchus sulcatus* in Europe (Zimmermann, 1982, 1986). The fungus is considered to be as effective as the use of chemicals. Other field tests in Australia have exploited this fungal species against termites (Hanel and Watson, 1983).

Verticillum lecanii is used in Europe for the control of aphids and whiteflies in greenhouses. The incubation period for the disease is about 10 days, so the pathogen must be introduced into the pest populations at an early stage, prior to significant damage to the crop. The fungus, once established, continues to give effective control of the pest for several weeks. It may be used to control aphids on food crops, such as cucumbers and egg plants (Hall, 1981; Hall and Papierok, 1982). Conidia and blastospores may be used with equal effectiveness but the latter are not so stable (Hall, 1981).

3.3.4 Problems with entomopathogenic fungi as control agents

In common with bacteria and viruses, the effectiveness of fungal biocontrol agents is determined by environmental conditions during and after pesticide application.

The mean temperature for fungi is normally between 20 and 25°C. For most relevant entomopathogenic fungi species the optimal temperature is between 15 and 25°C. Greenhouse temperatures above 35°C are often a limiting factor for fungi in the vegetation period. Zimmermann (1982) demonstrated that the resistance of *M. anisopliae* to higher temperatures increased as the relative humidity increased.

Relative humidity is often attributed the greatest influence on the potential of entomopathogenic fungi. Moist conditions favour the development of germinating condia. For example, *M. anisopliae* conidia cease germinating in relative humidities of <94% (Walstad *et al.*, 1970). These data suggest that unless relative humidity is high, entomopathogenic fungi may be ineffective in controlling insect pests.

UV light has a major role in determining the infectious half-life of fungal conidia, but no more so than for the entomopathogenic bacteria or viruses. In experiments to test the resistance of fungal, virus and bacterial pathogens, the conidia of *B. bassiana* were most resistant to UV light (Zimmermann, 1986).

Finally, the slow speed of kill of entomopathogenic fungi remains a critical factor in their future success as biological control agents. After the initial invasion of the insect host by the fungi, the pest continues to feed and cause damage. Unless this lag period in the fungal infection can be reduced, farmers are unlikely to select these agents in preference to chemicals.

3.3.5 Strategies to improve entomopathogenic fungal insecticides

There are two principal approaches to improving the effectiveness of entomopathogenic fungi. The first involves sexual and parasexual recombination, to exchange genetic traits favourable for insect control, between fungal strains. In the second method, it is proposed to use genetic engineering to modify candidate fungal entomopathogens.

Hybridization via the sexual cycle may be applicable for some fungi used as biocontrol agents, but *M. anisopliae* and *B. bassiana* do not have sexual stages. Recombination through the parasexual cycle has, however, been recorded. In the parasexual cycle, transient heterkaryons are formed by hyphal anastomosis or protoplast fusions. Thereafter, rare nuclear fusions produce a heterozygous diploid nucleus. The diploid nuclei are unstable, breaking down to a stable haploid state by chromosome loss during mitotic divisions. Genetic recombination occurs by mitotic crossing over between homologous chromosomes in diploid or aneuploid nuclei and by random segregation of chromosomes during haploidization. Parasexual hybridization between pathogenic *M. anisopliae* strains derived mutants with differing sporulation rates and pathogenicity towards *Nilaparvata lugens* (Heale *et al.*, 1989). Vegetative incompatibility restricts genetic exchange between unrelated fungal strains. The incompatibility may be overcome by utilizing the technique of protoplast fusion (Jackson and Heale, 1987).

The use of genetic engineering requires techniques for the introduction of exogenous DNA into the fungus genome. DNA-mediated transformation procedures for filamentous fungi are based on the addition of DNA to protoplasts in the presence of Ca^{2+} and polyethylene glycol. The frequency of transformation is low (0.1–10 transformants per microgram DNA). If the transforming DNA is capable of autonomous replication, transformation efficiencies of 100–10 000 transformants per microgram may be obtained. Benomyl resistance has been conferred on *M. anisopliae* by genetic transformation (Bernier *et al.*, 1989). The transformants were stable after removal of the selection pressure. Analysis of genomic DNA from the modified fungi revealed that most transformants contained tandemly repeated vector sequences inserted randomly in the chromosomes.

In the longer term, the use of genetic engineering offers the opportunity to introduce genes encoding potential pathogenicity factors, such as cuticle penetrating enzymes, into alternative fungal strains. For the present, there is a paucity of information as to the roles of enzymes or toxins in the gross pathology of fungal infection. To define these roles will require the development of a range of fungal mutants lacking one or more of the genes encoding pathogenicity factors. Since most deuteromycetes do not have a sexual stage, it is not possible to conduct conventional genetic analysis. Fortunately, the development of fungal transformation techniques for *M.*

anisopliae (Goettel *et al.*, 1990) will permit the isolation of many putative virulence genes. These may be altered *in vitro* before reintroduction into the fungal genome to monitor the effect.

3.4 ENTOMOPATHOGENIC VIRUSES

It has been demonstrated that insects are susceptible to a wide variety of virus infections (reviewed by King *et al.*, 1994). Viruses, unlike bacteria, are obligate parasites and depend on the availability of a permissive host cell for continued propagation. Many viruses are currently identified in insect hosts. These include: baculoviruses (large, covalently closed, circular DNA genomes); poxviruses (large, linear, covalently closed DNA genome); cytoplasmic polyhedrosis viruses (segmented, double-stranded RNA genome); picornaviruses (small, single-stranded RNA genome). It is obvious that many of the examples cited are related to similar viruses which cause disease in man and other mammals. The Baculoviridae are the one of the few virus families which are confined to invertebrate species. This makes them particularly useful as insect biocontrol agents, since the risk of cross-infection in mammals is considered to be vanishingly small. The advent of molecular biology has also made it relatively easy to manipulate the genomes of these viruses. In consequence, baculoviruses have been the main focus of recent efforts to develop cost-effective virus insecticides and will be the major topic of discussion in this section.

3.5 BACULOVIRUSES

3.5.1 Historical perspective

The symptoms of baculovirus infections in insects were first recognized in silkworms in the 16th century, although the cause of the disease was unknown at that time. In the 19th century the presence of refractile crystals or viral occlusion bodies was associated with the infection. Later Bergold (1947) provided definitive evidence of the viral nature of the disease. Many subsequent studies were recorded which demonstrated the potential of baculoviruses for the control of a variety of insect pathogens (Ignoffo, 1973). Unfortunately, as with many other biocontrol agents, the advent of relatively cheap, synthetic pesticides sounded the death knell for the future of baculoviruses as cost-effective products. Baculoviruses have been isolated from over 600 species of insects, including Lepidopteran, Hymenopteran and Dipteran orders.

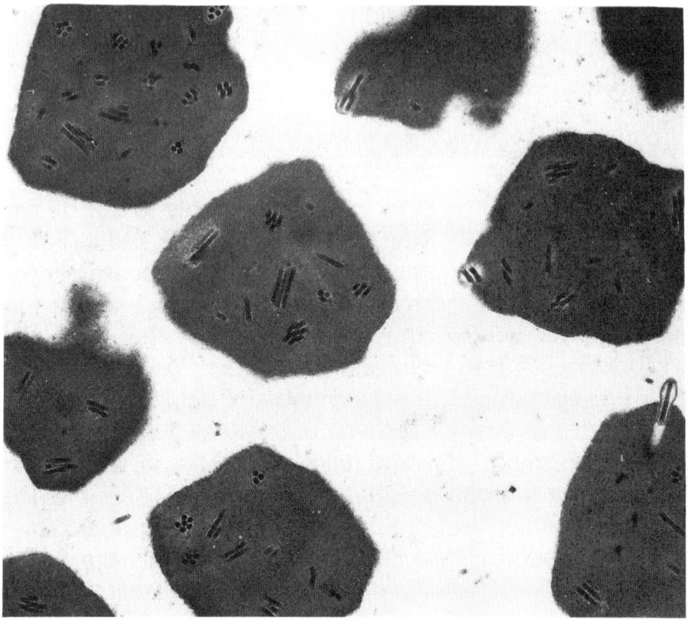

Figure 3.1 Transmission electron micrograph of polyhedra of AcNPV. Note the proteinaceous occlusion bodies containing bundles of nucleocapsids packaged within the virus particles. A polyhedron is about 1 μm in diameter.

3.5.2 Mode of action

Baculoviruses have a large, double-stranded, circular DNA genome (about 80–200 kilobase pairs). The DNA is packaged within a rod-shaped nucleocapsid which is further enclosed within a lipoprotein envelope and finally occluded by a crystalline matrix (occlusion body) largely comprising a single protein, polyhedrin or granulin (28 kDa), which serve to protect the virus in the environment (Figure 3.1). The Baculoviridae is subdivided into the Eubaculovirinae, including the nuclear polyhedrosis viruses (NPV) and granulosis viruses (GV) which occlude virus particles, and the Nudibaculovirinae including the non-occluded baculoviruses. The Nudibaculovirinae do not produce the viral inclusion bodies; rod-shaped virus particles are only protected by the lipoprotein envelope.

The peculiar structure of the members of the Eubaculovirinae is reflected in their mode of replication in insect larvae and individual cells. Occlusion bodies (polyhedra for NPVs and granules for GVs) are ingested by insect larvae. In the highly alkaline pH of the midgut, the occlusion body protein dissolves and is further degraded by host alkaline proteases. The virus particles are released from polyhedra and subsequently negotiate the peritrophic membrane lining the midgut. Thereafter, the lipoprotein membrane surrounding the virus fuses with plasma membrane of the gut wall cells and liberates nucleocapsids into the cytoplasm. The nucleocapsids transport virus DNA into the nucleus of the cell and virus gene expression begins.

Most studies on the molecular events occurring in baculovirus replication have been performed with the NPVs, and in particular, the *Autographa californica* (Ac) NPV. Virus genes are expressed in three temporally distinct phases; early, late and very late. In the early phase, virus gene products associated with the transactivation of other virus genes and virus DNA replication are produced. At 6 h post-infection, late gene expression begins and accounts for the production of various structural proteins associated with the formation of progeny virus nucleocapsids and particles. Late gene expression is also accompanied by the accumulation of an α-amanitin-resistant RNA polymerase in the virus-infected cells. By 12 h post-infection, progeny virus nucleocapsids emerge from the nucleus and bud through the plasma membrane of the infected cell, thus acquiring the lipoprotein envelope. These mature virus particles are infectious and proceed to invade other susceptible cells within the insect host. Virus replication, however, continues in the primary infected cells and enters the very late phase of virus gene expression at about 18 h post-infection. In this phase, the polyhedrin and p10 proteins are produced to levels which can account for 30–50% of the total virus-infected cell protein content. The polyhedrin protein occludes virus particles retained in the nucleus after the budding phase. The p10 protein appears to have some role in the process of occlusion body formation and in the lysis of virus-infected cells which aids release of the virus.

The replication of occluded baculoviruses can be considered a biphasic process. In the first phase, infectious virus particles bud from the cell and spread the infection throughout the insect host. While this process continues, occlusion bodies containing virus particles are formed in the nuclei of infected cells. In the latter stages of the infection of the insect, the host becomes packed with virus occlusion bodies. These structures are released from the larva when the integument ruptures. A milky suspension of virus is liberated and contaminates plant leaf surfaces which may be consumed by other insects.

Virus replication of occluded baculoviruses is only significant in the larval stage of insect development. As the larvae represent the most active feeding stage of the insect's life cycle, they are most likely to encounter infectious virus. Occlusion bodies cannot infect via the intact insect integument.

Non-occluded baculoviruses such as *Oryctes rhinoceros* NOV infect the rhinoceros beetle. In this host, virus replication can occur in both the larval and adult stages. Since the gut of the beetle is close to neutrality, occlusion bodies would not be dissolved. Naked virus particles are spread from host to host as the beetle voids virus with faecal material. The intra-cellular processes of virus replication, apart from the absence of the very late phase of virus gene expression, are the same as in the occluded viruses.

3.5.3 Examples of baculoviruses as biocontrol agents

There have been numerous examples of the use of baculoviruses to control insect pests. One of the most successful has been the application of *Anticarsia gemmatalis* NPV in Brazil, to control the velvet bean caterpillar (*A. gemmatalis*), a major pest of soybean (Moscardi and Sosa-Gomez, 1992). This programme has evolved so that over 1 million hectares of soybean are now sprayed annually with the virus. A further major success has been the control of the palm rhinoceros beetle (*Oryctes rhinoceros*) in the Pacific region with the non-occluded homologous baculovirus (Bedford, 1981; Zelanzy *et al.*, 1990). The control of forest pests has been particulary successful, primarily because the crop can tolerate a degree of damage before the insect succumbs to virus-infection. The pine sawfly (*Neodiprion sertifer*) is controlled with the *N. sertifer* (Ns) NPV (Doyle and Entwistle, 1988). In a similar manner, viruses from the Douglas fir tussock moth (*Orgyia pseudosugata*) (Otvos *et al.*, 1987), the gypsy moth (*Lymantria dispar*) (Lewis, 1981), the pine beauty moth (*Panolis flammea*) (Cory and Entwistle, 1990) and the red headed pine sawfly (*Neodiprion leconti*) (Wood and Granados, 1991) have also been used. Baculoviruses have been used in experimental field trials to control the cotton bollworm (*Heliothis zea*) in the USA. Their application has been limited, because of the competition from chemical agents. In China, however, larger areas of cotton crops have been sprayed with the *H. zea* NPV (Guangyu, 1989).

3.5.4 Problems with baculoviruses as biocontrol agents

Two major features of baculoviruses have limited their development as commerically viable biocontrol agents. The first concerns the relatively narrow host range of each baculovirus isolate. While this is an excellent safety factor, it does mean that a variety of baculovirus products must be used to control multiple insect pest species. This increases production costs to much higher levels than for chemical agents. The production of baculovirus insecticides is already complicated by the need to use insect larvae in large numbers. The maintenance of virus-free insects is not a facile matter and populations can crash at the most inconvenient times. The use of cell culture is increasingly being considered as an alternative, but synthetic media is also expensive. There are also concerns that the long-term propagation of baculoviruses in cell culture can reduce viability in the natural larval hosts via virus genome deletions and acquisition of host cell DNA.

The second limitation to the wider acceptance of baculoviruses concerns the time taken by the virus to kill the target insect after initial infection. Chemical agents are renowned for very rapid 'knock-down' in a matter of hours, while baculoviruses can take many days to kill the host. It is this last issue which will be considered in the remainder of this section.

3.5.5 Strategies to improve baculoviruses insecticides

The approach to the improvement of baculovirus insecticides favoured by most workers has been to insert genes encoding insect-specific hormones, enzymes or toxins into the virus genome. The expectation is that the synthesis of the foreign gene product in the virus-infected insect will reduce the time taken to kill the insect larvae or at least reduce the amount of feeding damage. The use of genes encoding hormones and enzymes, particularly when taken from the target pest insect species, offers the advantage that the virus will only be producing, albeit at a higher level, the same proteins that the insect larvae normally synthesizes during the natural course of development. The chances of the insect becoming resistant to its own gene products are considered remote, although as outlined below, it may not be quite so easy to overcome the tightly-regulated developmental processes. Furthermore, the use of insect genes present less of a problem in assessing the safety of these recombinant viruses, since the genes are not 'foreign' to the target insect.

The use of genes encoding toxins is particularly attractive since many of these proteins are very fast-acting in their native form and may be expected to have a similar effect when produced by the baculovirus. Conversely, there is probably a greater risk of the insect developing resistance to the foreign toxin. We consider this risk to be small, given that the insect would also have to become resistant to the effects of the virus infection.

It is only in recent years that it has been feasible to contemplate inserting foreign DNA into the baculovirus genome. The techniques of genetic manipulation documented elsewhere in this book and in other publications have been applied with as much vigour to baculovirology as to prokaryotic and other eukaryotic systems. The methodology involved in altering the baculovirus genome is outlined in Figure 3.2. Essentially, the approach depends on placing a copy of a foreign coding region under the transcriptional control of a duplicate of a strong virus gene promoter, such as polyhedrin or p10.

A particular problem encountered when performing the genetic manipulations with baculoviruses is the large size of the genome (e.g. 133 000 nucleotides for AcNPV). It is not possible to insert foreign DNA directly into a unique restriction enzyme site within the virus genome, as is readily achieved with bacterial plasmids. In order to circumvent this problem, a portion of the baculovirus genome is inserted into a bacterial plasmid, such as one of the pUC series and manipulated to achieve the desired genetic organization. Figure 3.2 shows that the portion of the virus genome spanning the polyhedrin gene is inserted in pUC8 and then tailored to produce a plasmid with a copy of the foreign gene coding region inserted upstream

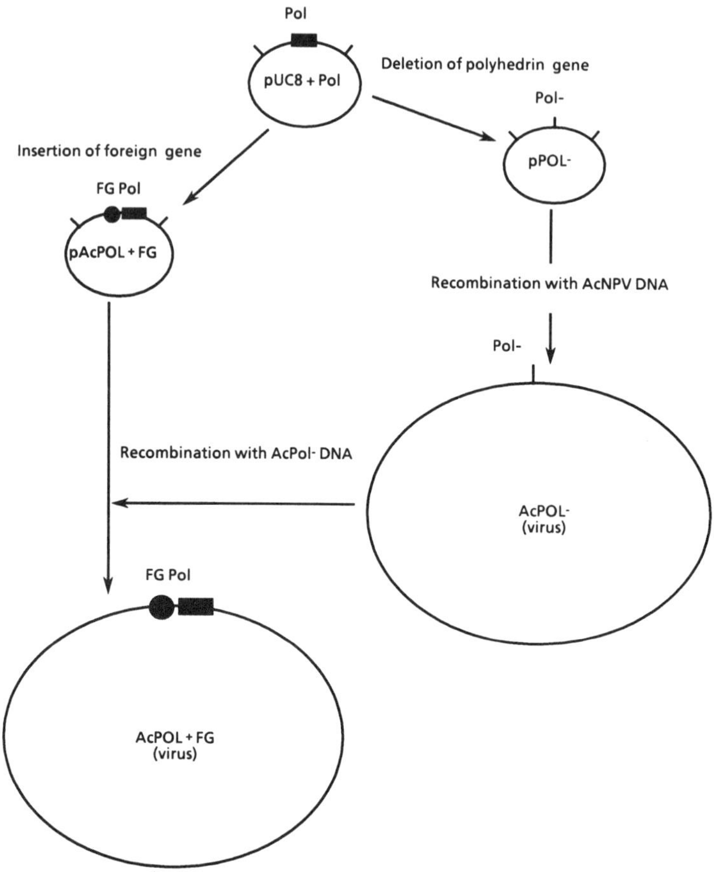

Figure 3.2 Construction of recombinant baculovirus insecticides. A portion of the virus genome spanning the polyhedrin gene is inserted into a plasmid vector such as pUC8. The polyhedrin coding region is removed from this plasmid to create pPOL–. The plasmid is co-transfected into insect cells with infectious baculovirus DNA. The homologous regions in each DNA recombine, replacing the polyhedrin gene in the virus with the defective polyhedrin gene from the plasmid to derive AcPOL–. Infectious DNA from this virus is co-transfected with pAcPOL+FG, containing a foreign gene (FG) inserted upstream of the intact polyhedrin gene, to produce AcPOL+FG.

of the polyhedrin gene. In other procedures, a similar plasmid is manipulated to remove the polyhedrin gene. This plasmid (pPOL–) is co-transfected with infectious virus DNA into insect cells to produce a virus infection. The homology between the baculovirus DNA in pPOL– results in recombination with the same sequences in the intact virus genome. The functional polyhedrin gene is replaced with the defective region of the virus genome from pPOL–. Although this event only occurs at a frequency of 2%, the polyhedrin-negative phenotype may be selected in a simple plaque assay, against a background of plaques that contain polyhedra. Viruses lacking the polyhedrin gene can still replicate efficiently in insect cells. After amplification of AcPOL–, the DNA from this virus is co-transfected with the recombinant plasmid, pAcPOL+.FG, to restore the polyhedrin gene and the foreign gene coding region into the baculovirus genome. This apparently circuitous route to making a recombinant baculovirus avoids the difficulties of selecting a polyhedrin-positive virus containing a foreign gene from a parental polyhedrin-positive background!

It is obvious that polyhedrin-negative viruses containing a foreign gene *in lieu* of the polyhedrin gene can also be readily produced. These will be discussed further below.

3.5.6 Baculoviruses producing insecticidal foreign gene products

The control of water balance in insects is influenced by the action of diuretic and anti-diuretic hormones (Maddrell, 1986). It has been proposed that the balance could be disrupted if a recombinant baculovirus produced elevated levels of either hormone in infected larvae. A diuretic hormone (DH) has been isolated from *Manduca sexta* (tobacco hornworm) and the protein sequence determined (Kataoka *et al.*, 1989). The hormone consists of a 41 amino acid protein, which is amidated at the C-terminus. This protein sequence was used to design and construct an appropriate synthetic oligonucleotide. A secretion peptide coding sequence from the *Drosophila melanogaster* (fruit fly) cuticle protein was added to the 5′ end of the DH coding region to facilitate the secretion of the hormone from virus-infected cells. This synthetic gene coding region was introduced into the *Bombyx mori* (silk worm) NPV genome, in place of the polyhedrin gene coding region, under the control of the polyhedrin gene promoter, thus producing a polyhedrin-negative virus (Maeda, 1989). The virus was propagated in *B. mori* cells in culture and behaved in a similar way to the wild-type, unmodified virus. When insect larvae were injected with the recombinant, polyhedrin-negative virus a decrease in haemolymph volume was noted during collection of samples. The reduction in haemolymph volume was estimated to be about 30% compared with mock-infected larvae or larvae infected with wild-type virus. Furthermore, insect larvae infected with the recombinant virus died 1 day earlier than larvae infected with the unmodified virus. The production of the DH in insect larvae was infered from analysis of mRNA from virus-infected insects. Measuring the levels of protein production in infected insects was complicated by the small size of the expected hormone. A reverse-phase column was used to isolate the DH, which was assayed in newly emerged adults of *Pieris rapae* (small white butterfly) (Kataoka *et al.*, 1989). However, direct detection of the hormone was not accomplished by standard techniques, such as protein gel electrophoresis.

The processes that regulate the metamorphosis of insect larvae into pupae are another suitable target for recombinant baculoviruses. In the last larval instar there is a decrease in the titre of juvenile hormone (JH) which has been shown to initiate metamorphosis and terminate feeding (deKort and Granger, 1981). This reduction is accompanied by a substantial increase in the levels of juvenile hormone esterase (JHE). The JHE hydrolyses the chemically-stable, conjugated methyl ester of JH to the JH acid (Hammock, 1985). Inhibition of the JHE results in the production of overgrown insect larvae, because the JH level remains sufficiently high to keep the insect in the feeding stage (Sparks and Hammock, 1980). Conversely, it was proposed that if sufficient JHE was produced early in the insect developmental stages, then JH would be inhibited, resulting in the abolition of feeding and possible premature pupation (Hammock *et al.*, 1990). The JHE from *Heliothis virescens* (tobacco budworm) was isolated by affinity purification and then partially sequenced (Abdel-Aal and Hammock, 1986; Hanzlik *et al.*, 1989). This information provided the basis for designing oligonucleotide probes which were employed to isolate clones from a complementary DNA library made from fat body mRNA (Hanzlik and Hammock, 1987). Thereafter, a full-length clone, encoding JHE, was inserted into a baculovirus transfer vector, utilizing the AcNPV polyhedrin promoter, and this was used to derive a polyhedrin-negative baculovirus (Hammock *et al.*, 1990). The recombinant virus, AcRP23.JHE, produced JHE in insect cell culture and in *Trichoplusia ni* (cabbage looper) larvae. When the virus was fed to first instar *T. ni* larvae a reduction in feeding activity was measured, in comparison with untreated control larvae or larvae treated with a virus lacking the JHE coding sequences. The levels of JHE measured in the haemolymph of the AcRP23.JHE-infected insects matched those normally recorded at ecdysis, but were only 10% of those normally seen in the last larval instar. The reduction in feeding was only observed when the recombinant virus was fed to first instar larvae. Later larval instars did not demonstrate a reduction in the rate of weight gain. Several explanations for this observation have been advanced. First, in later larval stages the levels of JHE produced may be unable to overcome hormone biosynthesis. Secondly, the production of a virus gene-encoded ecdysteroid UDP-glucosyl transferase (O'Reilly and Miller, 1989) may reduce the effects of JHE. Finally, the JHE is extremely unstable *in vivo* when produced by the recombinant virus or in its natural form.

The lack of effect on the later larval instars would appear to reduce the prospects for the successful use of this enzyme in a recombinant baculovirus insecticide. It is not likely that insect pests could consistently be targeted at the neonate stage. However, it is encouraging that an insect gene can be used to modify the efficacy of a baculovirus insecticide. Undoubtedly, progress will be made to increase the levels of JHE attained in virus-infected insects and to improve the stability of the enzyme *in vivo.* Mutation of the JHE coding region may offer a suitable approach.

A gene which is particularly attractive for expression in a baculovirus insecticide is the *B.t.* δ-endotoxin (see p. 70).The natural bacterial product has been used as an insecticide (Luthy *et al.*, 1982; Cunningham, 1988) and thus is already accepted as a safe product for use in the environment.

Two studies have reported the insertion of *B.t.* δ-endotoxin genes into the AcNPV genome (Martens *et al.*, 1990; Merryweather *et al.*, 1990) with essentially the same results recorded in both reports. A full-length copy of the endotoxin coding sequences was inserted into the baculovirus genome in place of the polyhedrin gene-coding region, to derive polyhedrin-negative viruses which produced the endotoxin. Martens *et al.* (1990) demonstrated that the protein produced in insect cells formed large crystals in the cytoplasm. These crystals had an ultrastructure similar to that of the native bacterial toxin. When insect larvae were fed recombinant virus-infected cell extracts there was an immediate reduction in feeding activity and the larvae eventually died. This mimicked the natural effect of the endotoxin.

Merryweather *et al.* (1990) also produced a polyhedrin-positive virus by inserting the endotoxin coding sequences upstream of the polyhedrin gene, under the control of a copy of the AcNPV p10 promoter. This facilitated the removal of active endotoxin from virus preparations and thus facilitated standard bioassays to be performed to assess the efficacy of the virus. However, when insects were given purified polyhedra in a bioassay, there was no improvement in the effectiveness of the virus insecticide. This was attributed to the fact that only the full-length protoxin was produced in insect cells. This must be cleaved to derive biologically-active toxin. Since this process normally occurs in the gut of insect larvae, it must be concluded that the protoxin produced by the virus does not enter the gut in the course of the infection. Furthermore, the baculovirus-derived toxin was only detected as an intracellular protein. To have a biological effect it must be exported from the cell. Future studies should address the question as to whether the addition of a suitable signal peptide sequence to the toxin would facilitate secretion of the recombinant product from the cell.

The use of an insect-specific toxin, such as the *B.t.* δ-endotoxin, which acts in the gut of the target insect appears to offer poor prospects for improving the efficacy of baculovirus insecticides at the present time. A more attractive proposition is the use of insect-specific neurotoxins. Two reports have recently been published which describe the production of a mite neurotoxin gene (Tomalski and Miller, 1991) and a scorpion neurotoxin gene (Stewart *et al.*, 1991) by recombinant baculoviruses in infected insects. Both examples clearly demonstrated biological activity of the neurotoxins and an improvement in the effectiveness of the baculovirus insecticide.

The North African (Algerian) scorpion, *Androctonus australis*, produces a venom containing an *insect-specific* neurotoxin (Zlotkin *et al.*, 1971). The neurotoxin affects the sodium conductance of neurons, producing a presynaptic excitatory effect which leads to paralysis and death (Walther *et al.*,

1976; Teitelbaum *et al.*, 1979). In the study reported by Stewart *et al.* (1991) a synthetic oligonucleotide was constructed, based on the protein sequence of the scorpion neurotoxin determined by Darbon *et al.* (1982). A copy of the AcNPV glycoprotein (gp) 67 signal peptide coding sequence (Whitford *et al.*, 1989) was added to the 5′ end of the coding region, to facilitate secretion of the neurotoxin by insect cells. This construction was inserted into the AcNPV genome, upstream of the polyhedrin gene, to derive a occlusion-positive virus. After infection of cells in culture with recombinant virus, a secreted form of the toxin was detected by injection of *Musca domestica* adults (flies) with toxin purified from cell culture medium by HPLC. The recombinant virus was assessed in bioassays with *T. ni* to determine the LD_{50} and ST_{50} values (LD_{50} is the dose required to give 50% kill, while ST_{50} is the time taken to achieve 50% kill). There was a small, but significant reduction in the LD_{50}, and a 25% reduction in the ST_{50}. When further tests were performed to assess the effect of the virus on the feeding behaviour of insect larvae, it was calculated that a 50% reduction in damage to cabbage leaves was achieved. If this result could be repeated in the field the recombinant baculovirus containing the scorpion toxin gene would offer a real advantage over the unmodified AcNPV. This remains to be tested.

In a similar study to that described by Stewart *et al.* (1991), Tomalski and Miller (1991) inserted a mite (*Pyemotes tritici*) neurotoxin gene (*TxP-1*) into the AcNPV genome. The toxin gene was isolated from a cDNA expression library produced with mite mRNA and identified with a polyclonal anti-TxP-1 serum. The full-length cDNA clone was sequenced to determine that the toxin probably consists of 291 amino acids with a molecular mass of 33 kDa. The toxin gene was inserted in place of the polyhedrin gene coding region, but under the control of the polyhedrin gene promoter. Proteins specific for the toxin were identified in cell lysates and the culture medium of cells infected with the recombinant virus. When fifth instar *T. ni* larvae were injected with the non-occluded virus the insects became paralysed two days after infection. Larvae injected with the unmodified AcNPV remained active. An occluded form of the virus was also produced, with the polyhedrin gene function retained to facilitate occlusion body formation. This virus, or AcNPV was fed to neonate *T. ni* and the insects monitored for signs of paralysis. At 3 days post-infection 60% of the larvae fed with the recombinant virus were paralysed or dead. In comparison, about 5% of the insects fed AcNPV showed similar symptoms. By day 4 the relative proportions were about 82 and 10%. All insects, in both groups, were dead by day 6. These results are very encouraging, but it has yet to be demonstrated that the virus can reduce the rate of larval feeding, although this may be infered since the insects are paralysed by the action of the toxin.

3.5.7 Field releases of genetically modified baculoviruses

The first field release of a genetically modified baculovirus was performed in Oxford, UK in 1986. A synthetic oligonucleotide marker was inserted within a non-essential region of the AcNPV genome to produce a polyhedrin-positive, genetically marked virus. Risk assessment analyses prior to the field experiment included examining the effect of the change on host range and physical stability of the virus. It was concluded that the genetic modification caused no changes in the phenotype of the baculovirus. The results from these tests were submitted to the appropriate regulatory authorities (see Bishop *et al.*, 1988 for full details) and permission was granted for a field trial. The marked AcNPV was released to the environment by feeding polyhedra to susceptible *Spodoptera exigua* (small mottled willow) larvae and placing these insects on sugar beet plants within an insect-proof enclosure. Uninfected insects were placed in a similar plot to serve as a control. The virus-infected larvae died within a 7 day period. Thereafter, marked virus was recovered from the soil in the enclosure until the site was disinfected with formalin 6 months later. Infectious virus could no longer be recovered from the site after this treatment. The trial demonstrated that it was possible to release a genetically altered virus into the environment, monitor its survival over a period of months and ultimately inactivate residual virus at the release site.

There are concerns that once a genetically modified virus is released into the environment, it will persist indefinitely. One approach to avoiding this problem is to render the virus insecticide less able to survive in the field by altering the structural integrity of the protective occlusion body. Therefore, in 1987, a second field trial utilized a baculovirus lacking the polyhedrin gene. The study was intended to determine whether the persistence of a baculovirus insecticide in the environment could be limited by removal of the protective polyhedrin protein from the virus particles. The virus was released after infecting *S. exigua* larvae in the laboratory with non-occluded virus and then placing them on the sugar beet plants in the enclosed facility used for the 1986 trial. After 1 week all of the larvae had died. Two weeks post-release, infectious virus could not be identified on the plant surfaces or in soil samples. This showed that the baculovirus insecticide could be modified to reduce its persistence in the environment.

In 1988 and 1989 two further field-release experiments were conducted using a recombinant AcNPV, containing a copy of the *Escherichia coli* β-galactosidase (*lac*Z) coding region, under the control of the polyhedrin promoter. The aim of these studies was to demonstrate that a polyhedrin-negative virus could be used to produce a foreign gene product in insect larvae in the field.

The recombinant virus was constructed by standard techniques, to insert a copy of the *E. coli lac*Z coding region *in lieu* of the polyhedrin gene coding sequences. Essentially, the virus was identical to the recombinant, non-occluded baculovirus used in the 1987 field study, with the addition of the foreign DNA sequences. It was expected to behave in the same way in the field-release experiment, that is, to kill the insect larvae and fail to persist on leaf surfaces or in the soil. The AcNPV polyhedrin promoter was preserved to enable high-level production of the β-galactosidase protein to be achieved in insect cells. The amount of β-galactosidase protein produced by the recombinant virus matched that of the normal polyhedrin protein synthesized by unmodified virus. After repeated passage of the virus in cell culture and insects, it was found to remain stable and to continue to produce the β-galactosidase gene product.

The virus was subjected to the routine risk assessment programme established in the previous studies (host-range testing, etc), before permission was sought and granted for the field-release experiment. This was conducted in a similar way to the previous two release experiments. About 400 *S. exigua* larvae were fed the virus in the laboratory, before they were placed onto sugar beet plants in the contained experimental area. A similar number of uninfected larvae were placed in another area within the contained facility. The fate of the laboratory virus-infected larvae was monitored by testing larvae, foliage and soil samples recovered from the site for the presence of the recombinant virus. Materials returned to the laboratory were fed to highly susceptible *T. ni* larvae to detect the recombinant virus.

To summarize, both the 1988 and 1989 field trials were unsatisfactory in their outcome. First, few of the virus-infected larvae placed in the facility died from virus infection. The majority of the larvae pupated before the end of the experimental period. Secondly, plants infested with virus-infected larvae or uninfected larvae were equally damaged. These results may have been caused by a cold weather spell during the post-release period, in contrast with the warm weather during the 1987 trial with the genetically marked, non-occluded baculovirus. Reduced temperatures will slow the rate of larval development and almost certainly have an effect on virus replication. However, bioassays prior to release indicated that a single dose of the same virus inoculum killed *S. exigua* larvae, subsequently reared on diet in the laboratory at 18°C. Any change in the course of infection such as low temperatures, can be expected to reduce the progress of infection, unless an overdose of virus was used. In fact, double infections were employed to provide as much virus as possible. Two successive doses of undiluted virus were fed to *S. exigua* in the laboratory before release. However, this proved inadequate in the subsequent field study. It was concluded that polyhedrin-negative virus, although suitable for laboratory analyses is not suitable for field work unless the species is highly susceptible.

Future field-release experiments should make use of the polyhedrin-positive baculovirus insecticides. The 1986 study used a marked, occluded AcNPV. It demonstrated that larval deaths occurred as the result of virus infection and that virus persisted *in situ* in cadavers, in the soil and on the plants. Virus was not detected outside the area of the barriered plot within the facility or in locations distant to the plot. Furthermore, at the end of the study the plot was successfully decontaminated by formaldehyde treatment. This confirms that it is safe to use the occluded form of the virus, since despite virus persistence, the site may be cleared of infectious virus at the end of the study, or if the study has to be prematurely terminated.

3.5.8 Risk assessment analyses of genetically modified baculoviruses

One of the most important aspects is whether the addition of foreign DNA affects the host range of the recombinant virus insecticide. The procedures whereby this information may be gathered are now well-established and involve the collection of wild insect species for challenge with the virus in the laboratory. Given that viruses are now being produced with improved characteristics, these host range tests remain important. For instance, a species which previously was only marginally susceptible to the unmodified virus, may be killed more efficiently by the recombinant baculovirus. However, in preliminary studies with the virus containing the scorpion toxin gene this has not been confirmed. The host range of the virus remained unchanged.

The persistence of the recombinant baculovirus in the environment is likely to be similar to that of the wild-type virus, given that the additional foreign DNA sequences do not affect the formation of mature polyhedra. Therefore, the results from earlier studies with genetically marked occlusion positive viruses may be used to predict the consequences of releasing a baculovirus containing a foreign gene. Recombination between baculoviruses may occur in nature, but probably only between related viruses after co-infection of the same host. The concern is whether the foreign gene in the recombinant baculovirus insecticide may be able to move to a virus with a different host range and thus affect a wider selection of species. Tests conducted in our laboratory showed that it was not possible to 'rescue' a polyhedrin-negative AcNPV with a UK baculovirus, *Panolis flammea* (pine beauty moth) (Pf) NPV polyhedrin gene. In these tests the chances for recombination were maximized by performing a co-transfection with AcNPV DNA and a plasmid containing the PfNPV polyhedrin gene. This ensured that cells were likely to receive both virus and plasmid DNA. In parallel experiments it was shown that if the PfNPV polyhedrin promoter is deliberately introduced into the AcNPV genome it would function efficiently (Cameron and Possee, 1989). This demonstrated that the failure to derive polyhedrin-positive virus was not a consequence of the inability of the heterologous promoter to direct transcription.

Probably the most serious concern of those undertaking the field release experiments, and of parties concerned with assessing the safety of these trials, is whether the recombinant protein produced in the virus-infected insect will present a hazard to other species, particularly mammals. This is particularly the case when a toxin gene has been inserted into the virus genome. It is important, therefore, to be sure that the toxin in question is insect-specific. This can be readily ascertained in laboratory studies by injection, ingestion and inhalation studies with small mammals. The LD_{50} values determined from these studies may be used to estimate the dose of toxin which would have an effect on humans. Given a knowledge of the yield of the toxin in virus-infected insects it will also be possible to estimate the toxin load in the environment, after release of the recombinant virus. This estimate, together with the LD_{50} information will permit the proposers and regulators alike to determine whether man or other mammals would be likely to receive a harmful dose of toxin in the environment.

A further question to be addressed is whether the toxin gene inserted into the baculovirus could mutate to yield a product which might be active against mammals. Analysis of protein sequence databases to identify closely-related proteins is the best way to investigate this possibility. Calculations may then be made to determine the number of nucleotide changes which would be required to derive a mammalian-active protein. Given that the rate of DNA mutation at a given nucleotide is constant, a reasonably accurate estimate may be made of the likelihood that these changes could occur.

3.6 CONCLUSIONS AND FUTURE PROSPECTS

For the present, biotic crop protection agents pose little threat to the supremacy of chemicals in the control of insect pests. The rapid action of chemicals remains a major attraction for the crop grower. Chemicals are easier to produce on a large scale, in comparison to biological agents. Storage conditions are also less critical for chemicals, while biological agents must be maintained in an infectious form. The problems faced by those wishing to exploit biological insecticides, while considerable, are not insurmountable. It is a fair comment that, had a fraction of the resources committed to the development of chemical agents been devoted to research on bacteria, viruses and fungi, there would be many more successful biological control agents in the marketplace.

The drive to improve biological control agents will come from two quarters. In the first, techniques for genetic manipulation will allow the modification of the genotype of the microorganism. These modifications may result in improved phenotypes. The introduction of insect-specific toxins, hormones or enzymes into the genomes of bacteria, viruses and fungi appears to offer the greatest opportunity for short term success. In the longer term, as we understand more about the natural history of the microorganisms, we may be able to make more subtle alterations to their genomes to affect their efficacy. For example, virus productivity in an infected insect may be sacrificed for a more rapid kill. Secondly, the increasing concern of the general public about the saturation of food stuffs with chemicals is prompting a reevaluation of the benefits of biological control agents.

REFERENCES

Abdel-Aal, Y.A.I. and Hammock, B.D. (1986) Transition state analogs as ligands for affinity purification of juvenile hormone esterase. *Science*, **233**, 1073–6.

Akiba, Y. (1986) Microbial ecology of *Bacillus thuringiensis* VI. Germination of *Bacillus thuringiensis* spores in the soil. *Applied Entomology and Zoology*, **21**, 76–80.

Angus, T.A. (1954) A bacterial toxin paralysing silkworm larvae. *Nature*, **173**, 545–6.

Barnes, A.C. and Cummings, S.E. (1987) US Patent No. 4695455.

Beach, R.M. (1990) Application for an experimental use permit to ship and use a pesticide for experimental purposes only. Permit number 58788-EUP-4 for InCideTM 586. Crop Genetics International, Hanover, MD.

Bedford, G.O. (1980) Biology, ecology, and control of palm rhinoceros beetles. *Annual Review of Entomology*, **25**, 309–39.

Bedford, G.O. (1981) Control of the rhinoceros beetle by baculovirus, in *Microbial Control of Pests and Diseases* 1970–1980, (ed. H.D. Burges), Academic Press, New York, pp. 409–26.

Bergold, G.H. (1947) Die Isolierung des Polyeder-virus und die Natur der Polyeder. *Zeitschrift fur Naturforsch Teil B*, **2**, 122–43.

Berliner, E. (1911) Uber die Schlaffsucht der Mehlmottenraupe. *Zeitschrift fur Gesamte Getreidewsen (Berlin)*, **3**, 63–70.

Berliner, E. (1915) Uber die Schlaffsucht der Mehlmottenraupe. *Zeitschrift fur Angewandte Entomologie*, **2**, 29–56.

Bernier, L., Cooper, R.M., Charnley, A.K. and Clarkson, J.M. (1989) Transformation of the entomopathogenic fungus *Metarhizium anisopliae* to benomyl resistance. *FEMS Microbiology Letters*, **60**, 261–6.

Bishop, D.H.L., Entwistle, P.F., Cameron, I.R., Allen, C.J. and Possee, R.D. (1988) Field trials of genetically engineered baculovirus insecticides, in *The Release of Genetically Engineered Micro-organisms*, (eds M. Sussman, C.H. Collins, F.A. Skinner and D.E. Stewart-Tull), Academic Press, New York, pp. 143–79.

Burges, H.D. (1973) Enzootic diseases of insects, in *Regulation of Insect Populations* by Microorganisms, (ed. L.A. Bulla Jr), *Annals of the New York Academy of Sciences*, **217**, 31–49.

Cameron, I.R. and Possee, R.D. (1989) Conservation of polyhedron gene promoter function between *Autographa californica* and *Mamestra brassicae* nuclear polyhedrosis viruses. *Virus Research*, **12**, 183–99.

Carlton, B.C., Gawron-Burke, C. and Johnson, T.B. (1990) Exploiting the genetic diversity of *Bacillus thuringiensis* for the creation of new biopesticides, in *Proceedings of the Fifth International Colloquium on Invertebrate Pathology and Microbial Control*, Society for Invertebrate Pathology, Adelaide, pp. 18–22.

Charnley, A.K. (1984) Physiological aspects of destructive pathogenesis in insects by fungi: a speculative review, in *Invertebrate–Microbe Interactions*, (eds J.M. Anderson, D.M. Rayner and D.W.H. Walton), British Mycological Society Symposium 6, Cambridge University Press, Cambridge, pp. 229– 70.

Charnley, A.K. and St Leger, R.J. (1991) The role of cuticle-degrading enzymes in fungal pathogenesis in insects, in *The Fungal Spore and Disease Initiation in Plants and Animals*, (eds G.T. Cole and H.C. Hoch), Plenum Publishing Co., London, pp. 267–86.

Coles, R.B. and Pinnock, D.E. (1982) Control of the pasture cockchafer with the fungal pathogen *Metarhizium anisopliae*, in *Proceedings of the 3rd Australian Conference on Grassland Invertebrate Ecology* (ed. K.E. Lee), CSIRO, Australia, pp. 191–98.

Cory, J.S. and Entwistle, P.F. (1990) The effect of time of spray application on infection of the pine beauty moth, *Panolis flammea* (Den. & Schiff). (Lep., Noctuidae), with nuclear polyhedrosis virus. *Journal of Applied Entomology*, **110**, 235–41.

Crickmore, N., Nicholls, C., Earp, D.J., Hogman, T.C. and Ellar, D.J. (1990) The construction of *Bacillus thuringiensis* strains expressing novel entomocidal delta-endotoxin combinations. *Biochemical Journal*, **270**, 133–6.

Cunningham, J.C. (1988). Baculoviruses: their status compared to *Bacillus thuringiensis* as microbial insecticides. *Outlook on Agriculture*, **17**, 10–7.

Darbon, H., Zlotkin, E., Kopeyen, E., van Rietschoten, J., and Rochat, H. (1982) Covalent structure of the insect toxin of the North African scorpion *Androctonus australis* Hector. *International Journal of Peptide and Protein Research*, **20**, 230–3.

Doyle, C.J. and Entwistle, P.F. (1988) Aerial application of mixed virus formulations to control joint infestations of *Panolis flammea* and *Neodiprion sertifer* on lodgepole pine. *Annals of Applied Biology*, **113**, 119–27.

Dulmage, H.T., Graham, H.M. and Martinez, E. (1978) Interactions between the tobacco budworm, *Heliothis virescens*, and the delta-endotoxin produced by the HD- 1 isolate of *Bacillus thuringiensis* var. *kurstaki*: relationship between length of exposure to the toxin and survival. *Journal of Invertebrate Pathology*, **32**, 40–50.

Ellar, D.J. (1989) Investigation of the molecular basis of *Bacillus thuringiensis* delta-endotoxin specificity and toxicity. *Abstracts of the Society for Invertebrate Pathology Meeting*, College Park, MD, 20–24 August.

Fargues, J. (1984). Adhesion of the fungal spore to the insect cuticle in relation to pathogenicity, in *Infection Processes of Fungi – Conference Report*, (eds D.W. Roberts and J.R. Aist), Rockefeller Foundation, New York, pp. 90–110

Ferron, P. (1978) Etiologie et epidemiolgie des muscardines. These Doctorate d'Etat. University Pasteur and Marie Curie, Paris.

Ferron, P. (1981) Pest Control by the fungi *Beauveria* and *Metarhizium*, in *Microbial Control of Pests and Plant Diseases* 1970–1980, (ed. H.D. Burges), Academic Press, New York, pp. 465–82.

Franz, J.M. and Krieg, A. (1980) Mikrobiologische Schadlingsbekampfung in China. Ein Reisebericht. *Forum Mikrobiologie*, **3**, 173–6.

Ge, A.Z., Rivers, D., Milne, R. and Dean, D.H. (1991) Functional domains of *Bacillus thuringiensis* insecticidal crystal proteins: refinement of the *Heliothis virescens* and *Trichoplusia ni* specificity domains on CryIA(c). *Journal of Biological Chemistry*, **266**, 17954–8.

Gelernter, W. and Schwab, G.E. (1993) Transgenic bacteria, viruses, algae and other microorganisms as *Bacillus thuringiensis* toxin delivery systems, in *Bacillus thuringiensis, An Environmental Pesticide: Theory and Practice*, (eds P.F. Entwistle, J.S. Cory, M.J. Bailey and S. Higgs), John Wiley, New York, pp. 89–104.

Georghiou, G.P. and Lagunes, A. (1988) *The Occurrence Of Resistance To Pesticides: Cases Of Resistance Reported Worldwide Through 1988*, FAO, Rome, p 325.

Goettel, M., St Leger, R.J., Bhairi, S., Jung, M., Oakley, B., Roberts, D. and Staples, R.C. (1990) Pathogenicity and growth of metarhizium-anisopliae stably transformed to benomyl resistance. *Current Genetics*, **17**, 129–32.

Gonzalez, J.M. and Carlton, B.C. (1982) Plasmid transfer in *Bacillus thuringiensis*, in *Genetic Exchange: A Celebration and a New Generation*, (eds U.N. Steips, S.H. Goodall, W.R. Guild and G.A. Wilson), Marcel Dekker, New York, pp. 85–95.

Guangyu, Z. (1989) The first commercial viral pesticide in China. *Integrated Pest Management Practice*, **11**, 13.

Hall, R.A. (1981) The fungus *Verticillum lecani* as a microbial insecticide against aphids and scales, in *Microbial Control of Pests and Plant Diseases 1970–1980*, (ed. H.D. Burges), Academic Press, New York, pp. 483–98.

Hall, R.A. and Papierok, B. (1982) Fungi as biological control agents of agricultural and medical importance. *Parasitology*, **84**, 205–40.

Hammock, B.D. (1985) Regulation of juvenile hormone titer: degradation, in *Comparative Insect Physiology Biochemistry and Pharmacology*, (eds G.A. Kerkut and L.I. Gilbert), Pergamon Press, New York, pp. 431–72.

Hammock, B.D., Bonning, B.C., Possee, R.D., Hanzlik, T.N. and Maeda, S. (1990) Expression and effects of the juvenile hormone esterase in a baculovirus vector. *Nature*, **344**, 458–61.

Hanel, H. and Watson, J.A.L. (1983) Preliminary field tests on the use of *Metarhizium anisopliae* for the control of *Nasutitermes exitiosus* (Hill) (Isoptera: Termitidae). *Bulletin of Entomological Research*, **73**, 305–13.

Hannay, C.L. (1953) Crystalline inclusions in aerobic spore-forming bacteria. *Nature*, **172**, 1004.

Hannay, C.L. and Fitz-James, P. (1955) The protein crystals of *Bacillus thuringiensis* Berliner. *Canadian Journal of Microbiology*, **1**, 694–709.

Hanzlik, T.N. and Hammock, B.D. (1987) Characterization of affinity-purified juvenile hormone esterase from *Trichoplusia ni*. *Journal of Biological Chemistry*, **262**, 13584–91.

Hanzlik, T.N., Abdel-Aal, Y.A.I., Harshman, L.G. and Hammock, B.D. (1989) Isolation and sequencing of cDNA clones coding for juvenile hormone esterase from *Heliothis virescens*. *Journal of Biological Chemistry*, **264**, 12419–25.

Heale, J.B., Isaac, J. and Chandler, D. (1989) Prospects for strain improvement in Entomopathogenic fungi. *Pesticide Science*, **26**, 79–92.

Hendrick, K., De Loof, A. and Van Mellaert, H. (1990) Effects of *Bacillus thuringiensis* isolates effective on resistant Indian meal moth, in *Genetics and Biotechnology of the Bacilli*, vol. 2,(eds A.T. Ganesan and J.A. Hoc), Academic Press, New York, pp. 233–8.

Honee, G., Vriezen, W. and Visser, B. (1990) A translation fusion product of two different insecticidal crystal proteins of *Bacillus thuringiensis* exhibits an enlarged insecticidal spectrum. *Applied and Environmental Microbiology*, **56**, 823–5.

Hussey, N.W. and Tinsley, T.W. (1980) Impressions of insect pathology in the People's Republic of China, in *Microbial Control of Pests and Plant Diseases* 1970–1980, (ed. H.D. Burges), Academic Press, New York, pp. 513–38.

Ignoffo, C.M. (1973) Development of a viral insecticide: Concept to commercialization. *Experimental Parasitology*, **33**, 380–406.

Ishawata, S. (1901) On a kind of severe flacherie (sotto disease). *Dainihon Sanshi Kaiho*, **114**, 1–5.

Jackson, C.W. and Heale, J.B. (1987) Parasexual crosses by hyphal anastomosis and protoplast fusion in the entomopathogen *Verticillium lecani*. *Journal of General Microbiology*, **133**, 3537– 47.

Katoaka, H., Troetschler, R.G., Li, J.P., Kramer, S.J., Carney, R.L. and Schooley, D.A. (1989) Isolation and identification of a diuretic hormone from the tobacco hornworm, *Manduca sexta*. *Proceedings of the National Academy of Science USA*, **86**, 2976–80.

Keller, S. (1983) Die mikrobiologische Bekampfung des Maikafers (Melolontha melolontha L.) mit dem Pilz Beauveria brongniarti. Mitt Schweiz. *Landwirtschaft*, **31**, 61–4.

King, L.A., Possee, R.D., Hughes, D.S., Atkinson, A.E., Palmer, C.P., Marlow, S.A., Pickering, J.M., Joyce, K.A., Lawrie, A.M., Miller, D.P. and Beadle, D.J. (1994) *Advances in insect virology*. Advances in Insect Physiology (in press).

Kinsinger, R.A. and McGaughey, W.H. (1979) Susceptibility of populations of Indian meal moth and almond moth to *Bacillus thuringiensis*. *Journal of Economic Entomology*, **72**, 346–9.

Klier, A., Bourgouin, C. and Rappoport, G. (1983) Mating between *Bacillus subtilis* and *Bacillus thuringiensis* strain aizawai 7-29 and comparison of the structural organization of genes from different serotypes, in *Molecular Biology of Microbial Differentiation*, (eds J.A. Hoch and P. Setlow), American Society for Microbiology, Washington, DC, pp. 217–224.

Knowles, B.H. and Ellar, D.J. (1986) Characterization and partial purification of a plasma membrane receptor for *Bacillus thuringiensis* var. *kurstaki* lepidopteran-specific delta-endotoxin. *Journal of Cell Science*, **83**, 89–99.

de Kort, C.A.D. and Granger, N.A. (1981) Regulation of juvenile hormone titre. *Annual Reviews of Entomology*, **26**, 1–28.

Krassilstschik, J. (1888). La production industrielle des parasites vegetaux pour la destruction des insectes nuisible. *Bulletin of Science, France, Belgium*, **19**, 461–72.

Lewis, F.B. (1981) Control of the gypsy moth by a baculovirus, in *Microbial Control of Pests and Diseases 1970–1980*, (ed. H.D. Burges), Academic Press, New York, pp. 363–77.

Luthy, P., Cordier, J.L. and Fischer, H.M. (1982) *Bacillus thuringiensis* as a bacterial insecticide: basic consideration and applications, in *Microbial and Viral Pesticides*, (ed. E. Kurstak), Marcel Dekker, New York, pp. 35–74.

Maddrell, S. (1986) Hormonal control of diuresis in insects, in *Insect Neurochemistry and Neurophysiology* (eds A.B. Borkovec and D.B. Gelman), Humana, Clifton, NJ, pp. 79–90.

Maeda, S. (1989) Increased insecticidal effect by a recombinant baculovirus carrying a synthetic diuretic hormone. *Biochemical and Biophysical Research Communications*, **165**, 1177–83.

Martens, J.W.M., Honee, G., Zuidema, D., van Lent, J.W.M., Visser, B. and Vlak, J.M. (1990) Insecticidal activity of a bacterial crystal protein expressed by a recombinant baculovirus in insect cells. *Applied and Environmental Microbiology*, **56**, 2764–70.

McGaughey, W.H. (1985) Insect resistance to the biological insecticide *Bacillus thuringiensis. Science* **229**, 193–5.

Merryweather, A.T., Weyer, U., Harris, M.P.G., Hirst, M., Booth, T. and Possee, R.D. (1990) Construction of genetically engineered baculovirus insecticides containing the *Bacillus thuringiensis* subsp. *kurstaki* HD-73 delta endotoxin. *Journal of General Virology*, **71**, 1535–44.

Metschnikoff, E. (1879) *Krankheiten der Larve des Getreidekafers*. Odessa, 32 pp.

Moscardi, F. and Sosa-Gomez, D.R. (1992) Use of viruses against soybean caterpillars in Brazil, in *Pest Management in Soybean*, (ed. L.G. Copping, M.B. Green and R.T. Rees), Elsevier, Amsterdam, pp. 98– 109.

Nambier, P.T.C., Ma, S.W. and Iyer, V.N. (1990) Limiting an insect infestation of nitrogen-fixing root nodules of the pigeon pea (*Canjanus cajan*) by engineering the expression of an entomcidal gene in its root nodules. *Applied and Environmental Microbiology*, **56**, 2866–9.

Obukowicz, M.G., Perlak, F.J., Bolten, S.L., Kusano-Kretzmer, K., Mayer, E.J. and Watrud, L.S. (1987) IS50L as a non-self- transposable vector used to integrate the *Bacillus thuringiensis* delta-endotoxin gene into the chromosome of root-colonizing pseudomonads. *Gene*, **51**, 91–6.

Otvos, I.S., Cunningham, J.C. and Friskie, L.M. (1987) Aerial application of nuclear polyhedrosis virus against Douglas-fir tussock moth, *Orgyia pseudosugata* (McDunnough) (Lepidoptera: Lymantriidae): impact in the year of application. *Canadian Entomologist*, **119**, 69–76.

O'Reilly, D.R. and Miller, L.K. (1989) A baculovirus blocks insect molting by producing ecdysteroid UDP-glucosyl transferase. *Science*, **245**, 1110–2.

Petras, S.T. and Casida, L.E. (1985) Survival of *Bacillus thuringiensis* spores in soil. *Applied and Environmental Microbiology*, **50**, 1496–501.

Risco, B.S.H. (1980) Biological control of the leaf froghopper, *Mahanarva posticata* Stal., with the fungus *Metarhizium anisopliae* in the state of Alagoas-Brazil. *Entomology Newsletter for the Society of Sugarcane Techniques*, **9**, 10.

Sacchi, V.F., Parenti, P., Hanozet, G.M., Giordana, B., Luthy, P. and Wolfersberger, M.G. (1986) *Bacillus thuringiensis* toxin inhibits K^+gradient-dependent amino acid transport across the brush border membrane of *Pieris brassicae* midgut cells. *FEBS Letters*, **204**, 213–8.

Salama, H.S. and Sharaby, A. (1985) Histopathological changes in *Heliothis armigera* infected with *Bacillus thuringiensis* as detected by electron microscopy. *Insect Science Applications*, **6**, 503–11.

Samuels, R.I., Charnley, A.K. and Reynolds, S.E. (1988) Role of destruxins in the pathogenicity of 3 strains of *Metarhizium anisopliae*. *Mycopathologica*, **104**, 51–8.

Sparks, T.C. and Hammock, B.D. (1980) Comparative inhibition of the juvenile hormone esterases from *Trichoplusia ni*, *Tenebrio monitor*, and *Musca domestica*. *Pesticide Biochemistry and Physiology*, **14**, 290–302.

Stewart, L.M.D., Hirst, M., Lopez-Ferber, M., Merryweather, A.T., Cayley, P.J. and Possee, R.D. (1991) Construction of an improved baculovirus insecticide containing an insect-specific toxin gene. *Nature*, **352**, 85–8.

Suzuki, A., Kawakami, K. and Tamura, S. (1971) Detection of destruxins in silkworm infested with *Metarhizium anisopliae*. *Agricultural and Biological Chemistry*, **35**, 1641–3.

Tabashnik, B.E., Cushing, N.L., Finson, N. and Johnson, M. (1990) Field development of resistance to *Bacillus thuringiensis* delta-endotoxins in diamond moth (Lepidoptera: Plutellidae). *Journal of Economic Entomology*, **83**, 49–55.

Teitelbaum, Z., Lazarovici, P. and Zlotkin, E. (1979) Selective binding of the scorpion venom insect toxin to insect nervous tissue. *Insect Biochemistry*, **9**, 343–6.

Thompson, W.T. (1989) *Agricultural Chemicals*, Book 1, Thompson Publications, Fresno, CA, pp. 64–70.

Tomalski, M.D. and Miller, L.K. (1991) Insect paralysis by baculovirus-mediated expression of a mite neurotoxin gene. *Nature*, **352**, 82–5.

Waalwijk, C., Dullemans, A. and Maat, C. (1991) Construction of a bioinsecticidal rhizosphere isolate of *Pseudomonas fluorescens*. *FEMS Microbiology Letters*, **77**, 257–64.

Walstad, J.D., Anderson, R.F. and Stambaugh, W.J. (1970) Effects of environmental conditions on two species of muscardine fungi (*Beauveria bassiana* and *Metarhizium anisopliae*). *Journal of Invertebrate Pathology*, **16**, 221–6.

Walther, C., Zlotkin, E. and Rathmayer, W. (1976) Action of different toxins from the scorpion *Androctonus australis* on a locust nerve-muscle preparation. *Journal of Insect Physiology*, **22**, 1187–94.

Whitford, M., Stewart, S., Kuzio, J. and Faulkner, P. (1989) Identification and sequence analysis of a gene encoding gp67, an abundant envelope glycoprotein of the baculovirus *Autographa californica* nuclear polyhedrosis virus. *Journal of Virology*, **63**, 1393–9.

Wood, H.A. and Granados, R.R. (1991) Genetically engineered baculoviruses as agents for pest control. *Annual Reviews of Microbiology*, **45**, 69–87.

Zelazny, B., Lolong, A. and Crawford, A.M. (1990) Introduction and field comparison of baculovirus strains against *Oryctes rhinoceros* (Coleoptera: Scarabaeidae) in the Maldives. *Environmental Entomology*, **19**, 1115–21.

Zimmermann, G. (1982) Effect of high temperatures and artificial sunlight on the viability of conidia of *Metarhizium anisopliae*. *Journal of Invertebrate Pathology*, **40**, 36–40.

Zimmermann, G. (1986) Insect pathogenic fungi as pest control agents. *Progress in Zoology*, **32**, 217–31.

Zlotkin, E., Rochat, H., Kopeyan, C., Miranda, F. and Lissitzky, S. (1971) Purification and properties of the insect toxin from the venom of the scorpion *Androctonus australis* Hector. *Biochemie*, **53**, 1073–7.

Molecular determinants for resistance to crop protection chemicals

D.W. Hollomon and J.A. Butters

4.1 INTRODUCTION

Crop protection chemicals play an important part in the production of high yielding quality food and fibre crops, and their failure can cause considerable losses. Biological material is seldom uniform, and wide variation in sensitivity to pesticides often exists in target populations of insects, pathogens and weeds. Rare resistant mutants favoured by selection increase in frequency so that chemical control is no longer effective. Resistance is a stable and heritable characteristic and although resistant mutants can often be generated and studied in the laboratory, their early and accurate detection in field populations is of major concern in practice.

The impact of resistance on growers stems from the loss of harvestable and saleable crops. If resistance spreads too rapidly, development costs will not be recovered by the agrochemical industry. Considerable human effort is invested in the successful marketing of a new pesticide and loss of a biochemical target site through resistance means a valuable resource is no longer available for further exploitation.

Resistance is not a new phenomenon and strategies of pesticide use have been developed in many cropping systems to combat it. To be effective, strategies must be based on sound information about the biochemical, genetic and epidemiological aspects of resistance. Much of this evidence is imprecise and fragmentary at present, and may lead to the introduction of inappropriate strategies. This chapter explores how advances in molecular techniques are improving the understanding of fundamental aspects of

Molecular Biology in Crop Protection. Edited by George Marshall and Dale Walters. Published in 1994 by Chapman & Hall, London. ISBN 0 412 54390 7(HB) 0 412 54400 8(PB).

pesticide resistance, allowing more precise and rapid identification, and contributing to more durable strategies to contain resistance.

4.2 BIOCHEMISTRY AND GENETICS OF RESISTANCE

To reach its target site a pesticide must pass through several barriers, any one of which can change, confering resistance. Penetration may be retarded (Sawicki, 1973) or in some cases, membrane changes lead to active excretion of fungicides such as fenarimol and other sterol 14 α-demethylation inhibitors (DMIs; De Waard and van Nistelrooy, 1982). Resistance to the herbicide paraquat in several weed species is due to reduced symplastic movement, apparently because paraquat is immobilized in some way in mesophyll cells (Fuerst *et al.*, 1985). Slower movement prolongs exposure to the mainly constitutive battery of hydrolytic and oxidative enzymes that detoxify pesticides and can cause resistance. Although enhanced detoxification is seldom a feature of fungicide resistance, it frequently is the mechanism involved in insecticide and herbicide resistance. Not only can the amount of these enzymes be dramatically increased so that organophosphorus insecticides, for instance, are sequestered by carboxylesterase before they reach their target site (Devonshire and Moores, 1982), but catalytic efficiency of hydrolases may also increase (Brown, 1990). Isozymic alteration leading to changes in specificity, coupled with inducible increases in the levels of cytochrome P450 monooxygenases, detoxify a wide range of insecticides through oxidation of the carbon of alkyl substituents, or even oxidation of the P=S bond of phosphorothioates (Brattsten, 1990). Enhanced cytochrome P450 activity in cell cultures of black grass (*Alepocurus myosuroides*) is also associated with resistance to substituted urea herbicides (Caseley *et al.*, 1990). Metabolism of herbicides such as atrazine (McCall *et al.*, 1986) and organophosphorus insecticides (Brattsten, 1990) to form non-toxic glutathione conjugates provides another route to detoxification and resistance. Conjugation is catalysed by glutathione-S-transferases, which generally possess low substrate specificity, and can simply act as binding proteins. As with other detoxifying enzymes, isozymic forms of glutathione-S-transferase allow changes in substrate specificity, which may be linked to resistance. This association of metabolic detoxification with resistance to both herbicides and insecticides opens the way for synergists, which inhibit detoxification enzymes, to be used in strategies to combat resistance.

Changes at the target site provide another important general mechanism of pesticide resistance. It is a common feature of fungicide resistance, where cross-resistance seldom extends to fungicides with another mode of action. The affinity of methylbenzimidazole carbamate (MBC, Davidse and Flach, 1977) and *N*-phenylcarbamate (Ishii *et al.*, 1990) fungicides to the target β-tubulin protein is reduced in resistant strains of several fungi, but increased in super-sensitive mutants. In this example, resistance to the two fungicide groups is negatively correlated, and increased affinity for *N*-phenylcarbamate is coupled with reduced binding of MBC fungicides. An acetylcholinesterase with reduced affinity for methyl carbamate insecticides accounted for resistance in rice green leaf hopper, although larger *N*-alkyl groups still bound to the enzyme and killed the leaf hoppers (Brown, 1990). Modification of the D1 target protein of chloroplast photosystem II accounts for resistance to a range of photosynthetic inhibiting herbicides, including atrazine (van Rensen and de Vos, 1992).

Although the biochemical mechanisms of resistance may be well understood, the genetic basis is generally less clear. Often, many genetic factors are associated with resistance (Crute, 1992), and very few genes confering resistance have been isolated and correlated with the biochemical mechanisms involved. The majority of genes that have been mapped involve changes at the target site, and several alleles, each confering different level resistance, are commonly found. Where only a single gene is involved further research is feasible to uncover the molecular mechanisms causing resistance.

4.3 PROCEDURES FOR CLONING TARGET SITE GENES

4.3.1 Herbicides

A combination of biochemical and genetic studies identified the site of action of sulphonylurea herbicides as the enzyme acetolactate synthase (ALS), which catalyses the first common step in the biosynthesis of the essential branched chain amino acids, valine, leucine and isoleucine (Ray, 1984). Resistant tobacco (*Nicotiana tabaccum* L.) plants possessed a sulphonylurea insensitive ALS which co-segregated with the herbicide-resistant whole plant phenotype. Sulphonylureas also inhibit ALS in many microorganisms, including yeast (*Saccharomyces cerevisiae*), suggesting that the enzyme is highly conserved. In what has now almost become a standard approach to cloning target site genes, the yeast ALS gene was cloned to provide a heterologous probe to screen plant genomic DNA libraries (Hartnett *et al.*, 1990).

A high copy number plasmid library of yeast DNA was screened in a wild-type strain by selecting for chlorosulfuron resistance, coupled with over-production of ALS. Location of the ALS gene was identified by deletion analysis of positive plasmid clones and by hybridizing them to another cloned ALS gene from *Salmonella typhimurium*. A series of deletion clones were sequenced and compared with other microbial ALS sequences available in data bases to identify conserved nucleotide regions, which might provide probes likely to hybridize to plant ALS sequences. A yeast

DNA fragment containing almost the complete ALS coding region was chosen and used to screen genomic phage libraries of *Arabidopsis thaliana* and tobacco. In this way, several plaques containing presumptive ALS genes were identified and the phages isolated.

Complete sequence analyses of *Arabidopsis* and tobacco genes revealed highly conserved proteins with 85% sequence identity at the amino acid level. Southern blots of *Arabidopsis* and tobacco genomic DNA were probed with isolated ALS genes to determine the number of ALS genes present in the two species. Two fragments of tobacco DNA digested with restricton endonucleases hybridized to the probe and one fragment from *Arabidopsis*, suggesting that tobacco, which is an allotetraploid, has two ALS genes and *Arabidopsis* one. This agrees with the genetic analysis of sulphonylurea resistance in tobacco, which is controlled by two distinct genes. Southern analysis of maize and soybean DNA also indicated the presence of multiple ALS genes.

Genomic DNA libraries were prepared from two sulphonylurea-resistant tobacco lines. One, mutated at the SuRA locus, showed only a 100-fold more resistance to chlorosulfuron than its wild-type parent; the second was a 1000-fold more resistant and mutated at a second locus, SuRB. Complete sequence analysis of the two genes revealed the identity of the point mutations confering herbicide resistance. In SuRA, a C to A mutation substituted glutamine for proline at amino acid position 191. In SuRB a C to G change subsitituted alanine for proline at the same amino acid position, whilst a second mutation of G to T substituted leucine for tryptophan at amino acid position 568. ALS activity of these two resistant mutants was not affected, indicating that changes to the catalytic site of the enzyme were apparently not involved. Although only two sites in conserved regions of the ALS gene have been associated with resistance in plants, in yeast substitutions at 10 separate sites in the protein have been connected with resistance to sulphonylureas (Hartnett *et al.*, 1990).

This example of ALS outlines many aspects of strategies commonly used to isolate target site genes and to understand the molecular basis of resistance (Figure 4.1). Confirmation of gene function often rests on sequence comparisons with known genes already present in data bases. Heterologous probes of β-tubulin genes from other sources have frequently been used to clone the same gene from many different fungi (May *et al.*, 1987). A similar approach has been used to clone the D1 protein of photosystem II reaction centre, and to identify the many point mutations predominantly between valine 217 and leucine 275 which confer resistance to a range of photosynthetic inhibiting herbicides (Trebst, 1991). Homologous probes were used to clone the sodium channel genes of pyrethroid susceptible and resistant *Drosophila* strains. Partial sequencing revealed that a single-base DNA change substituting asparagine for aspartic acid at amino acid 1172 was correlated with the 'knock down' (Kdr) resistance factor (Pauron *et al.*, 1992).

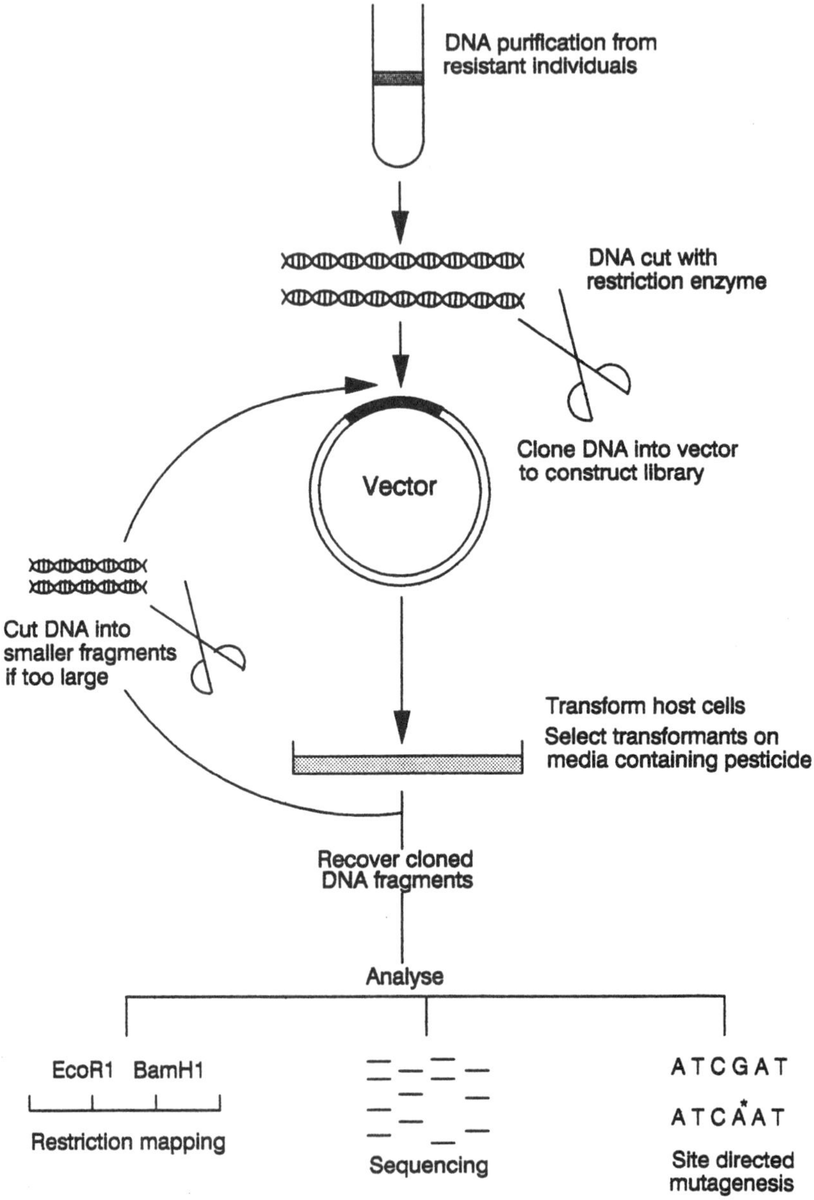

Figure 4.1 General strategy for cloning pesticide resistance genes.

4.3.2 Fungicides

Selection of plasmid libraries prepared from fungicide resistant strains has commonly been used to clone target site genes, especially where the resistant allele shows dominance and can be identified in the genomic background of the wild-type strain. Several MBC resistant β-tubulin alleles have been cloned and sequenced in this way. Recently, selection for carboxin resistance was used to isolate the *Cbx* gene from *Ustilago maydis* which appears to code for the iron–sulphur subunit of mitochondrial complex II (Keon *et al.*, 1992). In this case substitution of a histidine for a leucine in the third iron–sulphur cluster of the protein is associated with resistance. This arises through changes in two adjacent base pairs altering amino acid codon 752, suggesting that perhaps the *Cbx* allele is a double mutant. The sterol 14 α-demethylase gene from yeast, the target site for important triazole and other fungicides, was cloned by screening a genomic library in a multi-copy plasmid vector, selecting for resistance to ketoconazole and overexpression of cytochrome P450 (Kalb *et al.*, 1987).

Cloning of target site genes and identifying single DNA base changes causing resistance is not without its pitfalls. Genes unconnected with the target site, but which code for pesticide resistance, might equally well be cloned. DNA sequencing is not without error; the human genome project for instance is based on the expectation that 1:400 nucleotide base pairs will be incorrectly assigned. Although databases are increasing in size, they are by no means exhaustive and may not provide matching sequences to identify a possible function of a cloned DNA fragment. Where extensive gene families exist, such as with cytochrome P450 and β-tubulin genes, and several isozymes may be found in the same tissue, sequence homology may allow genes to be cloned which are unconnected with pesticide action. For instance, the *ben A* allele of an *Aspergillus nidulans* β-tubulin gene confers resistance to MBC fungicides even in heterologous transformation systems, whereas a second β-tubulin gene, tub C, does not. At least nine β-tubulin genes are expressed in *Arabidopsis* (Snustad *et al.*, 1992). To exploit molecular technology and fully understand the mechanisms of resistance, cloned genes clearly must be expressed to confirm their function.

Yeast can provide a good expression system, since incoming foreign DNA, which can be disrupted by insertion of another gene into the cloned DNA fragment, generally recombines with homologous resident sequences. By transforming diploid yeast with appropriate constructs, null mutants can be recovered as haploid progeny and the site of any biochemical lesion determined. Restoration of wild-type activity by complementation with the cloned heterologous gene allows its function to be confirmed. Unfortunately, yeast lacks the enzyme machinery necessary to correctly handle intron sequences so, unless cDNA clones are available, genomic DNA fragments containing introns are unlikely to be expressed. Although vectors are available which can be used to generate null mutations in a few other fungi, such as *Ustilago*, in most filamentous fungi transformation leads to insertion of incoming DNA at random loci involving unrelated sequences. Targeted null mutations can be produced in filamentous fungi such as *Neurospora crassa* using the phenomenon of Repeat Induced Point mutation (RIP; Selker, 1990) through extensive cytosine methylation of homologous introduced DNA and A–T mutations. Once appropriate strains are available, site-directed mutagenesis can be used to confirm the molecular basis of resistance.

4.4 POLYMERASE CHAIN REACTION (PCR) AND PESTICIDE RESISTANCE

PCR techniques greatly simplify the analysis of genetic variation and mutations associated with pesticide resistance (further details in Chapter 1). The abundance of amplified target sequence greatly reduces detection limits, and allows the practical use of non-isotopic detection systems. We have used PCR techniques to analyse the mutational changes associated with MBC

Figure 4.2 PCR amplification of a β-tubulin fragment from *R. secalis* genomic DNA. Phenol-chloroform purified genomic DNA was used as a template in a PCR reaction (100 μl) for 30 cycles under the following conditions. Annealing 65°C/1 min; primer extension 72°C/2 min; denaturation 94°C/1 min. Then 20 μl of each reaction was separated on a 3.0% 'Nusieve' agarose gel (containing ethidium bromide) at 4.7 v/cm. Lanes 1–3: complete reaction mixture; lanes 4 and 5: no template DNA; lanes 6 and 7: no *Taq* polymerase; lane 8: No [Mg^{2+}]; lanes 9 and 10: no primers.

Table 4.1 Amino acid substitutions in β-tubulin genes confering MBC resistance

Organism	*Amino acid substitution*	*position*	*Reference*
Laboratory mutants			
A. nidulans	His → Tyr	6	Osmani and Oakley (1991)
	His → Leu	6	Osmani and Oakley (1991)
	Tyr → Asn	50	Koenraadt *et al.* (1992)
	Tyr → Ser	50	Koenraadt *et al.* (1992)
	Gln → Lys	134	Koenraadt *et al.* (1992)
	Ala → Val	165	Osmani and Oakley (1991)
	Glu → Asp	198	Osmani and Oakley (1991)
	Glu → Gln	198	Osmani and Oakley (1991)
	Glu → Lys	198	Osmani and Oakley (1991)
	Phe → Tyr	200	Osmani and Oakley (1991)
	Met → Leu	257	Koenraadt *et al.* (1992)
N. crassa	Phe → Tyr	167	Orbach *et al.* (1986)
	Glu → Gly	198	Fujimura *et al.* (1990)
R. secalis	Glu → Lys	198	Hollomon and Butters (1992)
S. cerevisiae	Arg → His	241	Thomas *et al.* (1985)
S. nodorum	His → Tyr	6	Cooley and Caten (1993)
Field isolates			
B. cinerea	Glu → Ala	198	Martin *et al.* (1992)
R. secalis	Glu → Gly	198	Hollomon and Butters (1992)
S. homoeocarpa	Glu → Lys	200	Koenraadt *et al.* (1992)
V. inaequalis	Glu → Ala	198	Koenraadt *et al.* (1992)
	Glu → Gly	198	Koenraadt *et al.* (1992)
	Glu → Lys	198	Koenraadt *et al.* (1992)
	Phe → Tyr	200	Koenraadt *et al.* (1992)
V. pirina	Glu → Ala	198	Koenraadt *et al.* (1992)
	Phe → Tyr	200	Koenraadt *et al.* (1992)
M. fructicola	Glu → Lys	198	Koenraadt *et al.* (1992)
Penicillium spp.	Glu → Lys	198	Koenraadt *et al.* (1992)
	Glu → Ala	198	Koenraadt *et al.* (1992)
	Glu → Val	198	Koenraadt *et al.* (1992)
	Phe → Tyr	198	Koenraadt *et al.* (1992)

resistance in *Rhynchosporium secalis*, which causes leaf blotch of barley. Studies on several fungi have shown that amino acid substitutions involved in MBC resistance can occur throughout the β-tubulin gene (Table 4.1; Hollomon and Butters, 1992), although only mutations affecting amino acids at positions 198 and 200 seem important in filamentous plant pathogenic fungi. Upstream (5′) from these two amino acids there are many conserved nucleotide sequences in β-tubulin genes that might be exploited as PCR oligonucleotide primers, but downstream (3′) sequence homology is more limited. Table 4.2 shows conserved regions of nine fungal β-tubulin genes, and the consensus sequence used to synthesize the fully degenerate primers with a GC:AT ratio around 1. These two primers were used to amplify a

Table 4.2 Example of conserved regions in nine fungal β-tubulin genes for designing degenerate PCR primers

Organism	*Amino acid*	
	134 141	
	Q I T H S L G G	
Epichloe typhina	CAGATCACCCACTCTCTTGGTGGTGG	.....
Neurospora crassa	CAGATCACCCACTCCCTCGGTGGTGG	.....
Schizosaccharomyces pombe	CAATTAACTCACTCTTTGGGTGGTGG	.....
Erysiphe graminis	CAAATAACACATTCTCTTGGGGGTGG	.....
Candida albicans	CAAATCACCCATTCTTTGGGTGGTGG	.....
Aspergillus nidulans	CAGATCACCCACTCTCTCGGTGGTGG	.....
Aspergillus flavus	CAGATTACCCACTCCCTCGGTGGTGG	.....
Colletotrichum graminicola	CAGATTACCCACTCCCTCGGTGGAGG	.....
Saccharomyces cerevisiae	CAGATCACACATTCTCTTGGTGGTGG	.....
Degenerate primer 5′	CAAATCACCCACTCCCTCGGTGGTGG	3′
	G T T T T	

		Amino acid	
		300 306	
		M A A S D F R	
Epichloe typhina	.....	TGATGGCTGCTTCTGATTTCAGAAA	
Neurospora crassa	.....	TGATGGCTGCTTCTGACTTCCGCAA	
Schizosaccharomyces pombe	.....	TGATGGTCGCCGCTGATCCTCGCCA	
Erysiphe graminis	.....	TGATGGCAGCATCCGACTTCCGAAA	
Candida albicans	.....	TGATGGCTGCTTCTGATCCAAGAAA	
Aspergillus nidulans	.....	TGATGGCTGCCTCTGACTTCCGCAA	
Aspergillus flavus	.....	TGATGGCTGCTTCTGACTTCCGTAA	
Colletotrichum graminicola	.....	TGATGGCTGCCTCTGACTTCCGCAA	
Saccharomyces cerevisiae	.....	TGATGGCTGCTGCCGATCCAAGAAA	
Degenerate primer (reverse)3′		ACTACCGACGGAGACTGAAGGCTTT	5′
		A A	

520 bp fragment from *R. secalis* genomic DNA coding for 173 amino acids. Direct sequencing of the PCR products allowed identification of two internal priming sites which were not only totally complementary to the *R. secalis* β-tubulin genomic sequence, but which could be used as specific primers to reduce the probability of amplication of non-target DNA (Figure 4.2).

PCR techniques allowed rapid accumulation of sequence data for several *R. secalis* strains which differed in sensitivity to MBC fungicides. These revealed changes only at amino acid position 198, with substitution of glutamic acid in the wild-type with either glycine or lysine (Table 4.3). These substitutions were all the result of single DNA base changes. A similar, but more extensive PCR analysis of MBC resistance in *Venturia inaequalis, Monilinia fructicola* and several *Penicillium* species has confirmed additional amino acid substitutions at position 198, as well as a change of phenylalanine for tyrosine at the neighbouring amino acid position 200 (Koenraadt *et al.*, 1992).

Table 4.3 DNA changes and MBC resistance in *R. secalis*

	Amino acids						*Fungicide sensitivity*
Isolate	*196 (Ser)*	*197 (Asp)*	*198 (Glu)*	*199 (Thr)*	*200 (Phe)*	*201 (Cys)*	*(MBC)*
K1124	TCT	GAT	GAG	ACC	TTC	TGT	S
554	TCT	GAT	GAG	ACC	TTC	TGT	S
666.02	TCT	GAT	*Gly* GGG	ACC	TTC	TGT	R
Ben 22.	TCT	GAT	*Lys* AAG	ACC	TTC	TGT	R

4.5 GENE AMPLIFICATION AND RESISTANCE

In many natural populations, resistance to drugs, pesticides and other toxic chemicals often increases in a stepwise geometric progression, becoming unstable and reverting to susceptibility when selection is relaxed. Studies with microorganisms link these changes with gene amplification, and increased levels of target proteins or detoxifying enzymes may reach 10% of total cellular protein. A similar pattern of development was observed for resistance to a wide range of insecticides in the peach-potato aphid (*Myzus persicae*), and this was linked with a stepwise increase in levels of esterases (E4 and FE4) which both hydrolyse and sequester insecticidal esters (Devonshire and Sawicki, 1979). Several powerful molecular techniques have been employed to confirm the amplification hypothesis and to rule out the possibilty that selection of independent genes with purely additive effects might account for how insecticide resistance developed in aphids.

E4 protein was purified from a resistant aphid clone and used to immunize rabbits to produce polyclonal antisera to the E4 esterase. Fractions of poly A RNA (mRNA) enriched with E4 mRNA were identified in a cell-free rabbit reticulocyte protein synthesizing system by specific immunoprecipitation with E4 antibodies. Reverse transcriptase generated double-stranded DNA copies (cDNA) from the mRNA fractions and these were cloned into a plasmid vector. To identify clones that contained at least part of the E4 coding region, those that hybridized strongly to E4 mRNA were treated with RNA-ase H, which destroys DNA–RNA hybrids but not single-stranded nucleic acid sequences. cDNA clones complementary to E4 mRNA arrested the synthesis of E4 esterase in the reticulocyte protein synthesizing system (Field *et al.*, 1988). E4 specific cDNA probes isolated in this way were hybridized by dot-blot assay to total denatured DNA from sensitive (S), R1 (×4), R2 (×16) and R3 (×64) aphid clones. Intensity of the probe signal increased in line with resistance, clearly indicating that E4 esterase levels and gene copy number were linked (Figure 4.3). Southern blots of *Eco*RI digested DNA showed that the amplified E4 genes were

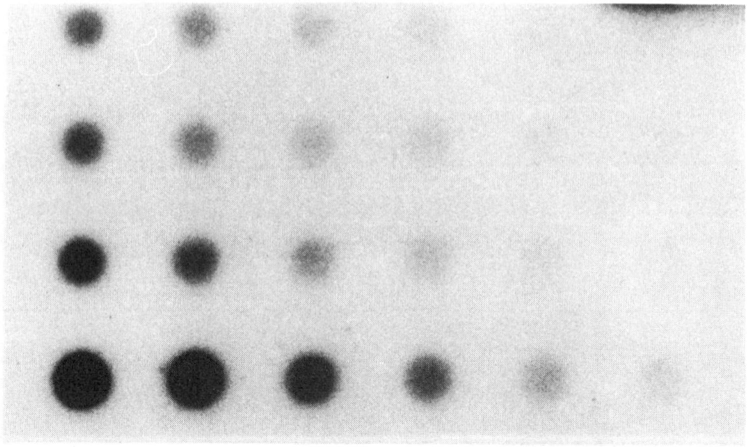

Figure 4.3 Binding of E4 esterase cDNA to genomic DNA extracted from susceptible and three insecticide resistant clones of *M. persicae* (R_1, R_2 and R_3). Dot blots of denatured aphid DNA (8, 4, 2, 1, 0.5 and 0.25 μg) made up to 8 μg with herring sperm DNA as necessary. From Field and Devonshire (1992).

present on an 8 kb fragment, whereas the FE4 gene, which codes for a slightly larger esterase associated with lower levels of resistance (R1), was located on a 4 kb fragment (Field and Devonshire, 1992). Cloning and sequencing these genomic fragments suggest that the two genes have a common 5′ coding region spanning the enzyme active site, but that the E4 gene is smaller perhaps as a consequence of rearrangements caused by translocations.

Since highly resistant aphid clones can spontaneously loose both their resistance and elevated esterase levels, it is of particular interest to understand the molecular mechanisms regulating amplification events. A fully susceptible revertant glasshouse aphid clone was found to have normal levels of E4 mRNA, but the amplified E4 gene remained still located on the 8 kb *Eco*RI fragment, and without any obvious rearrangement. One difference between expressed and silent E4 sequences was revealed by restriction digests of genomic DNA using the isoschizomers *Msp*I and *Hpa*II, which both cleave CCGG sequences, but only *Msp*I cuts if the internal cytosine is methylated. DNA from resistant clones was cut differently by the two restriction enzymes, whereas for the revertant clone banding patterns were the same (Figure 4.4). It seems that decreased methylation suppressed transcription of both E4 and FE4 genes and this accompanied a loss of resistance in *M. persicae*. This is in complete contrast with the majority of other organisms where demethylation of DNA is a signal to initiate transcription. None of the *Msp*I sites in the cloned E4 DNA fragments are upstream of the coding region, but if methylation sites can be identified in

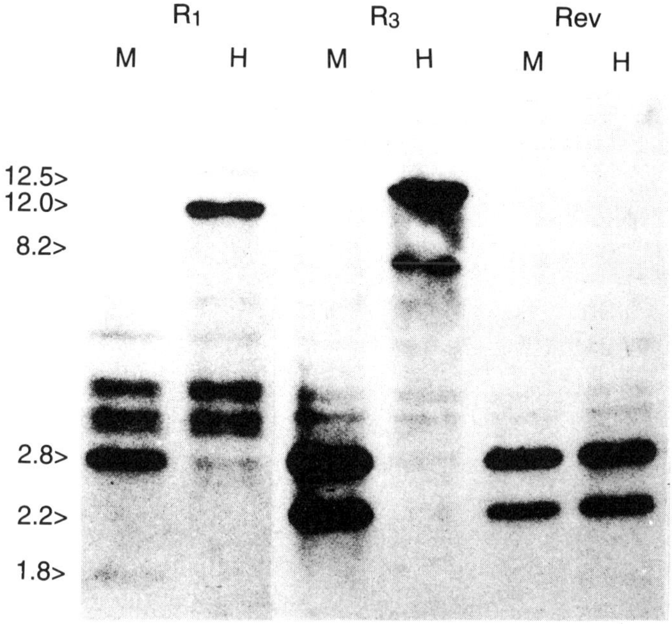

Figure 4.4 Southern blot of DNA extracted from insecticide resistant (R_1 and R_3) and a susceptible revertant (Rev) clone and probed with esterase E4cDNA. DNA was restricted with either *Msp*I (M) or *Hpa*II (H). Arrows indicate size of major bands. From Field and Devonshire (1992).

the promoter region, this may open the way to controlling methylation and combatting resistance.

4.6 IMPACT OF MOLECULAR TECHNIQUES ON PRACTICAL MEASURES TO COMBAT RESISTANCE

The examples discussed so far provide an insight into how various techniques have been used to expand understanding of resistance mechanisms at a molecular level. How are these findings likely to improve strategies to combat resistance? Already, herbicide resistant varieties have been engineered in some crops following recognition and isolation of herbicide resistance genes from different plants and microorganisms to broad spectrum weedkillers. Progress towards introducing these novel varieties into cropping systems is discussed in Chapter 8 and hopefully they will be seen as a way to increase the range of alternative herbicides available for use in strategies to combat existing herbicide resistance problems.

Although it may seem that a novel compound has a different mode of action from existing products, it is often difficult to identify its precise mode of action using conventional biochemical approaches. Selecting for single-gene resistance to the novel compound, and then cloning and sequencing the gene involved, can point towards the target enzyme involved, if sufficient sequence homology exists with known genes in databases. This approach has already been used to identify the resistance mechanism in *A. nidulans* mutants selected by a new class of fungicides (Gustaffson *et al.*, 1990) and led directly to establishing a mode of action. Cloning the same target gene from fungus, plant and insect, and expressing these all in yeast, or some other well understood organism, allows for rapid screening of new chemistry, identifying the most promising chemicals for more extensive examination using target organisms. By expressing the corresponding mammalian gene in the same way, early information on the toxicology of a new compound can be obtained. Molecular biology offers more effective ways to screen for novel activity and to improve the quality of structure/activity data if the target genes are from actual pathogen, pest or weed targets. Coupled with these screening methods, the impact of molecular techniques in expanding basic biochemical understanding of metabolic pathways in fungi, including the biosynthesis of natural toxins, should help in the search for novel target sites. Any novel products that emerge will certainly play an important part in strategies to combat resistance.

Genetic manipulation of biological control agents may help improve their effectiveness, allowing extension of their use, especially to outdoor cropping systems. Transfer of fungicide resistance genes for instance, to fungi which offer potential as control agents of weeds, pests or diseases, will allow biological and chemical control to be integrated into sustainable control measures.

4.6.1 Use of biochemical and serological assays to detect resistance

Improved understanding of pesticide resistance mechanisms introduces opportunities for developing biochemical assays which rapidly and accurately diagnose resistance. Large-scale rapid throughput tests for insecticide resistance are in routine use based on micro-titre plate technology, and measurement of inhibition kinetics of acetylcholinesterase (Devonshire, 1990). Where this enzyme substantially increases in resistant insects, these tests may reliably detect resistance in individual insects and even distinguish between esterase variants coded for by different alleles (Moores *et al.*, 1988). However, much depends on which model substrate is used in assays and organophosphorus and carbamate resistance may well result from changed enzyme activities not detected in the assay. Complications may also arise when assaying insects directly from treated crops, where overproduced enzyme in resistant insects may already be saturated with insecticide. Only by removing insects to an insecticide-free environment, where the enzyme recovers, can total esterase activity be measured accurately. Nevertheless, these simple assays have allowed insecticide resistance to be followed closely in several important insect pests (Byrne *et al.*, 1992).

Attempts to use serology to detect pesticide resistance have been less successful. Monoclonal antibodies (mAbs) specific for short synthetic peptides based on the sequence of *N. crassa* β-tubulin surrounding amino acid 167 did distinguish between extracts from MBC-resistant and -sensitive *N. crassa* conidia on Western blots (Martin *et al.*, 1993). As expected, perhaps, these mAbs did not diagnose resistance in *B. cinerea* where amino acid 167 is unchanged in resistant strains (Table 4.1). Unfortunately, these immunoassays failed to work in an ELISA format.

4.6.2 Use of PCR and associated technologies to detect resistance

PCR and associated technologies have opened up a vast new field of detection systems, and combined with allele specific oligonucleotides (ASOs), have already had a significant impact on the detection of point mutations, especially in the diagnosis of human genetic disorders.

Allele specific amplification of either resistant or sensitive gene segments may be achieved using ASOs directly as PCR primers (Williams *et al.*, 1990). In this system the mis-matched base pair is at the 3′ end of the PCR primer and, under appropriate annealing conditions, only when there is a perfect match between primer and genomic template will a PCR product of the expected size be produced, which can then be visualized by gel electrophoresis. Allele specific amplification of MBC resistant and sensitive β-tubulin sequences directed from genomic DNA of *B. cinerea* using this approach was unsuccessful. Only by first amplifying a non-specific 579 bp DNA fragment, and using this as template in a second 'nested' ASO-PCR, was the expected 310 bp fragment amplified (Martin *et al.*, 1992). In addition, to obtain specificity, ASO primers were tailed at the 5′ end with sequence unrelated to the β-tubulin gene to reduce the non-specific priming ability of *Taq* polymerase (Jeffreys *et al.*, 1991). Uncertainties over the fidelity of PCR-priming reactions suggest that caution is needed when identifying PCR products solely on their electrophoretic mobility. However, six out of seven *B. cinerea* strains were correctly diagnosed in this way. Bioassay of the seventh strain revealed an intermediate level of MBC resistance, but unfortunately no DNA sequence data are available to identify the DNA change causing resistance in this strain.

An alternative approach used ASO probes to hybridize directly to genomic DNA dot-blotted onto nylon membranes. By washing the membranes under the right stringency conditions, only where a perfect match between ASO and target sequence exists, does hybridization occur. In practice, a PCR amplified fragment of genomic DNA is used containing the expected sequence rather than total genomic DNA, to ensure that sufficient target DNA is present, especially where small samples, such as single disease lesions, are involved. Using this approach, and amplifying a 1.2 kb fragment of the β-tubulin gene of *V. inaequalis*, 25 apple scab isolates from around the world were correctly diagnosed as either sensitive, moderately (MR), highly (HR) or very highly (VHR) resistant to MBC fungicides (Figure 4.5; Koenraadt and Jones, 1992). Four labelled ASOs, each with a different nucleotide in the middle of the probe, corresponding to the known changes in either amino acid codon 198 or 200 were needed to achieve these results. Only one MR single spore isolate was incorrectly diagnosed where an alternate codon was used for tyrosine at amino acid codon 200. Low level (LR) resistant strains were unchanged in this region of the β-tubulin DNA and could not be distinguished from sensitive ones. Diagnosis of MBC resistance in 1.2 kb PCR products from apple scab lesions rather than *V. inaequalis* cultures was also possible and all 12 lesions examined were correctly assigned as either sensitive or VHR to MBC. We have used an identical approach in our laboratory to diagnose MBC resistance in a 381 bp PCR amplified β-tubulin fragment of *R. secalis*, where only two ASO probes (15mer) were needed to identify the single base change so far encountered in MBC-resistant field strains.

Irrespective of how ASOs are used to detect point mutations, the target DNA sequence must be known before they can be designed as probes. Nowadays, development of any novel pesticide is accompanied by some

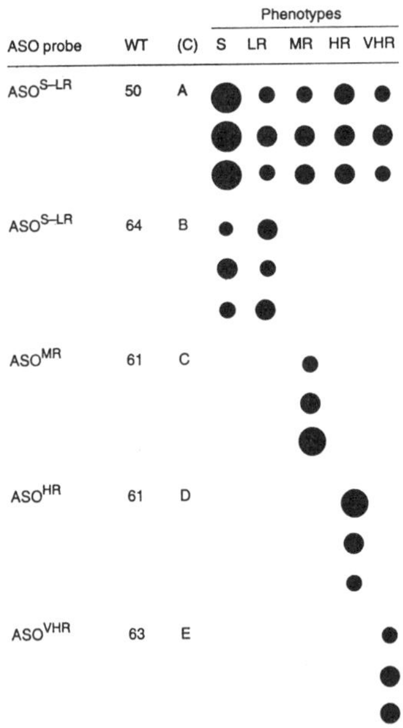

Figure 4.5 Detection of MBC resistance in *V. inaequalis*. PCR amplified β-tublin DNA from S, LR, MR, HR, and VHR strains was applied in columns to each panel of blots. Panels A and B were probed with $ASO^{S\text{-}LR}$ and washed at low stringency (A) or optimum stringency (B). ASO^{MR}, ASO^{HR} and ASO^{VHR} were applied to panels C, D and E, respectively, and washed at optimum stringency. Data from Koenraadt and Jones (1992).

early assessment of resistance risk, which involves studies on resistant mutants where these can either be generated in the laboratory or isolated as rare individuals from field populations. Where biochemical, genetic and molecular analysis of these mutants reveals a single DNA base change as the cause of resistance, ASO probes can be designed even before a new product is introduced. This should help in checking the appearance and spread of at least these point mutations in field populations and evaluating early predictions of resistance risk.

Degeneracy of the genetic code appears to limit ASOs to species specific diagnosis of resistance. None of the ASOs specific for *V. inaequalis* hybridized to amplified β-tubulin DNA from even the closely related *V. pirina*, on account of differences within these codons of the probe sequence (Koenraadt and Jones, 1992). Careful design of the primers might allow diagnosis to be extended across species boundaries, especially if a set of 'generic' ASOs aimed at each base pair position could be used in refinement of the 'nested' PCR primer approach. However, short ASO probes may offer a distinct advantage where visual diagnosis of pest, disease or weed is difficult. DNA probe techniques are rapid and with present technology should provide a result within 24–48 h, without requiring growth of the target species. Consequently, they offer considerable advantages in monitoring resistance in organisms that are either slow growing, are difficult to isolate, or indeed are obligate and cannot be maintained away from host material. Although ASO-PCR techniques are potentially very sensitive and can identify resistance even in mixtures, how quantitative they will be in monitoring mixed infections or heterokaryotic organisms remains to be established. Calibration against a standard set of known mixtures of template DNA may be one approach to overcoming this problem, which is equally important in other biochemical and conventional bioassays.

PCR-ASO techniques should detect resistance within 24–48 h, allowing results to be incorporated into spraying decisions. However, both electrophoretic separation of PCR products and membrane-based hybridization methods are not particularly quantitative, and there is a need to improve these aspects of assays if they are to have a practical value outside research laboratories. PCR can be carried out in micro-titre plates, and it should be possible to include the hybridization recognition event in the same wells of the plate. Linking this recognition event to an ELISA-based detection system should provide a rapid, robust, assay format allowing a large throughput of samples. Much research effort is currently directed towards refining ASO-PCR technology along these lines, especially in connection with detecting human genetic disorders, and approaches to adapting this methodology to the detection of pesticide resistance have been discussed elsewhere (Hollomon and Butters, 1992). Improvements in the accuracy of resistance monitoring, both in terms of the alleles involved and the numbers of samples tested, should not only improve early detection and help evaluate strategies aimed at combatting the problem, but it should open the way to exploit more fully the potential of mathematical modelling to predict how resistance will behave.

4.7 CONCLUSIONS AND FUTURE PROSPECTS

Today, few aspects of biology are not exposed to molecular analysis. As the examples discussed here reveal, molecular techniques play an important part in research on pesticide resistance, and it has undoubtedly expanded understanding of resistance mechanisms and how they are regulated. Molecular techniques inevitably shift the emphasis of work to the DNA level and it is not surprising, therefore, that a major thrust of the results obtained so far relates resistance to the DNA changes involved. Many allelic mutations are possible in a resistance gene, but fitness penalties severely restrict this number in field populations of target pests, weeds or diseases. As a result, better diagnosis of resistance seems possible, providing more rapid and accurate information about the field situation, and better management of the problem. Coupled with biochemistry, genetics and immunology, molecular techniques are helping to increase the availability of novel, alternative, chemical and biological pesticides, which should lessen selection pressure on any single product group. Common threads linking resistance problems across pest, disease and weed species are rare, but the involvement of cytochrome P450s, either through enhanced detoxification of insecticides and herbicides, or perhaps through altered target site in the case of important DMI fungicides, suggests that a concerted study of cytochrome P450 at a molecular level in relation to pesticide resistance may be worthwhile. No practical impact of molecular techniques on crop protection measures has yet reached the grower, but it will not be long before the benefits of better diagnosis, and herbicide resistant crops, will be tested on a wide scale. What is clear, however, is that the pace of molecular research is quickening, and we can expect more developments and techniques to find an application in crop protection at the farm level.

ACKNOWLEDGEMENTS

The authors wish to thank the many research collaborators and colleagues who have contributed to the thinking on which this chapter is based. We also wish to thank HGCA and MAFF for financial support towards the *Rhynchosporium* work briefly mentioned in the text.

REFERENCES

Brattsten, L.B. (1990) Resistance mechanisms to carbamate and organophosphate insecticides, in *Managing Resistance to Agrochemicals: From Fundamental Research to Practical Strategies*, (eds M.B. Green, H.M. le Baron and W.K. Moberg), ACS Symposium Series **421**, ACS, Washington, DC, pp. 42–60.

Brown, T.M. (1990) Biochemical and genetic mechanisms of insecticide resistance, in *Managing Resistance to Agrochemicals: From Fundamental Research to Practical Strategies*, (eds M.B. Green, H.M. le Baron and W.K. Moberg), ACS Symposium Series **421**, ACS, Washington, DC, pp. 61–76.

Byrne, F.J., Denholm, I., Birnie, L.C., Devonshire, A.L. and Rowland, M.W. (1992) Analysis of insecticide resistance in the white fly, *Bemisia tabaci*, in *Resistance 91: Achievements and Developments in Combatting Pesticide Resistance*, (eds I. Denholm, A.L. Devonshire and D.W. Hollomon), Elsevier, London, pp. 165–78.

Caseley, J.C., Keuh, J., Jones, O.T.G., Hedden, P. and Cross, A.R. (1990) Mechanism of chlorotoluron resistance in *Alepocurus myosuroides*, in *Proceedings of the 7th International Congress of Pesticide Chemistry*, (eds E. Kessler-Schnitz and S. Conway), IUPAC, Hamburg, p. 417.

Cooley, R.N. and Caton, C.L. (1993) Molecular analysis of the *Septoria nodorum* β-tubulin gene and characterisation of a benomyl-resistance mutation. *Molecular General Genetics*, **287**, 58–61.

Crute, I.R. (1992) The contribution of genetic studies to understanding fungicide resistance, in *Resistance 91: Achievements and Developments in combatting Pesticide Resistance*, (eds I. Denholm, A.L. Devonshire and D.W. Hollomon), Elsevier, London, pp. 190–202.

Davidse, L.C. and Flach, W. (1977) Differential binding of methyl benzimidazol-2-yl carbamate to fungal tubulin as a mechanism of resistance to this anti-mitotic agent in mutant strains of *Aspergillus nidulans*. *Journal of Cell Biology*, **72**, 174–93.

De Waard, M.A. and van Nistelrooy, J.G.M. (1982) Accumulation of SBI fungicides in a wild-type and fenarimol-resistant isolates of *Penicillum italicum*. *Pesticide Science*, **22**, 371–82.

Devonshire, A.L. (1990) Biochemical and DNA probing techniques for monitoring insecticide resistance genes, in *Pesticides and Alternatives: Innovative Chemical and Biological Approaches to Pest Control*, (ed. J. Casida), Elsevier, Amsterdam, pp. 421–31.

Devonshire, A.L. and Moores, G.D. (1982) A carboxylesterase with broad substrate specificity causes organophosphorus, carbamate, and pyrethroid resistance in peach-potato aphid (*Myzus persicae*). *Pesticide Biochemistry and Physiology*, **18**, 235–46.

Devonshire, A.L. and Sawicki, R.M. (1979) Insecticide-resistant *Myzus persicae* as an example of evolution by gene duplication. *Nature*, **270**, 140–1.

Field, L.M., Devonshire, A.L. and Forde, B.G. (1988) Molecular evidence that insectide resistance in peach-potato aphids (*Myzus persicae* Sulz,) results from amplification of an esterase gene. *Biochemical Journal*, **251**, 309–12.

Field, L.M. and Devonshire, A.L. (1992) Insecticide resistance by gene amplification in *Myzus persicae*, in *Resistance 91: Achievements and Developments in combatting Pesticide Resistance*, (eds I. Denholm, A.L. Devonshire and D.W. Hollomon), Elsevier, London, pp. 240–50.

Fuerst, E.P., Nakatami, H.Y., Dodge, A.D., Penner, D. and Arntzen, C.J. (1985) Paraquat resistance in Conyza. *Plant Physiology*, **77**, 984–9.

Fujimura, M., Oeda, K., Inove, H. and Kato, T. (1990) Mechanism of action of N-phenylcarbamates in benzimidazole-resistant *Neurospora* strains, in

Managing Resistance to Agrochemicals: From Fundamental Research to Practical Strategies, (eds M.B. Green, H.M. Le Baron and W.K. Moberg), ACS Symposium Series 421, Washington DC, pp. 224–36.
Gustaffson, G.D., Waldron, C. and Davis, G.E. (1990) Isolation of a DNA fragment that transforms *Aspergillus nidulans* to resistance to the antifungal compound LY 214352, presented at the *Fourth International Mycological Conference*, Regensburg, Abstract IIF-282/1.
Hartnett, M.E., Chui, C.F., Mauvais, C.J., McDevitt, R.E., Knowlton, S., Smith, J.K., Falco, S.C. and Mazur, B.J. (1990) Herbicide-resistant plants carrying mutated acetolactate synthase genes, in *Managing Resistance to Agrochemicals: From fundamental research to practical strategies*, (eds M.B. Green, H.M. le Baron and W.K. Moberg), ACS Symposium Series **421**, ACS, Washington, DC, pp. 459–73.
Hollomon, D.W. and Butters, J.A. (1992) New approaches to detecting fungicide resistance. *Agro-Food Hi-tech*, **3**, 20–3.
Ishii, H., Iwasaki, S., Sato, Z. and Inoue, I. (1990) Binding of cellular protein from *Venturia nashicola* isolates to carbendazim. Its relationship with sensitivity to N-phenylcarbamates, N phenyl-formadoximes and Rhizoxin, in *Managing Resistance to Agrochemicals: From Fundamental Research to Practical Strategies*, (eds M.B. Green, H.M. le Baron, and W.K. Moberg), ACS Symposium Series **421**, ACS, Washington, DC, pp. 237–48.
Jeffreys, A.J., Macleod, A., Tamaki, K., Neil, D.L. and Monckton, D.G. (1991) Minisallelite repeat coding as a digital approach to DNA typing. *Nature*, **354**, 204–9.
Kalb, V.F., Woods, C.W., Turi, T.G., Dey, C.R., Sulter, T.R. and Loper, J.C. (1987) Primary structure of the P_{450} lanosterol demethylase gene from *Saccharomyces cerevisiae*. *DNA*, **6**, 529–37.
Keon, J.P.R., Broomfield, P.L.E., Hargreaves, J.A., and White, G.A. (1992) Molecular genetic analysis of carboxin resistance in *Ustilago maydis*, in *Proceedings of the Brighton Crop Protection Conference – Pests and Diseases-1992*, The British Crop Protection Council, Farnham, UK, pp. 221–6.
Koenraadt, H. and Jones, A.L. (1992) The use of allele-specific oligonucleotide probes to characterize resistance to benomyl in field strains of *Venturia inaequalis*. *Phytopathology*, **82**, 1354–8.
Koenraadt, H., Somerville, S.C. and Jones, A.L. (1992) Characterization of mutations in the beta-tubulin gene of benomyl-resistant field strains of *Venturia inaequalis* and other plant pathogenic fungi. *Phytopathology*, **82**, 1348–54.
Martin, L.A., Fox, R.T.V., Baldwin, B.C. and Connerton, I.F. (1992) Use of polymerase chain reaction for the diagnosis of MBC resistance in *Botrytis cinerea*, in *Proceedings of the Brighton Crop Protection Conference – Pests and Diseases-1992*, The British Crop Protection Council, Farnham, UK, pp. 207–14.
Martin, L.A., Fox, R.T.V. and Baldwin, B.C. (1993) Rapid methods for detection of MBC resistance in fungi: 1. Immunological approach, in *Proceedings of the 10th International Symposium on Systemic Fungicides and Antifungal Compounds*, (eds H. Lyr and C. Polter), Reinhardsbrunn, 209–18.
May, G.S., Tsang, L.-S., Smith, H., Fifel, S. and Morris, N.R. (1987) *Aspergillus nidulans* beta-tubulin genes are unusually divergent. *Gene*, **55**, 231–43.
McCall, P.J., Stafford, L.E., Zorner, P.S. and Gavit, P.D. (1986) Modelling the foliar behaviour of atrazine with and without crop oil concentrate on giant foxtail and the effect of tridiphane on the model rate constants. *Journal of Agricultural Food and Chemistry*, **34**, 235–8.
Moores, G.D., Devonshire, A.L. and Denholm, I. (1988) A microtitre plate assay for characterising insensitive acetylcholinesterase genotypes of insecticide-resistant insects. *Bulletin Entomological Research*, **78**, 537–44.

Orbach, H.J., Porro, E.B., and Yanofsky, C. (1986) Cloning and characterization of the gene for β-tubulin from a benomyl resistant mutant of *Neurospora crassa*, and its use as a dominant selectable marker. *Molecular and Cellular Biology*, **6**, 2452–61.

Osmani, S.A. and Oakley, B.R. (1991) Cell cycle and tubulin mutations in filamentous fungi, in *More gene manipulations in fungi*, (eds J.W. Benaett and L.L. Lasure), Academic Press, London, pp. 107–25.

Pauron, D., Amichot, M. and Berge, J.-B. (1992) Voltage-dependent sodium channels in susceptible and pyrethroid-resistant *Drosophila* strains, in *Resistance 91: Achievements and Developments in combatting Pesticide Resistance*, (eds I. Denholm, A.L. Devonshire and D.W. Hollomon), Elsevier, London, pp. 228–39.

Ray, T.B. (1984) Site of action of chlorsulfuron: inhibition of valine and isoleucine biosynthesis in plants. *Plant Physiology*, **75**, 827–31.

Sawicki, R.M. (1973) Recent advances in the study of the genetics of resistance in the house fly, *Musca domestica*. *Pesticide Science*, **4**, 501–12.

Selker, E.U. (1990) Pre-meiotic instability of repeated sequences in *Neurospora crassa*. *Annual Review of Genetics*, **24**, 579–613.

Snustad, D.P., Haas, N.A., Kopczak, S.D. and Silflow, C.D. (1992) The small genome of *Arabidopsis* contains at least nine expressed β-tubulin genes. *Plant Cell*, **4**, 549–56.

Thomas, J.H., Neff, N.F. and Botstein, D. (1985) Isolation and characterisation of mutations in the β-tubulin gene of *Saccharomyces cerevisiae*. *Genetics*, **112**, 715–34.

Trebst, A. (1991) The molecular basis of resistance of phytosystem II herbicides, in *Proceedings of the 11th Long Ashton International Symposium 'Herbicide Resistance'*, (ed. J.C. Caseley), Butterworths, London, pp. 52–61.

van Rensen, J.J.S. and de Vos, O.J. (1992) Biochemical mechanisms of resistance to photosystem II herbicides, in *Resistance 91: Achievements and Developments in Combatting Pesticide Resistance*, (eds I. Denholm, A.L. Devonshire and D.W. Hollomon), Elsevier, London, pp. 251–61.

Williams, J.G.K., Kubelik, A.R., Livak, K.J., Rafalski, J.A. and Tingey, S.V. (1990) DNA polymorphisms amplified by arbitary primers are useful as genetic markers. *Nucleic Acids Research*, **12**, 6531–5.

PART THREE

Engineering Resistance to Diseases, Herbicides and Pests

Molecular biology and genetic engineering to improve plant disease resistance

E. Ward, S. Uknes and J. Ryals

5.1 INTRODUCTION

One of the main limiting factors in food and fibre production is plant disease. The exact worldwide losses to disease are hard to estimate relative to other agricultural problems such as herbivory, weeds, drought and other stresses. However, it is generally accepted that diseases are responsible for a considerable fraction of crop loss. The diseases that are most damaging are caused by fungi, bacteria and viruses. Currently, the main methods of controlling disease involve the use of chemical pesticides, resistant cultivars and agricultural practice. These methods have helped in managing many diseases, but there are limitations to the current technologies available. For instance, while chemical solutions are very effective, there are problems with resistance which could affect the sustainability of the solution and thus jeopardise the farmer's livelihood in the long term. Likewise, the use of genetically resistant cultivars has limitations in that the resistance is usually effective only against subspecies of certain pathogens. In some cases, resistant varieties have been grown for many years, providing successful disease control. However, as with chemical solutions, there are also resistance problems that are sometimes so severe that a resistant variety will have only a short practical lifetime. An example of this is resistance to rust in wheat (*Triticum aestivum* L.), where a resistant variety may only be

Molecular Biology in Crop Protection. Edited by George Marshall and Dale Walters. Published in 1994 by Chapman & Hall, London. ISBN 0 412 54390 7(HB) 0 412 54400 8(PB).

effective for two to three seasons before a new virulent isolate of the pathogen arrives. Furthermore, genetic resistance can only be used in crops where a particular resistance source is known. For example, genetic resistance to important diseases, such as take-all in wheat, is simply not available. Thus, a need exists for improved solutions to disease control.

Over the past 10 years there has been an explosion of knowledge in the field of genetic engineering. In plants and animals, genes from one organism have been moved to another and expressed to give new and important traits. Because of this renaissance, it seems likely that the tools of genetic engineering might be used to increase our options for disease-tolerant germplasm. For instance, it may prove feasible to isolate disease resistance genes and express them in a variety of crops at will. Other strategies might employ the expression of antibiotic genes from sources such as viruses (e.g. bacteriophage T4 lysozyme), birds (e.g. hen egg white lysozyme), or even frog skin (e.g. magainins), and in this way there could be a large increase in available diversity. Even more elegant strategies might involve the harnessing of inducible disease control systems in plants, to effectively manage pathogen growth. These types of solutions were previously unavailable to plant breeders.

Genetic engineering solutions will undoubtedly have an impact on agriculture by providing a greater diversity of disease-resistant plants. These new varieties will then be used to provide sustainable solutions to plant disease in the future. This chapter is designed to give a perspective on the possibilities that will be available for disease control over the next decade. The future has never been more exciting; frontiers of knowledge extend in all directions and are limited only by our vision and our dogma.

5.2 AN OVERVIEW OF PLANT–PATHOGEN INTERACTIONS

Most interactions between plants and microbes do not result in disease. Thus, given the vast number of organisms with pathogenic potential in the environment, it seems reasonable to concentrate not only on why plants are resistant to infection, but also on how a disease-causing interaction develops. Thus, although it is not within the scope of this chapter to deal with plant–pathogen interactions in detail, a brief look at pathogen attack and plant defence is important.

5.2.1 Pathogen attack

In order for a pathogen to cause disease on a plant, it must:

1. Find and recognize the appropriate host (e.g. by sensing stomata on leaf surfaces).

2. Attach and penetrate the surface of the host (e.g. by using hydrolytic enzymes).
3. Grow and develop utilizing nutrients provided by the host.
4. Fend off host defences (e.g. by suppressing inducible defences or detoxifying a phytoalexin).
5. Reproduce.
6. Induce disease symptoms by directly or indirectly impairing or altering the vital functions of the plant (e.g. by producing a toxin).

These aspects of pathogen attack have been covered by a number of reviews (see Further Reading).

5.2.2 Plant defence

In order to protect itself from disease a plant must do one of the following:

1. Fail to be recognized as a potential host by the pathogen.
2. Prevent or inhibit attachment and entrance of the pathogen.
3. Inhibit growth and development of the pathogen (e.g. using inducible defences; see below).
4. Contain or prevent the spread of disease symptoms.

Good accounts of plant defence can be found in several recent reviews (see Further Reading).

5.2.3 A brief look at inducible plant defences

Several lines of evidence indicate that the timing of induction of active defence responses appears to be a critical factor in whether or not a plant becomes diseased (de Wit and Van der Meer, 1986; Slusarenko and Longland, 1986; Collinge and Slusarenko, 1987). Therefore, considerable research has focused on defence mechanisms that are induced by pathogens and factors thought to be associated with pathogenesis.

(a) Systemic acquired resistance (SAR)

Induced resistance, or SAR, can result from infection with a pathogen that causes necrosis. About the time that the necrotic lesions appear on the lower leaves of the plant, the rest of the plant becomes noticeably resistant to a variety of bacterial, fungal, and viral pathogens. This phenomenon has been observed in dozens of plant species; indeed, in any species in which a systematic attempt has been made to document SAR, the response has been demonstrated (Figure 5.1).

Recent work has begun to shed some light on how SAR works. First, at the same time that resistance appears systemically in the plant, expression of at least nine sets of genes is induced (Ward *et al.*, 1991) (Figure 5.2). Some

Figure 5.1 Systemic acquired resistance, demonstrated in *A. thaliana*. Both plants have been infected with *P. syringae* pv. *tomato*, the symptoms of which are easily seen in the plant on the left. The plant on the right had previously been inoculated on lower leaves with turnip crinkle virus, which induced resistance to the subsequent bacterial infection. After Uknes *et al.* (1993).

of these genes encode so-called pathogenesis-related proteins, which are extracellular, acidic polypeptides, long associated with pathogen infection. The products of these induced genes may be involved in establishing SAR; in one case, this has been shown directly (see section on genetic engineering below). Second, while the role of the downstream protein products is now coming into focus, their expression must lie at the end of a signal transduction pathway that begins with the initial contact between host plant and necrotizing pathogen. Very little is understood about the nature of this pathway. In the past few years, several groups have shown that salicylic acid (2-hydroxybenzoic acid) increases transiently in the phloem of inoculated plants and in tissues that are about to become resistant (Malamy *et al.*, 1990; Métraux *et al.*, 1990). In addition, salicylic acid has been shown to induce expression of the same set of genes that are systemically induced as a result of pathogen infection (Ward *et al.*, 1991). These results are consistent with salicylic acid playing a role in signalling to the rest of the plant that a leaf has contracted a necrotic lesion from pathogen infection.

A recent experiment from our laboratory has now shown definitively that salicylic acid participates in the SAR signalling pathway. The bacterium *Pseudomonas putida* contains an enzyme, salicylate hydroxylase, that

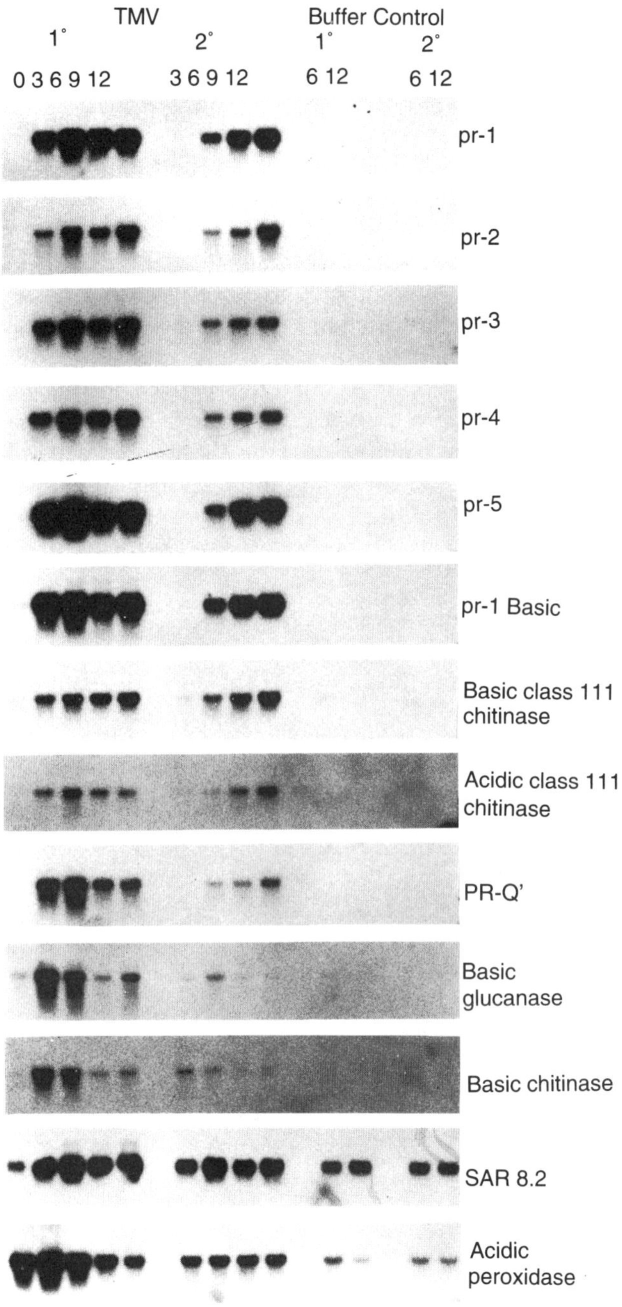

Figure 5.2 Coordinate gene induction correlated with SAR. Gel blots containing RNA from leaves infected with TMV (1°) or from uninfected leaves on infected plants (2°) were probed with the various genes named at the right of the figure. mRNAs detected by the first nine probes are all induced with similar kinetics in 2° tissue, and first appear at 6 days after infection, the same time at which SAR is first detectable. From Ward *et al.* (1991).

converts salicylate to catechol. The gene for this enzyme (*nahG*) was engineered for expression in plants and transformed into tobacco (*Nicotiana tabacum* L.). The resulting plants produced virtually no salicylate after pathogen infection, and did not show an SAR response (Figure 5.3). (Catechol has no resistance-inducing effect on plants, and does not trigger SAR-related gene expression.) This is the first example of manipulation of a signalling pathway involved in plant–pathogen interactions. This experiment illustrates the power of using plant genetic engineering methods to explore basic biology in ways that would be virtually impossible by other techniques.

Early work on SAR had shown that a transmissible signal emanates from the initially infected leaf and travels through the phloem to distal organs. This signal could move through a graft junction from an infected rootstock to a healthy scion. To address whether SA is this long distance signal for SAR, *nahG* transgenic and non-transgenic (Xanthi) tobacco plants were grafted. When Xanthi was used as a induced root stock and a *nahG* plant was used as the scion (in which SAR was measured) no resistance was observed. However when *nahG* plants were used as the induced root stock

Figure 5.3 Expression of the bacterial *nah*G (salicylate hydroxylase) gene in transgenic tobacco abolishes SAR. Both plants were infected on lower leaves with TMV and then challenge-inoculated 7 days later. The plant on the right (Xanthi wild-type) shows the reduction in lesion size characteristic of SAR; the *nah*G transgenic on the left shows no SAR.

and Xanthi was the scion, normal levels of SAR were detected. Thus, *nahG* rootstocks can generate the long distance signal despite their inability to develop SAR or accumulate SA. These results indicate that SA is required for the development of resistance but is not the long distance signal for SAR.

5.3 GENETICS OF PLANT–PATHOGEN INTERACTIONS

Because a great deal of research is being invested in isolation and characterization of so-called 'major' resistance genes, we will present here some of the basic genetic models that account for how plants and pathogens associate. When a plant and pathogen come into contact one of three possibilities exist: disease, a resistance interaction, or no interaction at all. If the pathogen recognizes the plant as a host and the plant does not have a defence against this particular pathogen, the interaction is said to be compatible. A compatible interaction results in disease of the host plant. If the plant can recognize the pathogen rapidly, generally by virtue of a specific recognition factor called a resistance (R) gene that interacts with a specific pathogen factor called an avirulence (*avr*) gene, then the interaction is said to be incompatible. An incompatible interaction often results in the rapid host cell death known as a hypersensitive response (HR). Whether the HR causes the restriction of the disease lesion or whether it is a downstream effect of resistance is a hotly debated topic. Whatever the case, a plant producing an HR is able to effectively limit the extent of the disease. In the most extreme cases, a single-cell HR occurs on the host plant in response to a single pathogen, resulting in no macroscopically visible symptoms. If the pathogen does not recognize the host, or the plant defences are impervious to any attack by this particular pathogen, the interaction is called non-host resistance and typically results in no observable reaction by the plant and no growth of the pathogen.

The first two interactions can be most easily understood by a model developed by Flor (1956; Figure 5.4). In the gene-for-gene model of plant–pathogen interactions, a dominant resistance gene in the host (R) interacts with a dominant avirulence gene in the pathogen (A) to cause an incompatible interaction (I). Absence of either the dominant resistance allele or the avirulence gene results in a compatible interaction (C). One of the many examples of a gene-for-gene interaction is resistance to *Pseudomonas syringae* pv. *tomato*, the causal agent of bacterial leaf speck disease in tomato (*Lycopersicon esculentum* L.). The *Pto* gene is a dominant resistance gene bred into tomato from the sexually compatible species *Lycopersicon pimpinellifolium*. Plants with *Pto* are resistant to any *Pseudomonas syringae* that contains the avirulence gene *avrPto* (Ronald *et al.*, 1992; Martin *et al.*, 1993). This type of dominant resistance to specific races of pathogens has been widely used to develop resistant cultivars of crops. For a few of these R genes, the resistance has worked season after season with little virulence developing

		Plant genotype	
		R–	rr
Pathogen genotype	A–	I	C
	aa	C	C

Figure 5.4 Gene-for-gene model of plant–pathogen interactions. I, incompatible interaction (no disease phenotype); C, compatible interaction (disease phenotype). R, resistance gene; r, not resistant. A, avirulence gene; a, no avirulence gene.

in the pathogen. In these cases the R gene may be recognizing an *avr* gene that is required for survival of the pathogen (e.g. a virulence factor) (Kearney and Staskawicz, 1990). More commonly, the R gene recognizes an *avr* gene in the pathogen that is not absolutely required for viability and pathogenicity. In these cases, the resistance does not last for many seasons in the field because elimination of the *avr* gene in the pathogen (a single, apparently non-deleterious change) will allow virulence. A good example of this is the breeding of rust resistance in wheat, where the pathogen keeps pace with each newly bred R gene by evolving back toward virulence. More than 90 different resistance alleles have now been identified for rust fungi and powdery mildew in wheat. The protein products of many *avr* genes have been determined, but in no case is their function known. In bacteria, one of the best elucidated cases is the *avrD* gene from *Pseudomonas syringae* pv. *tomato*. This gene encodes a protein involved in production of two low molecular weight γ-lactones, called syringolides 1 and 2 (Keen *et al.*, 1993). These small molecules cause necrosis only on soybean cultivars containing the resistance gene *Rpg4*, which is responsible for resistance to *Pseudomonas syringae* harboring *avrD*.

Interestingly, in an interaction between a single host and single pathogen species, specific gene-for-gene interactions do not always result in the same plant symptoms. For example, the incompatible interaction of an R gene, (R1) and an *avr* gene (A1) may result in a classic single-cell HR, while the interaction of a different R gene (R2) with its corresponding *avr* gene (A2) may result in a much greater degree of leaf necrosis. Although both interactions effectively prevent full-blown disease symptoms, they are morphologically quite different. Moreover, combinations of some R genes within the same host can interact with each other, providing altered resistance phenotypes compared to those seen when the combination is not present.

Thus, within any plant–pathogen interaction, a continuum of disease possibilities exists, even if discrete R/*avr* gene combinations govern the interaction (Crute, 1985). It has been argued by many that these interactions indicate that the plant R gene encodes a receptor and the *avr* gene directs synthesis of an elicitor which directly interacts with the receptor (Gabriel and Rolfe, 1990). However, many other possibilities exist that are completely consistent with all available data. Because only a few *avr* genes have been cloned and characterized, and no plant R genes (of the classic 'Flor' type) have been cloned, the exact nature of these interactions remains a mystery.

		Plant genotype	
		S–	ss
Pathogen genotype	V–	D	N
	vv	N	N

Figure 5.5 Model of dominant plant sensitivity, dominant pathogen virulence. D, disease phenotype; N, no disease phenotype. V, gene for production of virulence factor, v, no virulence factor. S, gene for sensitivity to the virulence factor; s, insensitive to the virulence factor.

Two other genetically determined plant–pathogen interactions that fall outside of the classic gene-for-gene paradigm have been characterized at the molecular level in sufficient detail to reveal the fundamental nature of the interaction. The first example is that of dominant plant sensitivity and dominant pathogen virulence (Figure 5.5). In this interaction a virulence gene in the pathogen is required for disease to occur. However, disease can only occur in plants that carry the dominant sensitivity gene. A classic example of this type of interaction is between Texas male-sterile cytoplasm (*cms-T*) maize (*Zea mays* L.) and *Bipolaris maydis* race T (Levings and Brown, 1989). The second example is that of dominant plant resistance and dominant pathogen virulence (Figure 5.6). In this interaction, the plant gene can encode a protein responsible for detoxifying a pathogen factor that is required for virulence. A classic example of this type of interaction is between *Cochliobolus carbonum* race 1 and *Hm1* maize (Meeley *et al.*, 1992).

5.4 STRATEGIES FOR ENGINEERING RESISTANCE TO PATHOGENS

Now that we have covered some of the basic principles of the ways in which plants and pathogens interact, this section will consider some applications

		Plant genotype	
		R–	rr
Pathogen genotype	V–	N	D
	vv	N	N

Figure 5.6 Model of dominant plant resistance, dominant pathogen virulence. N, no disease phenotype; D, disease. R, gene for resistance; r, not resistant. V, gene for a required virulence factor; v, no virulence factor.

of this knowledge. Viral diseases will be dealt with first, followed by bacterial and fungal pathogens.

5.4.1 Viruses and pathogen-derived resistance

Pathogen-derived resistance is a key concept in the engineering of plants resistant to viral infection. This idea consists of transferring functions from the viral genome to the host in order to interfere with the normal biogenesis of the virus. Host expression of these pathogen-derived genes can alter the life cycle of the virus in several ways: the gene product can be produced in large excess to its normal concentration, the gene product can be defective in some way that dominantly inhibits normal function of the cognate gene in an infecting virus, the gene product be modified to possess a new activity that is detrimental to viral replication or maturation, or the gene product can be produced at an inappropriate stage in viral development.

Because viruses are so simple in genomic organization and their genomes have been well characterized, a considerable amount is known about the various functions encoded and their roles in the viral life cycle. This is the case for many of the more than 30 virus families that have been described to date. A typical virus produces as few as half a dozen gene products, which is about 1000 times fewer than are produced by a bacterium or fungus. Although knowing the full range of proteins produced by a pathogen does not give a complete description of its mode of pathogenesis, it clearly provides a major advantage in figuring out how the pathogen grows. The extent of this detailed understanding is a key factor in the ability to design effective biotechnological strategies for controlling plant diseases.

(a) Coat protein protection

The best and most widely exploited example of pathogen-derived resistance is the use of coat protein genes to protect against viral infection and systemic movement. The idea for this method of genetically engineered resistance came from a phenomenon called cross protection. Cross protection results when infection with a less virulent strain of a virus reduces disease symptoms caused by a more virulent form the same virus. Although the biochemical basis for cross protection is poorly understood, it probably results from interference between virus strains during virus particle disassembly or later during replication of the viral genome. Early studies implicated viral coat protein as a key mediator of cross protection, making it a natural target for engineering into plants.

In a seminal experiment, Beachy and coworkers used this strategy to genetically engineer tobacco plants with resistance to tobacco mosaic virus (TMV) (Powell-Abel *et al.*, 1986). They created a chimeric gene consisting of a cDNA encoding the TMV coat protein gene linked to a strong constitutive plant promoter and, with collaborators at the Monsanto company, inserted this gene into tobacco using *Agrobacterium*. The resulting plants manifested resistance to TMV in three ways. First, fewer viral infection sites were visible after hand inoculation. Second, plants showed much less of the yellowish 'mosaic' pattern characteristic of systemic disease development. Third, and consistent with the first two observations, the plants accumulated less virus. Interestingly, coat protein protection can be effective against viruses other than the one from which the transgene is derived. Tobacco transformed with a chimeric gene that expresses the coat protein from soybean mosaic virus is resistant to its relatives, tobacco etch virus and potato virus Y. These viruses share approximately 60% amino acid sequence homology with the coat protein of SMV. A thorough study has shown that ability of TMV coat protein to protect against different viruses or strains of TMV falls off somewhere between 45 and 60% homology (Nejidat and Beachy, 1990).

TMV coat protein has now been expressed in tomato, where it protects against both TMV and the closely related tomato mosaic virus. The performance of transgenic tomatoes expressing high levels of coat protein was first tested in the field in 1987 (see Chapter 8). After mechanical inoculation with TMV, essentially all of the non-transformed control plants showed virus symptoms, while only 5% of transgenic coat protein expressors showed symptoms by harvest time. Moreover, yield of fruit from the infected transgenic plants was equal to uninfected controls and about 25% higher than infected controls. Thus, expression of coat protein gave high levels of protection from the disease, but did not adversely affect fruit production. This is the first of many cases now recorded where high levels of expression of a foreign protein do not seem to drain the plant's nitrogen or energy reserves, as had frequently been assumed in the early days of

Figure 5.7 An example of transgenic coat protein protection against virus infection in the field. Both rows were inoculated with potato viruses X and Y; control potato plants are in the front row and transgenic potato plants expressing the coat proteins of both viruses are in the back row. From Kaniewski *et al.* (1990).

agricultural genetic engineering. Within the next few years, transgenic tomatoes resistant to TMV may be sold as one of the first commercial products of plant genetic engineering (Figure 5.7).

(b) Protection derived from other viral genes

Other viral proteins can be effective in protecting plants against viral infection. Recently, several groups have shown that intact viral replicase genes, or portions of these genes, are able to confer high levels of protection when expressed as chimeric genes in plants. For instance, the *ORF 1* gene of potato virus X (PVX) is thought to encode a replicase function, based on sequence homology with known viral replicases. Transgenic tobacco expressing either the intact *ORF 1* gene or the 5′ half show protection against PVX infection approximately 10-fold greater than that seen with CP-mediated protection (Braun and Hemenway, 1992). Interestingly, 3′ portions of the gene do not provide protection, despite containing the conserved glycine–aspartate–aspartate (GDD) motif known to be functionally important in related replicases. A recent report has appeared, however, which shows that transgenic plants containing a PVX replicase gene in which the GDD motif has been mutated to alanine–aspartate–aspartate (ADD) causes potent inhibition of viral accumulation (Longstaff *et al.*, 1993). This result confirms that the replicase protein *per se* is responsible for protection and that missense mutations in viral functions can act as dominant negative mutations capable of interfering with viral multiplication.

(c) *Protection resulting from non-structural components*

Other viral components that do not comprise functioning genes can be used to provide protection. For example, satellite RNAs are small nucleic acid molecules that are found in association with some viruses. In most cases, the virus can be propagated free of satellites; thus, the satellite is thought to be a 'molecular parasite' that uses the virus as an agent to replicate itself and promote its own transfer from plant-to-plant. The genetic engineer's interest in satellites comes from the observation that satellites can, in some cases, dramatically decrease the symptom formation normally caused by certain viruses. Thus, providing the host plant genome with its own satellite RNA gene could conceivably protect against virus symptoms. In a few cases, this strategy has been shown to work. Satellites of tobacco ringspot virus and cucumber mosaic virus, driven by the strong 35S promoter in transgenic plants, could be amplified to high copy number by infection with the associated virus (Gerlach *et al.*, 1987; Harrison *et al.*, 1987). In both cases, replication of the virus was reduced and symptom formation was significantly alleviated.

An intriguing variation on protection using satellite RNA is the use of short, defined sequences from satellites that can cleave RNA in a site-specific fashion. These so-called ribozymes adopt a tertiary structure that allows them to catalyse intramolecular cleavage at a specific sequence motif. By constructing a synthetic ribozyme which contains the catalytic domain attached to 'arms' that are complementary to a novel target sequence, it is possible to cleave that target sequence efficiently *in vitro*. In theory, it should be possible to inactivate any transcript *in vivo* by expressing an appropriately designed ribozyme transgene. To date, however, reported success has been limited. Very large ratios of ribozyme to target are apparently required to obtain even reasonable decreases in the steady state levels of a transcript.

(d) *Future prospects for transgenic virus resistance*

One curious result that appears repeatedly in the transgenic virus resistance literature is the observation that high expression levels of the gene product of interest, be it coat protein or replicase, do not always correlate with high levels of resistance. The exact basis for this seemingly anomalous phenomenon is not known. In some cases, expression of RNA, but not protein derived from the transgene, may provide the resistance. Why this should be so is not evident, but the observation falls within the broad theory of pathogen-derived resistance, which states that perturbation of the viral life cycle, even by means not completely understood, can lead to significant levels of resistance. Because viruses are quite specialized pathogens composed of rather simple arrays of macromolecules, many of the strategies outlined here may not apply to more complex pathosystems. Indeed, the simplest forms of pathogen-derived resistance may only work for viruses. For example, it is difficult to imagine how engineering a structural component of a phytopathogenic bacterium or fungus into a plant would result in defective multiplication of that organism. These more complex pathogens provide commensurately more complex challenges for the biotechnologist wishing to bolster a plant's defences against them.

5.4.2 Antifungal and antibacterial strategies

Even with the rather basic knowledge of plant–pathogen interactions we now possess, several ways to affect a pathogen adversely can be envisioned. These include:

1. Engineering a plant to synthesize a novel protein with direct activity against the pathogen.
2. Prematurely activating expression of some endogenous gene normally induced in response to pathogen infection.
3. Introducing a gene (or genes) that directs the synthesis of a novel low molecular weight compound with antimicrobial activity.
4. Introducing a gene to deactivate some pathogen-produced factor required for pathogenesis.

The following sections will describe some of the ways that potential factors for engineering disease resistance can be identified, together with some of the antimicrobial activities that have been identified. Emphasis will be placed on the cases in which transgenic plants have been shown to have enhanced disease tolerance.

(a) Finding candidate genes to combat microbial pathogens

***In vitro* screening**

One way to find candidate disease resistance factors for expression in transgenic plants is to screen for them *in vitro* (Figure 5.8). This approach has the major advantage of being rapid and sensitive, requiring relatively small amounts of candidate protein. *In vitro* screening allows the use of a 'wide net' in looking for potential antifungal activities. Approaches similar to this, but directed toward finding small molecules rather than proteins, have been in use by agrochemical and pharmaceutical companies for many years. Most of the published work on antifungal proteins found by *in vitro* screening has used seeds as sources of material. The embryo is the result of considerable energy input, and requires special protection from microbial attack. Examples of this extend beyond the plant kingdom; bird eggs, for

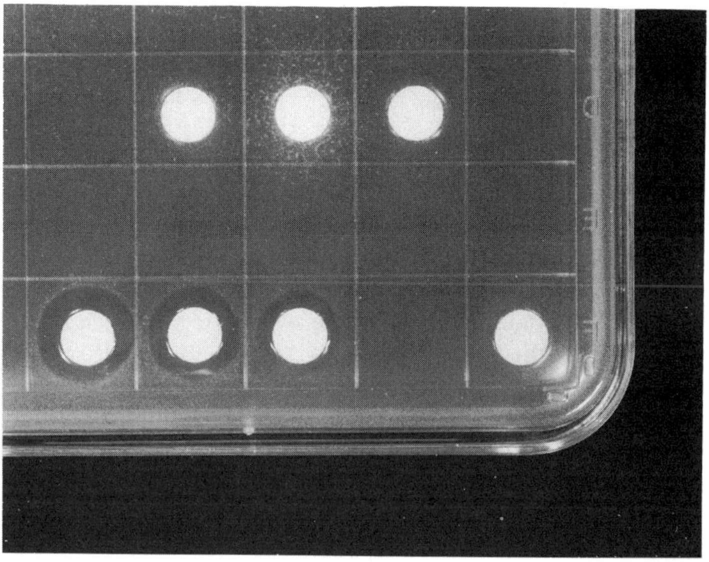

Figure 5.8 An *in vitro* assay for antifungal activity. Test compounds on filter discs are placed on an agar plate overlaid with spores of *D. maydis*, a maize stalk rot pathogen. Zones of growth inhibition are apparent around the lower row of disks. Photo courtesy of Dr T. Luntz, Ciba, Agricultural Biotechnology.

instance, contain high concentrations of lysozyme, an enzyme with hydrolytic activity on bacterial and fungal cell walls.

Constitutive expression of inducible host defence genes

Many pathogen-inducible genes are likely to play a key role in effecting or modulating various defence responses, directly inhibiting fungal growth or combatting fungal invasion of plant tissues. There are two especially convincing lines of evidence that genes expressed in response to pathogen attack may play a direct role in the disease resistance response. First, the types of plant proteins that have been demonstrated to possess antifungal activity *in vitro* can be similar to gene products induced by pathogen infection. Chitinase and osmotin (see below) are two such examples. Second, expression of many gene products induced systemically as a result of infection are tightly correlated with SAR (see section on host defences above), which argues that they may be causally involved in the resistance response. The rationale for expressing these inducible genes constitutively is that constant, high-level expression may give the plant a head start in its race against the invading pathogen. We know that in the case of SAR, induction by initial pathogen infection can dramatically decrease the extent of subsequent infections. Chemical inducers of resistance, which induce the same genes, have an equally dramatic effect. Thus, simply pre-inducing a set of genes that a pathogen normally induces during infection, appears to heighten host defences. By extension, constitutive expression of single inducible genes or groups of genes could have a similarly beneficial effect.

(b) Activities directed against the cell wall and membrane

Among the simplest strategies for combatting a pathogen is direct targeting of its structural components. This approach requires only a cursory knowledge of the basic make-up of the cells in the pathogen, and works on the idea that interfering with assembly of those components, or degrading them after they are assembled, will make the pathogen less able to cope with host defences. The attraction of this approach is that no detailed information about host/pathogen recognition is required. In some cases, mixing different proteins can synergistically increase their antifungal activity. For instance, early *in vitro* data showed that β1–3-glucanase and chitinase (known to degrade certain fungal cell wall components), while weakly active on their own, make a potent antifungal cocktail (Mauch *et al.*, 1988).

Chitinase

Broglie and colleagues at DuPont reported the antifungal effect of chitinase expression in transgenic plants (Broglie *et al.*, 1991). They took a bean (*Phaseolus vulgaris* L.) chitinase, which is locally inducible by pathogen infection in its native state, engineered it for constitutive high-level expression, and transformed it into both tobacco and oilseed rape (*Brassica napus* L.). The resulting plants showed levels of chitinase activity greater than non-transformed controls. These high expressors were significantly resistant to infection by the soil-borne, chitin containing fungus *Rhizoctonia solani*, which causes a seedling rot called post-emergent damping-off (Figure 5.9). This result has been carefully documented at the microscopic level, where lysis of hyphal tips has been observed in the tissues of transgenic plants.

Thionins

Among the earliest proteins shown to have antifungal activity *in vitro* were the thionins, which are present in the seeds and leaves of small grain cereals. Although the precise basis for the antimicrobial activity of the thionins is unclear, they are known to cause leakage from membranes. Recently, expression of barley α-thionin in tobacco has been shown to result in decreased lesion formation by both virulent and avirulent pathovars of *Pseudomonas syringae* (Carmona *et al.*, 1993).

Permatins

Looking for antifungal proteins from maize seeds, Roberts and Selitrennikoff (1990) found a 25 kDa protein they called zeamatin, which

Figure 5.9 Control tobacco plants (A) and transgenic plants constitutively expressing the bean basic chitinase (B) show differences in susceptibility to the damping-off disease caused by *R. solani*. From Broglie *et al.* (1991).

was able to inhibit the growth of the human pathogen *Candida albicans*, as well as filamentous fungi. Partial amino acid sequence of purified zeamatin showed that it was similar to a group of proteins that includes an intensely sweet-tasting protein called thaumatin. Although the activity of this group of proteins has not been well worked out, their antifungal activity is exerted by causing membrane leakage and cell lysis (Figure 5.10). The name 'permatins' has therefore been proposed for the group. Included in the group is osmotin, a protein inducible by osmotic or salt stress (Woloshuk *et al.*, 1991). Similar genes and proteins have been found in variety of plants. Osmotin is closely related to zeamatin, and to the extracellular pathogenesis-related protein PR-5, which is one of the set of genes that turn on during induction of SAR.

Lysozyme

Lysozyme occurs in various forms throughout the taxonomic spectrum of life. Organisms as diverse as bacteriophage and mammals possess the activity, which catalyzes degradation of peptidoglycan, a key component of bacterial cell walls. Lysozyme is both antibacterial and antifungal *in vitro*, and is therefore an obvious candidate to introduce into plants. Recently, a report appeared on engineering expression of the bacteriophage T4 lysozyme, an especially potent member of this class of enzymes, into transgenic potato plants (Düring *et al.*, 1993). The enzyme was targeted to the intercellular space by fusing the lysozyme coding sequence to the signal sequence of barley α-amylase. Although only minute amounts of foreign protein were detectable in the transgenic plants, a significant

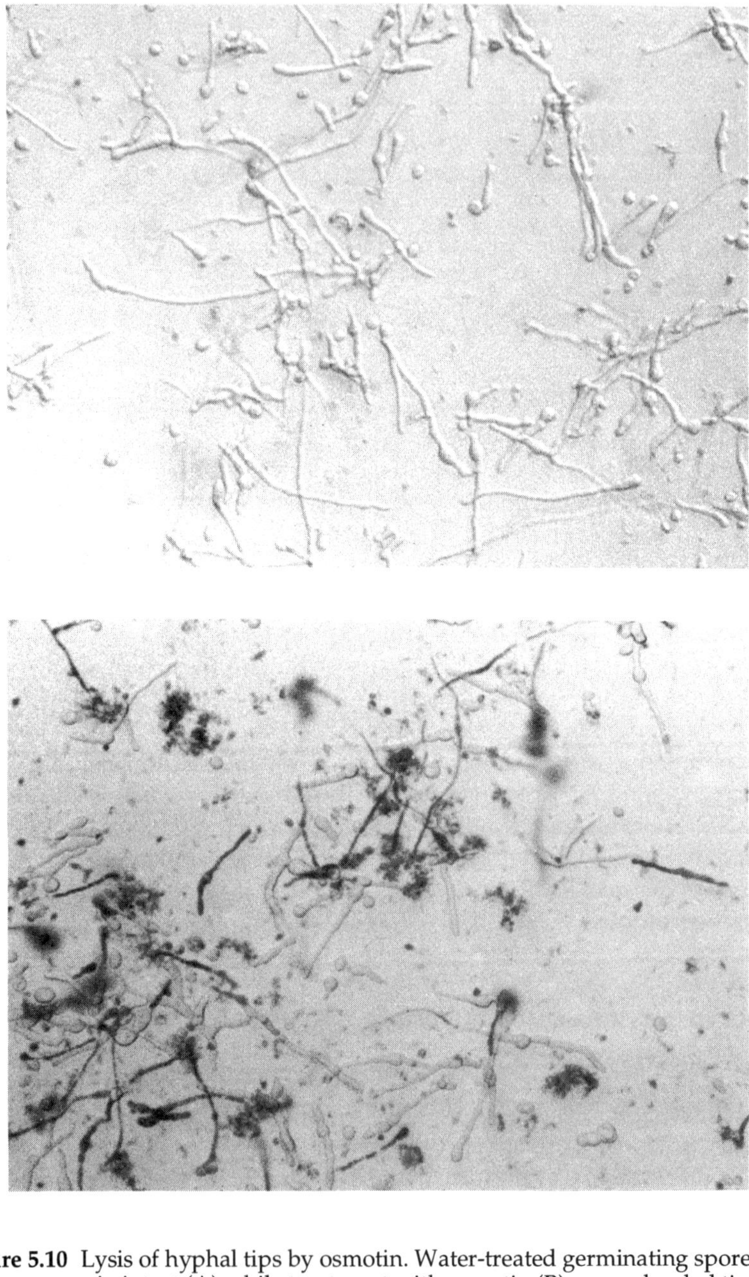

Figure 5.10 Lysis of hyphal tips by osmotin. Water-treated germinating spores of *N. crassa* remain intact (A) while treatment with osmotin (B) causes hyphal tip lysis (apparent as darkly-staining cytoplasm released upon rupture). From Vigers *et al.* (1992).

reduction in the growth of *Erwinia carotovora*, the causal agent of soft rot, was reported.

Lectins

Plants produce a variety of lectins, or carbohydrate-binding proteins. Several of these are known to bind chitin, including wheat germ agglutinin, which was shown many years ago to inhibit *in vitro* the growth of the fungus *Trichoderma viride* (Mirelman *et al.*, 1975). At the time these first experiments were reported, it was proposed that the antifungal effect of lectins might be to bind to or cross-link chitin polymers, thus disturbing the balance between synthesis and limited degradation that occurs at the growing hyphal tip. The stinging nettle plant (*Urtica dioica*) produces a chitin-binding lectin (UDA) that inhibits *in vitro* growth of a variety of chitin-containing fungi (Broekaert *et al.*, 1989). Unlike chitinase, which causes bursting of hyphal tips, UDA strongly inhibits hyphal growth without actually causing lysis. On some fungi, UDA is a more potent agent than chitinase, while on others, the order is reversed. Together, the two proteins act synergistically.

(c) Antifungal proteins with other functions

Most of the antimicrobial proteins studied to date for plant genetic engineering purposes are active against the cell wall or membrane. However, a few others with different or unknown modes of action are described below.

Ribosome-inactivating proteins (RIPs)

RIPs are plant proteins that potently inhibit protein synthesis by depurinating a specific residue near the 3′ end of 28S ribosomal RNA. Some RIPs are comprised of two polypeptide chains – one catalytic, the other a lectin domain that allows cell surface binding and may facilitate internalization. Such two-subunit RIPs are exceedingly potent cytotoxins that can be lethal in minute doses. Most RIPs, however, are single-subunit proteins lacking the lectin domain. RIPs are present in relatively high concentration in some seeds, and can have high levels of antifungal activity. Although most RIPs tested will inhibit protein synthesis in a variety of systems, they seem to possess selectivity, working at lower concentration on certain foreign ribosomes than on host ribosomes. The basis for the selective action of RIPs is unclear. A RIP from barley seed was recently reported to confer tolerance to *Rhizoctonia solani* when expressed in tobacco plants (Logemann *et al.*, 1992).

Pathogenesis-related protein 1 (PR-1a)

PR-1a is the most strongly inducible of the genes that are activated during SAR. Normally expressed at almost undetectable levels, it can reach 1–2% of the total mRNA population after pathogen infection or salicylic acid treatment. PR-1a is a member of a small, closely related gene family in tobacco and homologous genes occur in most plant species that have been examined. Despite its wide conservation and extreme induction both locally and systemically after pathogen infection, no function has yet been assigned to PR-1 nor does it resemble anything with known biochemical function in the available sequence databases. In order to investigate its function, several laboratories have expressed PR-1 protein at constitutive high levels in tobacco plants. No role for the protein in resistance to TMV or alfalfa mosaic virus was found, despite the tight correlation between induction of PR-1 expression and viral lesion formation. After extensive testing, however, PR-1 was found to significantly decrease disease development caused by two pathogens: *Peronospora tabacina* and *Phytophthora parasitica*, causal agents of blue mold and black shank diseases of tobacco (Figure 5.11). Both of these pathogens are oomycetes, which are not considered true fungi, and do not contain chitin in their cell walls. Thus, many of the cell wall hydrolytic activities that are effective against other fungal taxa are unlikely to inhibit oomycete growth. In this context, the activity of PR-1 is especially exciting.

(d) Modification of host metabolism

Beyond engineering plants with genes that directly affect an invading pathogen, lies the still mostly unexplored territory of modifying host metabolism. Presumably, conferring a novel biosynthetic capacity on a host

Figure 5.11 Protection against *P. parasitica* in transgenic plants constitutively expressing PR-1 (right) compared to control (left). Black shank disease is manifested as stunting and wilting in the controls. After Alexander *et al.* (1993).

would allow it to combat pathogens that are able to overcome the normal arsenal of defences. One recent report showed that such a strategy is possible in principle (Hain *et al.*, 1993). Stilbenes are a class of phytoalexins synthesized by a variety of plants, including grape (*Vitis vinifera* L.) and peanut (*Arachis hypogaea* L.). They are produced by the enzyme stilbene synthase from malonyl-CoA and *para*-comaroyl-CoA, both of which are ubiquitous in plants. Transformation of tobacco with a gene for stilbene synthase from grape, conferred the ability to synthesize the stilbene phytoalexin resveratrol in the transgenic tobacco. The transgenic lines varied in the speed with which they induced stilbene synthase expression after pathogen attack. One line that rapidly expressed the gene showed a 60–80% reduction in disease symptoms following infection by *Botrytis cinerea.*

(e) Interference with virulence mechanisms of pathogens

Other strategies for engineering disease resistance focus on key weapons used by the pathogen to invade or kill host tissue. A recent example from the literature illustrates the possibilities for overcoming the effects of bacterial toxins by genetic engineering. *Pseudomonas syringae* pv. *tabaci*, the causal agent of wildfire disease in tobacco, produces a toxic inhibitor of glutamine synthase called tabtoxin. The production of tabtoxin is solely responsible for the devastating systemic symptoms seen in plants infected by this bacterium. The bacterium avoids this toxicity by producing an enzyme that acetylates tabtoxin to an inactive form. Expression of this tabtoxin resistance gene in transgenic tobacco resulted in a strong reduction in symptom formation by bacterial infection even at high titers (Anzai *et al.*, 1989).

5.5 THE FUTURE: DEFINING GENES IMPORTANT IN HOST RESPONSES

Because so many of the basic ways in which plants and pathogens interact are still mysterious, biochemistry may not always be the best way to attack these problems. Modern molecular genetics has now made it possible to clone genes based only on knowledge of their mutant phenotype and genetic map position. Moreover, once the gene of interest is cloned, the sheer number of other genes whose sequences are available in data banks makes it increasingly likely to find a similar sequence with a known biochemical function (e.g. DNA binding, protein kinase, GTP hydrolysis, etc.). Given that information, it then becomes possible to design experiments to test specifically whether the gene of interest performs such a biochemical function. The plant species best-suited to this type of analysis is the crucifer *Arabidopsis thaliana*. Several laboratories are actively pursuing mutations that inactivate the ability of the plant to mount defence responses (e.g. Debener *et al.*, 1991).

5.6 CONCLUSIONS

Our existing methods for discovering genes that have activity against pathogens are limited by what we know about how plants and pathogens interact. Despite the fact that plant genetic engineering was impossible over a decade ago, the current state of affairs remains rather crude. Only by increasing our knowledge of the complex interactions between pathogens and plants at the most basic level will we be able to significantly increase our options for genetically engineered disease control. We hope that the information contained in this chapter will provide a foundation on which students interested in this area can build new and exciting methods for controlling plant diseases using molecular biology and genetic engineering.

FURTHER READING

Attachment to and recognition of the host

Hardham, A. (1992) Cell biology of pathogenesis. *Annual Review of Plant Physiology and Plant Molecular Biology*, **43**, 491–526.

Penetration of the host

Kolattukudy, P. (1985) Enzymatic penetration of the plant cuticle by fungal pathogens. *Annual Review of Phytopathology*, **23**, 223–50.

Bacterial pathogenicity

Long, S. and Staskawicz, B. (1993) Prokaryotic plant parasites. *Cell*, **73**, 921–35.

Host structural defences

Martin, J. (1964) The role of the cuticle in the defence against plant disease. *Annual Review of Phytopathology*, **2**, 81–100.

Inducible host defences

Bowles, D.J. (1990) Defence-related proteins in higher plants. *Annual Review of Biochemistry*, **59**, 873–907.

Systemic acquired resistance and the role of salicylic acid

Ryals, J., Uknes, S. and Ward, E. (1994) Systemic acquired resistance. *Plant Physiology* (in press).

Raskin, I. (1992) Role of salicylic acid in plants. *Annual Review of Plant Physiology and Plant Molecular Biology*, **43**, 439–63.

Viruses and pathogen-derived resistance

Wilson, T.M.A. (1993) Strategies to protect crop plants against viruses: pathogen-derived resistance blossoms. *Proceedings of the National Academy of Sciences USA*, **90**, 3134–41.

Arabidopsis genetics

Koncz, C., Chua, N.-H. and Schell, J. (1992) *Methods in* Arabidopsis *Research*, World Scientific Publishing Co., Singapore.

REFERENCES

Alexander, D., Goodman, R.M., Gut-Rella, M., Glascock, C., Weyman, K., Friedrich, L., Maddox, D., Ahl-Goy, P., Luntz, T., Ward, E. and Ryals, J. (1993) Increased tolerance to two oomycete pathogens in transgenic tobacco expressing pathogenesis-related protein 1a. *Proceedings of the National Academy of Sciences USA*, **90**, 7327–31.

Anzai, H., Yoneyama, K. and Yamaguchi, I. (1989) Transgenic tobacco resistant to a bacterial disease by the detoxification of a pathogenic toxin. *Molecular and General Genetics*, **219**, 492–94.

Braun, C. and Hemenway, C. (1992) Expression of amino-terminal portions or full-length viral replicase genes in transgenic plants confers resistance to potato virus X infection. *Plant Cell*, **4**, 735–44.

Broekaert, W.F., Parijs, J.V., Leyns, F., Joos, H. and Peumans, W.J. (1989) A chitin-binding lectin from stinging nettle rhizomes with antifungal properties. *Science*, **245**, 1100–2.

Broglie, K., Chet, I., Holliday, M., Cressman, R., Biddle, P., Knowlton, S., Mauvais, C.J. and Broglie, R. (1991) Transgenic plants with enhanced resistance to the fungal pathogen *Rhizoctonia solani*. *Science*, **254**, 1194–7.

Carmona, M.J., Molina, A., Fernandez, J.A., Lopez-Fando, J.J. and Garcéa-Olmedo, F. (1993) Expression of the a-thionin gene from barley in tobacco confers enhanced resistance to bacterial pathogens. *The Plant Journal*, **3**, 457–62.

Collinge, D. and Slusarenko, A. (1987) Plant gene expression in response to pathogens. *Plant Molecular Biology*, **9**, 389–410.

Crute, I. (1985) The genetic bases of relationships between microbial parasites and their hosts, in *Mechanisms of resistance to plant diseases*, (ed. R. Fraser), Kluwer, Boston, pp. 80–142.

Debener, T., Lehnackers, H., Arnold, M. and Dangl, J.L. (1991) Identification and molecular mapping of a single *Arabidopsis thaliana* locus determining resistance to a phytoapthogenic *Pseudomonas syringae* isolate. *The Plant Journal*, **1**, 289–302.

de Wit, P. and Van der Meer, F. (1986) Accumulation of the pathogenesis related tomato leaf protein P14 as an early indicator of incompatibility in the interaction between *Cladosporium fulvum* (syn *Fulvia fulva*) and tomato. *Physiological and Molecular Plant Pathology*, **28**, 203–14.

Düring, K., Porsch, P., Fladung, M. and Lörz, H. (1993) Transgenic potato plants resistant to the phytopathogenic bacterium *Erwinia carotovora*. *The Plant Journal*, **3**, 587–98.

Flor, H. (1956) The complementary genic systems in flax and flax rust. *Advances in Genetics*, **8**, 29–54.

Gabriel, D. and Rolfe, B. (1990) Working models of specific recognition in plant–microbe interactions. *Annual Review of Phytopathology*, **28**, 365–91.

Gerlach, W.L., Llewellyn, D. and Haseloff, J. (1987) Construction of a plant disease resistance gene from the satellite RNA of tobacco ringspot virus. *Nature*, **328**, 802–5.

Hain, R., Reif, H.-J., Krause, E., Langebartels, R., Kindl, H., Vornam, B., Wiese, W., Schmelzer, E., Schreier, P., Stöcker, R. and Stenzel, K. (1993) Disease resistance results from foreign phytoalexin expression in a novel plant. *Nature*, **361**, 153–6.

Harrison, B.D., Mayo, M.A. and Baulcombe, D.C. (1987) Virus resistance in transgenic plants that express cucumber mosaic virus satellite RNA. *Nature*, **328**, 799–802.

Kaniewski, W., Lawson, C., Sammons, B., Haley, L., Hart, J., Delannay, X. and Tumer, N.E. (1990) Field resistance of transgenic russet burbank potato to effects of infection by potato virus X and potato virus Y. *Bio/Technology*, **8**, 750–4.

Kearney, B. and Staskawicz, B. (1990) Widespread distribution and fitness contribution of *Xanthomonas campestris* avirulence gene *avrBs2*. *Nature*, **346**, 385–6.

Keen, N.T., Sims, J., Midland, S., Yoder, M., Jurnak, F., Shen, H., Boyd, C., Yucel, I., Lorang, J. and Murillo, J. (1993) Determinants of specificity in the interaction of plants with bacterial pathogens, in *Advances in Molecular Genetics of Plant–Microbe Interactions*, (eds E. W. Nester and D. P. S. Verma), Kluwer, Dordrecht, pp. 211–20.

Levings, C. and Brown, G. (1989) Molecular biology of plant mitochondria. *Cell*, **56**, 171–9.

Logemann, J., Jach, G., Tommerup, H., Mundy, J. and Schell, J. (1992) Expression of a barley ribosome-inactivating protein leads to increased fungal protection in transgenic tobacco plants. *Bio/Technology*, **10**, 305–8.

Longstaff, M., Brigneti, G., Boccard, F., Chapman, S. and Baulcombe, D. (1993) Extreme resistance to potato virus X infection in plants expressing a modified component of the putative viral replicase. *EMBO Journal*, **12**, 379–86.

Malamy, J., Carr, J.P., Klessig, D.F. and Raskin, I. (1990) Salicylic acid: a likely endogenous signal in the resistance response of tobacco to viral infection. *Science*, **250**, 1002–4.

Martin, G., Devicente, M. and Tanksley, S. (1993) High-resolution linkage analysis and physical characterization of the *Pto* bacterial resistance locus in tomato. *Molecular Plant Microbe Interactions*, **6**, 26–34.

Mauch, F., Mauch-Mani, B. and Boller, T. (1988) Antifungal hydrolases in pea tissue. II. Inhibition of fungal growth by combinations of chitinase and beta-1,3-glucanase. *Plant Physiology*, **88**, 936–42.

Meeley, R.B., Johal, G.S., Briggs, S.P. and Walton, J.D. (1992) A biochemical phenotype for a disease resistance gene of maize. *Plant Cell*, **4**, 71–7.

Métraux, J.-P., Signer, H., Ryals, J., Ward, E., Wyss-Benz, M., Gaudin, J., Raschdorf, K., Schmid, E., Blum, W. and Inverardi, B. (1990) Increase in salicylic acid at the onset of systemic acquired resistance in cucumber. *Science*, **250**, 1004–6.

Mirelman, D., Galun, E., Sharon, N. and Lotan, R. (1975) Inhibition of fungal growth by wheat germ agglutinin. *Nature*, **256**, 414.

Nejidat, A. and Beachy, R.N. (1990) Transgenic tobacco plants expressing a coat protein gene of tobacco mosaic virus are resistant to some other tobamoviruses. *Molecular Plant–Microbe Interactions*, **3**, 247–51.

Powell-Abel, P.A., Nelson, R.S., De, B., Hoffman, N., Rogers, S., Fraley, R.T. and Beachy, R.N. (1986) Delay of disease development in transgenic plants that express the tobacco mosaic virus coat protein gene. *Science*, **232**, 738–43.

Roberts, W.K. and Selitrennikoff, C.P. (1990) Zeamatin, an antifungal protein from maize with membrane-permeabilizing activity. *Journal of General Microbiology*, **136**, 1771–8.

Ronald, P., Salmeron, J., Carland, F. and Staskawicz, B. (1992) The cloned avirulence gene *avrPto* induces disease resistance in tomato cultivars containing the *Pto* resistance gene. *Journal of Bacteriology*, **174**, 1604–11.

Slusarenko, A. and Longland, A. (1986) Changes in gene activity during expression of the hypersensitive response in *Phaseolus vulgaris* cv. Red Mexican to an avirulent race 1 isolate of *Pseudomonas syringae* pv. *phaseolicola*. *Physiological and Molecular Plant Pathology*, **29**, 79–94.

Ukness, S., Winter, A.M., Delany, T., Vernooij, B., Morse, A., Friedrich, L., Nye, G., Potter, S., Ward, E. and Ryals, J. (1993) Biological induction of systemic acquired resistance in *Arabidopsis*. *Molecular Plant–Microbe Interactions*, **6**, 692–98.

Vigers, A.J., Wiedemann, S., Roberts, W.K., Legrand, M., Selitrennikoff, C.P. and Fritig, B. (1992) Thaumatin-like pathogenesis-related proteins are antifungal. *Plant Science*, **83**, 155–61.

Ward, E.R., Uknes, S.J., Williams, S.C., Dincher, S.D., Wiederhold, D.W., Alexander, D., Ahl-Goy, P., Métraux, J.-P. and Ryals, J.A. (1991) Coordinate gene activity in response to agents that induce systemic acquired resistance. *Plant Cell*, **3**, 1085–94.

Woloshuk, C.P., Meulenhoff, J.S., Sela-Buurlage, M., van den Elzen, P.J.M. and Cornelissen, B.J.C. (1991) Pathogen-induced proteins with inhibitory activity toward *Phytophthora infestans*. *Plant Cell*, **3**, 619–28.

Molecular mechanisms to confer herbicide resistance

D.J. Cole

6.1 INTRODUCTION

6.1.1 Weed control in agriculture

Modern agriculture has been shaped by several major innovations which include mechanization, intensive plant breeding and pest control. Not the least of these has been the use of chemical herbicides which in western agriculture have replaced traditional labour-intensive manual methods of removing weeds. The history of synthetic weed killers dates back to the second world war with the invention of the phenoxyalkanoic acid herbicides such as 2,4-dichlorophenoxyacetic acid (2,4-D). Eradication of weeds from a crop enables enhanced yields through removal of competition for resources and prevents undesirable contamination of the harvested crop with weed seeds. Weeds may additionally act as a reservoir for crop pests. Minimum tillage practices are facilitated by the use of broad-spectrum herbicides. Recently, the desirability of chemical weed control methods has been questioned in some quarters. However, there can be no doubt that the availability of inexpensive, plentiful and varied food in the developed countries has been due to agricultural intensification of which chemical weed control methods are an integral part.

Herbicides are therefore a key tool in the farmer's arsenal of crop management techniques. There is a constant flow of new products on to the market as older and less sophisticated products become obsolete. The trend is towards compounds which are effective at reduced application rates, have enhanced selectivity, possess low toxicity to man and wildlife and

Molecular Biology in Crop Protection. Edited by George Marshall and Dale Walters. Published in 1994 by Chapman & Hall, London. ISBN 0 412 54390 7(HB) 0 412 54400 8(PB).

dissipate rapidly in the environment. In these terms many contemporary herbicides are vastly more sophisticated than some of their antecedents.

6.1.2 Opportunities for herbicide-resistant crops

The farmer's first need is for agents which will selectively control economically important weeds in crops. Weed killers which can be well tolerated by the crop therefore find the greatest utility. Herbicides have normally been discovered by detection of a 'lead' chemical structure through random screening on plants. Once biological activity is established, potency and selectivity are maximized by an analogue chemistry programme in which variants are selected for the optimum combination of properties. Differential crop and weed behaviour is usually based on the serendipitous ability of tolerant crop species to metabolically degrade the herbicide to innocuous products more rapidly than the weed is able to (Cole *et al.*, 1987). Unfortunately the enzymes responsible can be highly species-specific and limited cross-tolerance between different crops can restrict the potential of an individual herbicide.

The greatest opportunity in herbicide usage presented by the advent of gene transfer technology is therefore to create new selectivities and optimize the use of the most benign herbicides. For example, glyphosate controls a wide spectrum of troublesome weeds but is a 'total' herbicide, being equally injurious to crops. Efforts to transform crops for glyphosate resistance have been driven by glyphosate's outstanding biological properties and environmental acceptability. Creating new herbicide resistance can assist in agronomic flexibility in other ways. For example, it may be desirable to enhance resistance in alternated crops to residual levels of herbicide, permitting more options in crop rotation. Reducing the number of applications, flexibility of application rate, resistance to spray drift from neighbouring crops or new approaches to controlling parasitic weeds are other instances that come to mind. Minor crops could particularly benefit from the availability of a wider range of modern herbicides enabled by transformation techniques.

6.1.3 Scientific background

Gene transfer techniques are of little use unless the biochemical basis of gene functionality is understood. Herbicides kill plants because they interact with enzymes or other sites involved in processes such as the synthesis of essential amino acids, lipids, plant pigments or vitamins, or with photosynthesis or developmental regulation (Devine *et al.*, 1992). Often these processes are unique to plants and microorganisms. Hitherto, studying the basis of herbicide action and selectivity has been rather a 'Cinderella' subject, of interest purely out of curiosity. In the 1980s it was discovered that compounds such as glyphosate and the sulphonylureas acted at single manipulatable sites. Plant or bacterial cells selected for resistance contained over-expressed or mutated target sites. It was therefore clear that resistant transgenic plants could be created if the target site was highly expressed in relation to the number of herbicide molecules encountered or was diluted by an insensitive, mutant enzyme from a donor species.

Of course, the target site may not be understood, may be multiple or not amenable to manipulation. In this case, one clear choice is to identify enzymes which attack the herbicidal structure, detoxifying it before it is able to reach the target site. The enzymes which have successfully been used have normally been specified by genes cloned from bacteria able to catabolize the herbicide. These advances have built upon the proven ability of bacterial antibiotic detoxification genes to impart antibiotic resistance to plant tissues, thus finding use as selectable markers (Bevan *et al.*, 1983; Herrera-Estrella *et al.*, 1983.). Low level expression of foreign detoxification genes can enable an impressive degree of herbicide resistance and has generally been found to be more effective than target site manipulation.

It is immediately clear that herbicide resistance and hence a trait of commercial potential can be conferred by introduction of a single new gene. In some cases this may need to be used in conjunction with a conventional breeding programme to obtain a value-added variety. In other cases, direct gene transfer to a commercial variety can be achieved. We can expect such varieties to be on the market in the next few years. This contrasts with the goal of manipulating crops for altered yield or quality. In the latter instances desirable traits are more likely to require the coordinated expression of multiple genes and this will be more exacting to achieve.

6.1.4 Techniques

Resistance to herbicides by recombinant techniques has so far been achieved for a handful of compounds which are all described within this article. A number of crops have been amenable to such transformation (Table 6.1). The major limitations to expanding this range are routine techniques for gene insertion and regeneration of the intact transgenic plants, notably for cereals (see Chapter 1). Unless otherwise mentioned, transformation has been carried out using the Ti plasmid as the gene vector mediated by *Agrobacterium tumefaciens*. Selection of transformants has routinely made use of identifying kanamycin resistance conferred by the NPTII gene (Bevan *et al.*, 1983) carried on the vector in tandem with the herbicide resistance gene. High level gene expression has normally been facilitated by use of the constitutive cauliflower mosaic virus 35S promoter (Odell *et al.*, 1985). This has been used widely to assist foreign gene expression in both dicot and monocot recipient plants. By and large, herbicide target sites are to be found within the chloroplast. Therefore it is necessary to ensure that constructs encoding manipulated target sites also carry sequences ensuring correct targeting of the gene product from the nucleus to the

Table 6.1 Transformation of crop species for herbicide resistance

Herbicide	*Novel gene product*	*Gene source*	*Gene function*	*Transformed crops*[1]
Sulphonyureas	acetolactate synthase	higher plants	mutated target site	rape, rice, flax, cotton, tomato, sugarbeet, lucerne, lettuce, melon, chicory, poplar
Imidazolinones	acetolactate synthase	higher plant	mutated target site	tobacco
Glyphosate	enolpyruvyl-shikimic acid phosphate synthase	soil and enteric bacteria, higher plant	mutated site	tomato, rape, soybean, poplar
	glyphosate oxido-reductase	soil bacterium	detoxification	maize, rape, soybean
Asulam	dihydropteroate synthase	enteric bacterium	mutated target site	tobacco
Atrazine	'D1' protein	higher plants	mutated target site	soybean
Glufosinate	N-acetyl transferase	bacteria	detoxification	maize, wheat, rice, cotton, rape, potato, tomato, sugarbeet, lucerne, aspen, tall fescue, belladonna
Glufosinate	glutamine synthetase	higher plant	over-expressed target-site	tobacco
Bromoxynil	nitrilase	soil bacterium	detoxification	rape, cotton potato, tomato
2, 4-D	monooxygenase	soil bacterium	detoxification	cotton
Dalapon	dehalogenase	soil bacterium	detoxification	tobacco
Cyanamide	nitrile hydratase	soil fungus	detoxification	tobacco

[1]Tobacco has been transformed as a model system in most cases.

chloroplast. In the following studies integration of the herbicide resistance gene into the host genome has generally been proven by Southern hybridization. Genetic analyses show that in most cases the herbicide resistance gene is expressed as a dominant character encoded by a determinant integrated at a single nuclear locus.

The ensuing examples show how successes have been achieved, how candidate genes have been identified and manipulated and the current technical limitations to be surmounted. Due to ease of transformation, tobacco (*Nicotiana tabacum*) is used as a model crop to study and optimize alien gene performance. This is a prelude to the introduction of these genes into crops of greater commercial value.

6.2 HERBICIDE RESISTANCE THROUGH TARGET SITE MANIPULATION

6.2.1 Sulphonylureas and imidazolinones

The approaches to developing herbicide-resistant crops through molecular means are well exemplified by the case of the sulphonylurea and imidazolinone families of herbicides. Products representing these two broad classes of herbicidal chemistry were launched in the 1980s. In particular the sulphonylureas aroused great interest since they represented a breakthrough in biological activity, being recommended for use at field application rates as low as 4 g/ha. Moreover, selectivity for broad-spectrum weed control in a number of major crops could be achieved by subtle variation in chemical substitution enabling rapid metabolic detoxification and thus tolerance in various crops (Brown *et al.*, 1991).

The potent weed control afforded by these families is due to a single site of action, namely inhibition of acetolactate synthase (ALS; also known as acetohydroxyacid synthase, AHAS; EC 4.1.3.18). This enzyme is the first common enzyme in the formation of the branched-chain amino acids valine, leucine and isoleucine and is ubiquitous in plants and microorganisms. Acetolactate synthase catalyses two distinct reactions (Figure 6.1), causing the condensation of two molecules of pyruvate leading specifically to the formation of valine and leucine, or the condensation of pyruvate with α-ketobutyrate, leading to the formation of isoleucine. Inhibition of ALS leads to starvation of branched-chain amino acids and inability to continue synthesizing proteins. This is most evident at the meristematic areas of the plant, where rates of macromolecule formation are highest.

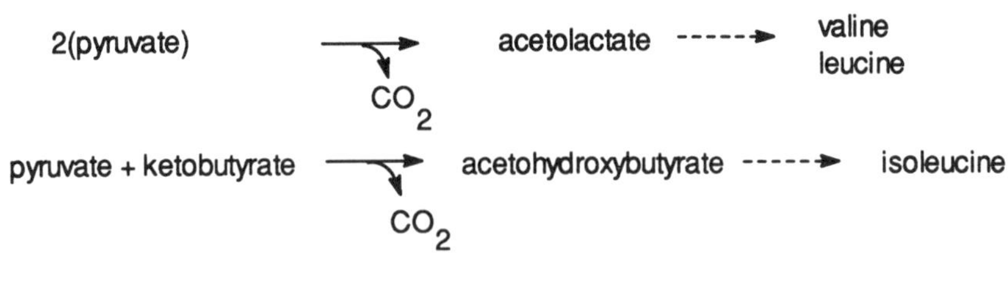

Figure 6.1 Reactions carried out by ALS, leading to formation of the branched-chain amino acids.

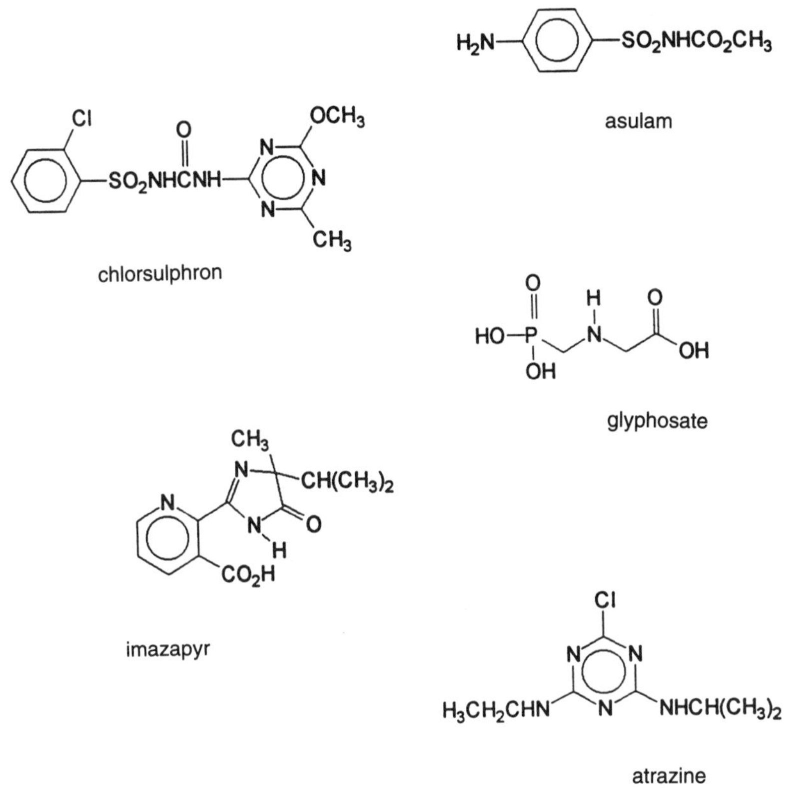

Figure 6.2 Structures of some herbicides mentioned in the text.

In addition to the sulphonylureas such as chlorsulfuron (Figure 6.2) and sulfometuron-methyl and the imidazolinones such as imazapyr (Figure 6.2) and imazaquin, ALS inhibition embraces other herbicidal chemistries such as the triazolopyrimidine sulphonamides, e.g. flumetsulam (Subraminian *et al.*, 1990). Herbicidal inhibitors do not compete with substrates or the co-factors FAD or thiamine pyrophosphate (La Rossa and Schloss, 1984; Shaner *et al.*, 1984). Evidence from resistant mutants yielding altered ALS enzymes suggests that the binding domains of different inhibitor classes overlap but are not identical (Saxena and King, 1988; Subraminian *et al.*, 1990; Hall and Devine, 1990; Hattori *et al.*, 1992).

Elucidation of the biochemical mode of action of ALS inhibitors relied on the use of microorganisms as models for plants in which simple nutrient complementation experiments were used to indicate where the point of inhibition was likely to lie. Amino acid feeding experiments in the presence of sulfometuron-methyl suggested that a common step in the synthesis of branched-chain amino acids was inhibited in *Salmonella typhimurium*. It was duly shown that sulfometuron-methyl was a potent and specific inhibitor of one particular isozyme of *S. typhimurium* ALS (La Rossa and Schloss, 1984). Subsequently it became clear that branched-chain amino acid supplementation could also ameliorate growth inhibition of plant tissue by chlorsulfuron (Ray, 1984) and that plant ALS was indeed highly sensitive to inhibition by sulphonylureas (Ray, 1984; Chaleff and Mauvais, 1984).

Once a novel herbicidal mode of action has been identified, this tends to create new interest in characterizing the unique site of action and the associated metabolic pathway. Certainly the advent of ALS inhibitors prompted new interest in understanding more precisely the regulation of branched-chain amino acid formation. Early on, it was identified that mutagenesis and herbicidal selection pressure could readily bring about ALS inhibitor-resistant bacterial and yeast mutants and indeed, also of plants. Such mutants proved to survive these herbicides due to possession of altered and insensitive ALS. Together, these realizations showed the potential to isolate and characterize mutant, insensitive ALS enzymes and to artificially create herbicide-resistant crops by the transfer and heterologous expression of the mutant target site protein.

In bacteria such as *Escherichia coli* and *S. typhimurium* the three ALS isozymes I, II and III each consist of two subunits, specified by one operon for each of the isozymes (Lawther *et al.*, 1979; Squires *et al.*, 1981; Newman *et al.*, 1982). By contrast, the yeast *Saccharomyces cerevisiae* contains only one ALS gene (Falco and Dumas, 1985) encoding a single polypeptide. The presence of multiple forms of ALS in plants is much less certain. However, plant ALS can be specified at more than one genetic locus, depending on the species (Mazur *et al.*, 1987; Lee *et al.*, 1988). Yadav *et al.* (1986) isolated spontaneous mutants of *E. coli* and *S. cerevisiae* which displayed resistance to sulfometuron-methyl. Analysis of ALS showed that in both cases, single amino acid changes occurred in the conserved amino-terminal area of the proteins, both possessing reduced sensitivity *in vitro* to sulfometuron-methyl. In the case of *E. coli*, an Ala → Val substitution occurred in the large subunit of the normally herbicide sensitive ALS II, whereas the yeast enzyme contained a Pro → Ser substitution. The correlation of reduced sensitivity *in vitro* with herbicide resistance was further supporting evidence for a single mode of action of the herbicide at this locus. Further mutations arising in yeast spontaneously or by site-directed mutagenesis revealed that there are at least 10 sites in the protein where substitution giving rise to sulphonylurea resistance can occur (Hartnett *et al.*, 1990).

The yeast ALS gene was used successfully to probe for the ALS genes in *Arabidopsis thaliana* and tobacco (Mazur *et al.*, 1987). Sequencing of the isolated plant genes revealed that the coding region of the *A. thaliana* gene specified 670 amino acid residues. For tobacco, the gene encoded 667 amino acid residues, therefore the proteins were of molecular weight 73 kDa. Excluding the chloroplast transit sequences, there is a high degree of homology between the proteins of the two species with 84% of the amino acid residues being conserved. There are several areas of the sequences showing good homology with areas of the yeast gene.

The subsequent identification of mutant plant ALS genes has enabled the complementation of a number of crop species with these to impart high level herbicide resistance. Probing of a sulphonylurea-resistant mutant of *A. thaliana* with the native wild-type gene resulted in the isolation of a gene encoding an inhibitor-insensitive ALS with a single predicted Pro → Ser substitution at position 196 (Haughn *et al.*, 1988). This is analogous to the single point mutation in the *S. cerevisiae* ALS gene (Yadav *et al.*, 1986). A 5.8 kb fragment containing the coding region of the mutant gene, termed *csr 1-1* was used to transform tobacco by conventional *Agrobacterium* infection. Under the control of its native regulatory sequences the gene conferred over 100-fold resistance to chlorsulfuron in regenerated plants. The mutant gene accounted for 20% of the ALS protein found in the transformants and it was clear that herbicide resistance was due to the presence of this fraction rather than over-expression of ALS as a whole. Superior resistance phenotype in tobacco using the *csr 1-1* gene could be obtained by placing the gene under the control of the cauliflower mosaic virus 35S promoter (Odell *et al.*, 1990; Charest *et al.*, 1990). Over-expression of the wild-type ALS gene from *A. thaliana* using 35S was not sufficient to confer useful chlorsulfuron tolerance to tobacco, even though gene transcription was enhanced 25-fold (Odell *et al.*, 1990).

The *csr 1-1* gene has also been used to confer sulphonylurea resistance to rapeseed (*Brassica napus*), rice (*Oryza sativa*) and oil-flax (*Linum usitatissimum*). In rape, *csr 1-1* governed by the 35S promoter imparted 30-fold tolerance to chlorsulfuron in progeny of transformants (Miki *et al.*, 1990). Mutant ALS expression was of a similar level to that of the host genes and the levels of mutant ALS transcripts and protein correlated well with resistance. Interestingly, the transformants could be selected directly with chlorsulfuron rather than using kanamycin. A similar construct imparted 200-fold tolerance to chlorsulfuron in rice callus developed from protoplasts transformed with the aid of polyethyleneglycol. Fertile plants were regenerated from the transgenic callus but were not independently tested for altered phenotype. Again the gene functioned well as a selectable marker (Li *et al.*, 1992). The unreconstructed *csr 1-1* gene imparted high level tolerance to chlorsulfuron in flax regenerated from hypocotyls treated with *Agrobacterium*, although the presence of the gene in the regenerated tissue was not directly proven (McHughen, 1989). A number of commercially important flax lines transformed with *csr 1-1* have now been trialled in the field and have similar agronomic performance to the non-transformed lines. Good resistance has occurred in soils containing residual levels of sulphonylureas which would have otherwise damaged the crop (McHughen and Holm, 1991). Minor crops transformed with *csr 1-1* include poplar (*Populus* spp.) (Brasileiro *et al.*, 1992) and chicory (*Cichorium intybus*) (Vermeulen *et al.*, 1992).

Tobacco has also served as a source of effective resistance genes for sulphonylureas, identified by hybridization to the *A. thaliana* ALS gene. A number of sulphonylurea-resistant mutants have been obtained from tobacco protoplasts, demonstrating that there are two genetic loci which can determine independently sulphonylurea resistance. Two specific mutants have been identified (Lee *et al.*, 1988), a mutant at the *Sur A* locus specifying a single Pro → Gly substitution at amino acid residue 196, and a triple mutation at the *Sur B* locus, two of which cause substitutions of Pro → Ala at position 196 and Try → Leu at position 573. One of the mutations is silent. The mutations at 196 are positionally analogous to the yeast and *A. thaliana* sulphonylurea resistance mutations. The multiple mutation at *Sur B* enabled 1000-fold resistance over the wild-type to sulphonylureas.

Tobacco plants transformed independently with either of these mutant genes displayed a level of herbicide tolerance similar to that of the original mutants. The doubly mutated gene conferred an 10-fold increase in resistance over the single mutant. Tobacco and tomato (*Lycopersicon escutentum*) lines transformed with this gene were completely resistant to up to four and eight times, respectively, the normal field application rates of sulphonylureas in field trials (32 g/ha). The gene has also been used to introduce sulphonylurea resistance to tomato, sugarbeet (*Beta vulgaris*), lettuce (*Lactuca sativa*), rape (*Brassica napus*), lucerne (*Medicago sativa*) and melon tissues (*Cucumis melo*) (Hartnett *et al.*, 1990).

The mutation in the *A. thaliana csr 1-1* gene allows resistance to sulphonylureas but not to imidazolinone herbicides. An *A. thaliana* mutant selected for 100-fold enhanced tolerance to imazapyr has yielded an imazapyr-insensitive ALS with a single predicted amino acid change of Ser → Asn at residue 653 near the carboxyl terminus of the protein (Sathasivan *et al.*, 1991). This is an area perfectly conserved in all plant ALSs. The mutant gene, termed *imr 1* was ligated with the *tml* promoter isolated from *Agrobacterium*. Tobacco plants transformed with *imr 1* possessed up to 100-fold enhanced resistance to imazapyr and ALS extracted from the transformants was highly insensitive to the inhibitor *in vitro*. In a logical and elegant study, Hattori *et al*. (1992) found that by combining the appropriate sections of the *csr 1-1* and *imr 1* genes to create a hybrid mutant gene, transgenic tobacco could be created which possessed high level resistance both to sulphonylureas and to imidazolinones. Location of determinants for sulphonylurea and imidazolinone resistance at different areas of the ALS polypeptide does not necessarily indicate separate binding sites in wild-type enzyme. It is likely that these areas will be in proximity in the tertiary protein structure, giving a degree of overlapping binding between the two classes of inhibitors.

Introduction of sulphonylurea resistance to plants by the alternative method of identifying detoxification genes has not so far met with success. Microbial oxygenases capable of degrading sulphonylureas have been found, but it has not proved possible to express these correctly in plant tissues. Plants with innate tolerance to sulphonylureas possess specific hydroxylases which rapidly detoxify the herbicides. Cloning and manipulation of plant genes encoding such enzymes has so far had limited success, but rapid progress is now being made (Bozak *et al.*, 1992).

6.2.2 Glyphosate

Glyphosate is a remarkable, non-selective herbicide, first marketed nearly 20 years ago and in worldwide use. Its broad action, particularly against perennial weeds together with highly favourable ecotoxicological properties has made it an attractive idea to extend usage by engineering resistance in major crops. Progress in understanding the biochemical mode of action of this unique herbicide provided the foundation for strategies to create artificial herbicide resistance. As a result, it was announced in 1985 that glyphosate resistance had been achieved in a transgenic plant, the first instance for any herbicide.

Glyphosate (Figure 6.2) is a potent and specific inhibitor of 5-enol pyruvylshikimic acid-3-phosphate synthase (EPSPS, EC 2.5.1.19), catalysing one of the steps in the shikimic acid pathway (Figure 6.3). This is a key pathway linking primary and secondary metabolism leading to the synthesis of, *inter alia* the three aromatic amino acids phenylalanine, tyrosine and

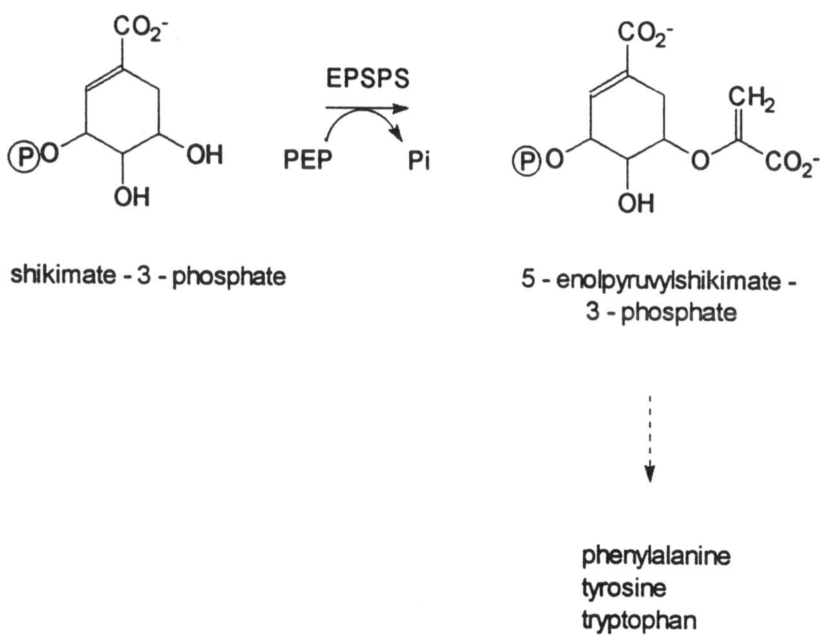

Figure 6.3 Reaction carried out by EPSPS leading to formation of the aromatic amino acids phenylalanine, tyrosine and tryptophan.

tryptophan. The pathway is restricted to plants and microorganisms, helping to account for the negligible animal toxicity of glyphosate. Curiously and despite much endeavour no other type of herbicide structure has been found which inhibits EPSPS or indeed any other reaction in this lengthy pathway. Glyphosate inhibits EPSPS from microbial and plant sources (Steinrücken and Amrhein, 1980; Amrhein *et al.*, 1980), competing with phosphoenolpyruvate (Mousedale and Coggins, 1984). Action at this site results in a massive accumulation of shikimic acid (Amrhein *et al.*, 1980) and the lethal depletion of end-product aromatic amino acids (Hoagland *et al.*, 1978).

Identification of a single target site for glyphosate suggested two alternative routes to engineering crop resistance, i.e. the insertion of either a mutant or over-expressed EPSPS. As was later to be the case with ALS inhibitors, prokaryotic organisms proved to be amenable to manipulation and were the first sources of altered EPSPS. Mutagenization of *S. typhimurium* yielded a glyphosate-resistant mutant possessing an insensitive EPSPS encoded by the *aro A* gene (Comai *et al.*, 1983; Stalker *et al.*, 1985). Sequence comparison with that of the native gene indicated that the mutant enzyme would contain a single Ser → Pro substitution at position 101. Splicing of the mutant *aro A* gene with the octopine or mannopine synthase promoters derived from *Agrobacterium* enabled the transformation of tobacco and tomato with recovery of plants possessing significant tolerance to glyphosate (Comai *et al.*, 1985; Fillatti *et al.*, 1987). Molecular analyses showed integration of the *aro A* gene and the presence of the gene product, yet an incomplete level of herbicide resistance was achieved. Similar partial resistance was found in transformed poplar (Riemenschneider *et al.*, 1988).

In view of the normal subcellular location of EPSPS within the chloroplast (Mousedale and Coggins, 1985) it was reasoned that the failure to achieve high level herbicide resistance was due to incorrect targeting of the novel gene product. Ligation of a mutant *E. coli aro A* with the sequence encoding a *Petunia* chloroplast transit peptide enabled uptake of the hybrid translation product by isolated chloroplasts, with subsequent cleavage to a mature EPSPS of the expected size (della-Cioppa *et al.*, 1987). Tobacco plants transformed with a similar adaption of the *S. typhimurium* gene duly showed full resistance to glyphosate at a field application rate of 1.3 kg/ha (Larson-Kelly *et al.*, 1988). Enhanced expression of a mutant *aro A* gene was also obtained by splicing with a sequence encoding the transit peptide and part of the *N*-terminal region of the RUBISCO small subunit. Tomatoes transformed in this way have given normal fruit yields following treatment with 1 kg/ha glyphosate. Some initial stunting was still seen, illustrating the continuing difficulties in achieving sufficient gene expression at the growing points (Stalker, 1991) which are the primary areas of glyphosate accumulation. Unsatisfactory yields of glyphosate-tolerant rape seed have been ascribed to inadequate gene expression in floral meristems (Kishore *et al.*, 1992).

A complementary approach based on over-expression of plant EPSPS has also been extensively evaluated. Plant cell suspension cultures can be used to develop variants under continuous selection pressure. An EPSPS gene has been isolated from glyphosate-tolerant *Petunia* suspension cells in which EPSPS is overproduced due to a 20-fold gene amplification. It was therefore clear that over-expression of native EPSPS could be a viable strategy to engineering herbicide resistance. Splicing the full length EPSPS cDNA from *Petunia* with the 35S promoter and reintroduction into transgenic *Petunia* conferred several fold tolerance to regenerated plants in greenhouse tests (Shah *et al.*, 1986). Site-specific mutation of petunia EPSPS has also resulted in glyphosate-insensitive enzymes which have been used to transform other species, e.g. oil-flax (*L. usitatissimum*) and soybean (*Glycine max*) (Jordan and McHughen, 1988; Hinchee *et al.*, 1988). However, high level expression of mutant plant EPSPS does not guarantee a practical degree of resistance to glyphosate in field-grown transgenic crops (Fraley, 1988; Kishore *et al.*, 1992).

Many of the mutant EPSPS enzymes that have been isolated have their glyphosate insensitivity coupled to a lower affinity for phosphoenolpyruvate and are thus less efficient enzymes. This may be a contributory factor in the difficulties of achieving complete glyphosate resistance in transgenic plants. More recent searches for mutant EPSPSs have led to the identification of new EPSPS genes which encode enzymes insensitive to glyphosate but retain a low K_m for phosphoenolpyruvate. These new EPSPS genes show limited homology to those previously isolated. One such novel EPSPS gene, isolated from *Agrobacterium* ssp. strain CP4 gives superior glyphosate resistance in field-trialled rapeseed and soybean (Barry *et al.*, 1992), when the enzyme is directed to the chloroplast.

More recently, glyphosate-resistant crops have been engineered by introducing a glyphosate metabolizing gene rather than altering the target site. This is described in section 6.3.3.

6.2.3 Asulam

Cloning of an altered target site in the folic acid synthesis pathway has provided a means of conferring crop resistance to asulam. Asulam (Figure 6.2) is a broad-spectrum herbicide conventionally selective in a number of crops, notably sugar cane and is also effective in non-crop situations. Uniquely, its herbicidal action is due to interference with the formation of folates, specifically by inhibition of the enzyme dihydropteroate synthase (DHPS, EC 2.5.1.15). This occurs competitively with respect to one of its substrates, 4-aminobenzoic acid (Veerasekaran *et al.*, 1981; Figure 6.4). Formyl, methylene and methyl derivatives of tetrahydrofolate act as one carbon donors in the biosynthesis of nucleotides and some amino acids (Cossins *et al.*, 1987). Therefore inhibition of folic acid formation will affect a diverse number of further biosynthetic processes (Kidd *et al.*, 1982). The

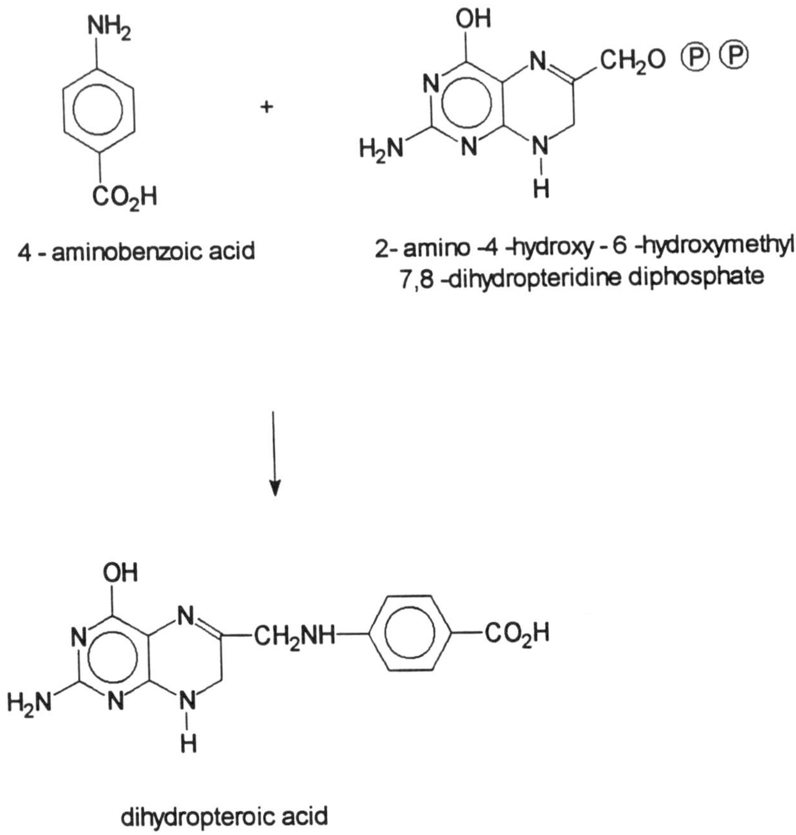

Figure 6.4 Reaction carried out by DHPS, leading to the formation of folic acid.

ability to convert guanasine triphosphate through to dihydrofolate is restricted to plants and microorganisms. Therefore as is the case for amino acid synthesis inhibitor herbicides, this site of action allows herbicidal efficacy and low mammalian toxicity.

Asulam is closely related to the sulphonamide antibiotics which possess an identical mode of action (Brown, 1962; Roland *et al.*, 1979). Because sulphonamide resistance in *E. coli* mutants was known to be conferred by plasmid-borne genes encoding altered DHPSs which are drug insensitive, it was realized that their availability could enable the construction of a novel selectable marker in plant transformation. A sulphonamide resistance gene termed *sul 1* was isolated from *E. coli* containing a single nucleotide change. From this the mutant DHPS was predicted to possess a Phe → Ile substitution accounting for sulphonamide insensitivity (Guerineau and Mullineaux, 1989).

Figure 6.5 Transgenic tobacco plants containing the *sul 1* gene, 21 days following post-emergence treatment with 8 kg/ha asulam. The *sul 1* gene is expressed in the left hand plant but not in the right hand plant. The resistant plants were identical to unsprayed control plants. (D.J. Cole, previously unpublished.)

Tobacco has been transformed with a construct containing *sul 1* ligated with the 35S promoter and a sequence encoding the pea RUBISCO small subunit chloroplast transit peptide (Guerineau *et al.*, 1990). Seedlings from selfed transformants possessed at least 10-fold tolerance to the antibiotic sulphadiazine *in vitro*. The presence of the chloroplast targeting sequence proved critical for gene expression, confirming the predominant location of DHPS within the chloroplast. In isolated chloroplasts the DHPS/transit peptide fusion protein was imported into the stroma and processed into a mature protein with a similar size to DHPS. Plants grown from F_2 seed collected in this study were tested under glasshouse conditions for tolerance to foliar applications of asulam. By visual inspection, plants proved to be completely resistant to asulam at rates up to at least 16 kg/ha which is eight-fold higher than that which causes significant phytotoxicity to tobacco (Figure 6.5). The resistance and sensitivity phenotypes segregated clearly in this generation in a 3:1 ratio demonstrating again a dominant character determined by gene integration at a single nuclear locus. Where resistance was expressed this was total at the herbicide rates used and normal flowering and viable seed set occurred. It is therefore likely that these plants had high level expression of the mutant DHPS, particularly at the apices where asulam is most active.

6.2.4 Atrazine

Many of the herbicides which have been in use since the 1950s and 1960s act by inhibiting photosynthetic electron transport. In most cases, as is exemplified by the s-triazines, this is achieved because the herbicide binds to a cleft in the chloroplastic 'D1' protein, displacing an electron-accepting quinone (Dodge, 1991). Atrazine (Figure 6.2) is the most widely used of the triazines, in part due to its good selective weed control in maize (*Zea mays*) and sorghum (*Sorghum bicolor*) crops. However, atrazine can degrade slowly in soil and this has contributed to the appearance of resistant biotypes of many different weed species (Le Baron, 1991). The D1 protein is encoded by the chloroplastic *psb A* gene specifying a 34 kDa polypeptide which is processed to yield a mature 32 kDa protein. Photosynthetic microorganisms can also be readily mutated for atrazine resistance. Analysis of the *psb A* coding sequences from many resistant plants and microorganisms reveals predicted single amino acid substitutions, normally of serine 264 (Hirschberg *et al.*, 1987).

An insensitive target site has therefore been readily available; however, the lack of technology to transform chloroplasts has posed a hurdle to engineering atrazine resistance. This was surmounted in a fascinating way by Cheung *et al.* (1988) who were able to target the mutant *psb A* gene from *Amaranthus hybridus* (Hirschberg and McIntosh, 1983) to the nucleus of recipient tobacco. This was achieved by fusing *psb A* with a nuclear promoter and chloroplast transit sequence, both derived from the pea RUBISCO small subunit gene. Low level tolerance to atrazine was exhibited by the regenerated transformants. The mutant D1 protein was detected in the tobacco thylakoids but it is clear that there was only a limited degree of expression.

Using a more novel approach to plant transformation, Liu *et al.*, (1990) have successfully delivered a mutant *psb A* gene from *Solanum nigrum* to soybean by direct microinjection of naked DNA to the fertilized ovary (see Chapter 1). The gene retained its own promoter sequence. A small proportion of the progeny had clear tolerance to foliarly applied atrazine and chloroplast DNA was shown to have integrated the vector DNA. While this is a rather exacting route of transformation, the potential for direct targeting of genes to the chloroplast is clear.

The ready detoxification of atrazine by a specific glutathione *S*-transferase (GST) in maize (Edwards and Owen, 1986) suggests an alternative strategy for engineering resistance to atrazine and indeed other herbicides conjugated by GSTs. Some work along these lines has been carried out but with limited success. The success of transferring specific GSTs from one organism to another depends upon satisfactory levels of the glutathione co-substrate in the recipient organism. Atrazine is likely to be supplanted by more modern herbicides in the near future, so the desirability of further efforts in this direction is questionable.

6.3 HERBICIDE RESISTANCE THROUGH METABOLIC DETOXIFICATION

6.3.1 Glufosinate

Engineering of crop resistance to the broad spectrum herbicide glufosinate by the introduction of a bacterial detoxification gene demonstrates vividly the success of a metabolism-based approach to resistance. Glufosinate, or L-phosphinothricin, is unusual in being a commercial herbicide based on a natural product. Glufosinate is a breakdown product of the antibiotic tripeptide bialaphos which is produced by certain *Streptomyces* species (Figure 6.6). Both glufosinate as its ammonium salt and bialaphos are in commercial use as herbicides, the latter undergoing peptide cleavage to form glufosinate within the plant.

As an analogue of L-glutamate, glufosinate is a specific, potent and competitive inhibitor of glutamine synthetase (Leason *et al.*, 1982), a key enzyme in nitrogen metabolism (Figure 6.7) and in the assimilation of toxic ammonia. While such action in the plant will deplete the level of glutamine and disrupt nitrogen assimilation the major biochemical lesion causing cell death through this action is the rapid accumulation of ammonia (Tachibana *et al.*, 1986). Glutamate synthetase exists in the plant in multiple forms, each

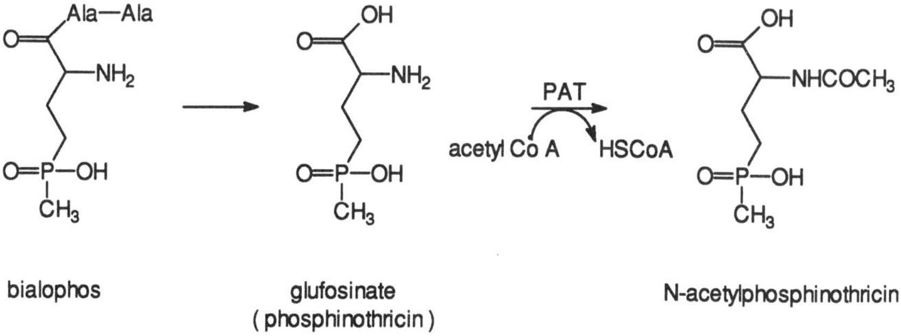

Figure 6.6 Derivation of glufosinate from bialaphos and the acetylation of glufosinate by PAT from *Streptomyces* spp.

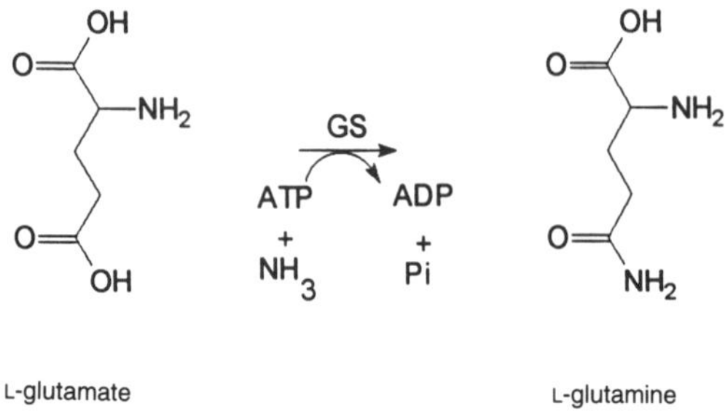

Figure 6.7 Reaction carried out by glutamine synthetase (GS).

comprising an octomer of catalytic subunits. The enzymes from a variety of plant species are inhibited similarly by glufosinate (Acaster and Weitzmann, 1985). Mammalian glutamine synthetase is also inhibited by glufosinate *in vitro*. Fortuitously however, glufosinate fails to cross the blood–brain barrier and therefore has low acute toxicity.

In antibiotic-producing bacteria, genes controlling detoxication of antibiotics and therefore protection of the organism against its own products, are often found closely linked to the genes responsible for synthesis of the antibiotic. In the case of glufosinate, genes have been identified encoding a specific *N*-acetylating enzyme (phosphinothricin *N*-acetylase, PAT) which inactivates the antibiotic (Figure 6.6). In *Streptomyces hygroscopicus* a gene conferring resistance to bialaphos (*bar*) was identified by complementation of susceptible bacteria (Murakami *et al.*, 1986). By subcloning into other bacterial hosts, the gene was found to encode a 22 kDa protein which acetylated glufosinate but not the bialaphos tripeptide (Thompson *et al.*, 1987). Similarly an analogous gene from *S. viridochromogenes* termed *pat* (Strauch *et al.*, 1988) encodes a PAT enzyme of 20.6 kDa showing 85% predicted amino acid sequence homology with that of the *bar* product (Wohlleben *et al.*, 1988).

Subsequently the *bar* gene has been extensively used to transform a range of important crop species to confer high-level and agronomically useful resistance to both glufosinate and bialaphos. The *bar* gene appears to function well as a selectable marker in the regeneration of transformants. Significantly, *bar* has featured in the seminal announcements of reproducible transformation methods for some key crops including sugarbeet and lucerne but notably the hitherto recalcitrant maize and wheat.

Expression of *bar* in plants required replacement of the bacterial initiation codon GTG with ATG and in most instances the gene has been coupled to the 35S promoter. Transgenic tobacco harbouring *bar* were completely tolerant in the greenhouse to repeated applications of both glufosinate and bialaphos, with normal flowering and seed setting (De Block *et al.*, 1987).

Immunoblotting and Northern hybridization showed that PAT enzyme and PAT-specific transcript levels were highly correlated, though contents showed wide plant-to-plant variation. Transgenic tomato and potato (De Block, 1988) have also been produced. Field evaluation of the resistance phenotype indicated that tobacco and three commercial potato lines containing *bar* possessed complete resistance to post-emergence treatment with 4 kg/ha glufosinate or four times the normal application rate (De Greef *et al.*, 1989). The lines showed variable levels of PAT expression, i.e. less than 0.01–0.1% total protein, but complete tolerance occurred in all cases. In the case of potato, PAT expression was driven by the TR2′ promoter isolated from *Agrobacterium* T-DNA. Further confirmation of the effectiveness of the *bar* gene was provided by the failure to detect intracellular ammonia over control levels in the transgenic tobacco, whereas glufosinate-treated non-transformed plants accumulated massive amounts of ammonia (De Block *et al.*, 1987).

Glufosinate tolerance has been successfully transferred to maize and wheat by microprojectile bombardment of embryogenic tissue culture followed by regeneration of fertile plants. This is currently the transformation method of choice for graminaceous crop species which are not amenable to transformation through the use of *Agrobacterium* techniques. Through this method, regenerants were recovered of maize (Gordon-Kamm *et al.*, 1990) and wheat (Vasil *et al.*, 1992) which harboured *bar* and expressed PAT. Fertilization of regenerated transformants with wild-type pollen resulted in progeny where clear co-segregation of the resistance phenotype and *bar* integration occurred. In these cases tolerance to glufosinate was determined by means of topical application to foliage. Similarly transformation of rice with *bar* has been carried out by microprojectile bombardment of immature embryos with recovery of plants resistant to bialaphos (Christou *et al.*, 1992). Rice has also been transformed with *bar* by direct gene transfer to protoplasts followed by regeneration of resistant plants (Datta *et al.*, 1992). A further success among graminaceous plants has been the transformation of the forage grass *Festuca arundinacea* with *bar* by a similar method (Wang *et al.*, 1992).

Several other dicotyledonous crops have been successfully transformed for glufosinate resistance with *bar* using *Agrobacterium* infection, with gene expression controlled either by the 35S or TR2′ promoters. These include sugarbeet (D'Halluin *et al.*, 1992), spring and winter rapeseed (*B. napus*) varieties (De Block *et al.*, 1989), lucerne (D'Halluin *et al.*, 1990) and aspen (*Populus* spp.) (Devillard, 1992). In all cases, the presence of the *bar* structural gene and enhanced PAT activity were shown in the transgenic plants, though some rape plants were positive for Southern hybridization (see Chapter 1) but did not possess the resistance phenotype. The 35S and TR2′ promoters were equally effective in aiding *bar* expression in sugar beet but 35S performance was superior in lucerne, where satisfactory resistance to glufosinate under field conditions was shown.

The analogous glufosinate resistance gene *pat* has also been used to impart resistance to tobacco plantlets *in vitro* (Wohlleben *et al.*, 1988) and also to carrot (*Daucus carota*) via direct treatment of protoplasts with a high copy number plasmid vector (Dröge *et al.*, 1992). In neither case were the molecular studies undertaken to show integration of the gene or presence of its product. However, in the glufosinate-tolerant plants, amino terminal acetylation of glufosinate took place, in contrast to other routes of metabolism in untransformed plants.

As an alternative means to the generation of glufosinate-resistant plants, over-expression of lucerne glutamine synthetase in tobacco has been evaluated. Control by the 35S promoter resulted in a several-fold increase in glutamine synthetase specific activity and a 20-fold increase in resistance to glufosinate of tobacco plantlets *in vitro* (Eckes *et al.*, 1989). However, this means of resistance has not as yet been rigorously evaluated. It may be speculated that in view of the multiple forms of glutamine synthetase existing in plants, over-expression of a single form may not be sufficient to guarantee high-level herbicide resistance. Over-expression of plant glutamine synthetase in a suitable *E. coli* host lacking its native enzyme may result in a system for generating mutant plant enzyme which is more likely to confer high level herbicide resistance upon reintroduction into plants (Das Sarma *et al.*, 1986).

6.3.2 Bromoxynil

Bromoxynil is another widely used herbicide which acts as a photosynthetic electron transport inhibitor and additionally possesses some uncoupling activity (Sanders and Pallett, 1985). It has conventional utility in the control of a range of broad-leaved weeds in cereals. Its use could be further extended if certain dicotyledonous crops could be rendered resistant.

Bromoxynil is rapidly degraded by soil microorganisms and in the

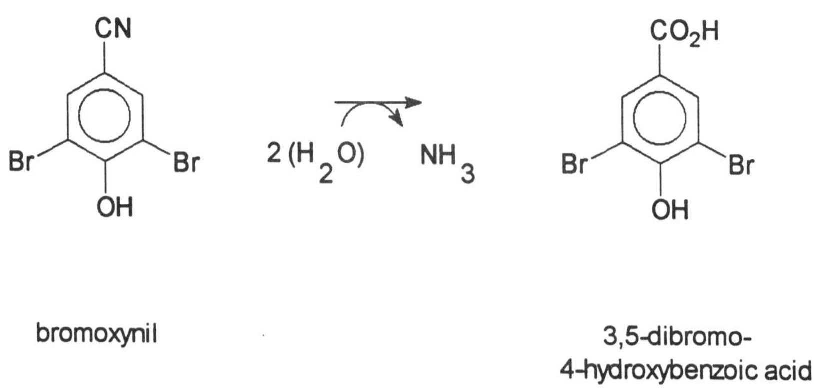

Figure 6.8 Detoxification of bromoxynil by the nitrilase from *K. ozaenae*.

search for a suitable detoxification gene a highly resistant isolate of *Klebsiella ozaenae* was found. This strain is enriched in a bromoxynil-specific nitrilase able to hydrolyse the cyano group of the herbicide (Figure 6.8), resulting in production of its benzoic acid derivative (McBride *et al.*, 1986). This nitrilase enables the bacterium to utilize bromoxynil as a sole nitrogen source. The *bxn* gene encoding the nitrilase is carried on a 82 kb plasmid and was subcloned in *E. coli* as a 2.6 kb fragment (Stalker and McBride, 1987). Analysis of the gene and enzyme purification indicated specification of a 38 kDa protein (Stalker *et al.*, 1988b).

When transferred to tobacco, the *bxn* gene confers excellent resistance to bromoxynil both in the laboratory and in the field, sufficient to enable satisfactory selective weed control (Stalker *et al.*, 1988a; Freyssinet *et al.*, 1989). The RUBISCO small subunit promoter was used to drive gene expression, showing that light-specific expression was sufficient to confer protection. Northern and Western blotting (see Chapter 1) of leaf material showed the presence of a 38 kDa protein and gene-specific transcripts, the level of which correlated to the degree of resistance. The enzyme was deduced to be located in the cytoplasm rather than the chloroplast, thus demonstrating effective inactivation of the herbicide prior to reaching the subcellular site of action (Stalker *et al.*, 1988a). In this way photosynthesis *in planta* was fully protected from bromoxynil even though chloroplasts isolated from resistant transgenic plants retained full biochemical sensitivity *in vitro* (Freyssinet *et al.*, 1989). From these studies, the *bxn* gene appears to show promise as a selectable marker. Bromoxynil resistance has now been introduced to a number of other crops. Tomato plants containing *bxn* under the control of the mannopine synthase promoter show good resistance to 1.65 kg/ha bromoxynil in the field (Stalker, 1991).

6.3.3 Glyphosate

Although most efforts towards obtaining glyphosate-resistant species have focussed on target site manipulation, a metabolism-based approach has also recently yielded success. In soil microorganisms, glyphosate (Figure 6.2) can be degraded by cleavage of the C–N bond to form aminomethylphosphonic acid and glyoxylate. An *Achromobacter* species has been isolated from a waste stream facility which is able to utilize glyphosate as a sole carbon or phosphate source.

The glyphosate oxidoreductase catalysing this process is a flavoprotein requiring FAD as a cofactor. Isolation of the encoding gene has enabled transformation to generate glyphosate resistance in transgenic maize, soybean and rape (Barry *et al.*, 1992). Gene expression was optimum when a modified structural coding sequence was used, together with a chloroplast targeting sequence. Rapeseed transformed in this way shows very good glyphosate resistance in field trials. So far, little published information is available to compare the performance of these crops with those modified at the target site.

6.3.4 2,4-D

The phenoxyacetic acid family of herbicides was introduced after the second world war, heralding the advent of weed control by artificial chemicals. The most prominent of these is 2,4-D, which is used in many situations to control a wide spectrum of broad-leaved weeds. Accidental spray drift damage to sensitive dicotyledonous crops has encouraged interest in engineering resistance to 2,4-D, particularly in cotton.

The phenoxyacetic acids exert their herbicidal effects by overly mimicking the action of the endogenous auxin plant hormones (Devine *et al.*, 1992). Engineering enhanced herbicide degradation in the plant is therefore a more realistic approach to creating resistance than modifying the receptor site(s). Numerous soil bacteria are able to degrade 2,4-D to less herbicidal products. One species, *Alcaligenes eutrophus*, has proved to be a practical source of a 2,4-D detoxifying gene enabling the creation of 2,4-D resistance in transgenic cotton and tobacco. The strain *A. eutrophus* JMP 134 contains a 75 kb plasmid pJP4 harbouring six genes encoding enzymes which sequentially degrade 2,4-D (Don *et al.*, 1985). The initial modification involves cleavage of the acetic acid side chain from 2,4-D by a mono-oxygenase-type reaction, forming the less phytotoxic 2,4-dichlorophenol (Figure 6.9). This enzyme is specified by the *tfd A* gene which has been isolated and characterized by Streber *et al.*, (1987).

As with other prokaryotic genes encoding detoxification enzymes, preparation for expression in plants required replacement of the GTG start codon by ATG. In the studies on tobacco transformation a consensus sequence for plant translation was placed immediately upstream of the start codon, and a number of promoter arrangements were investigated (Streber and Willmitzer, 1989; Lyon *et al.*, 1989). Initial over-expression of *tfd A* in *E. coli* verified the catalytic nature of the gene product (Lyon *et al.*, 1989). Expression of *tfd A* in tobacco enabled a 20- to 30-fold resistance in regenerants *in vitro* and at least 10-fold enhanced resistance in greenhouse trials. Hybridization showed integration of the *tfd A* gene and the presence

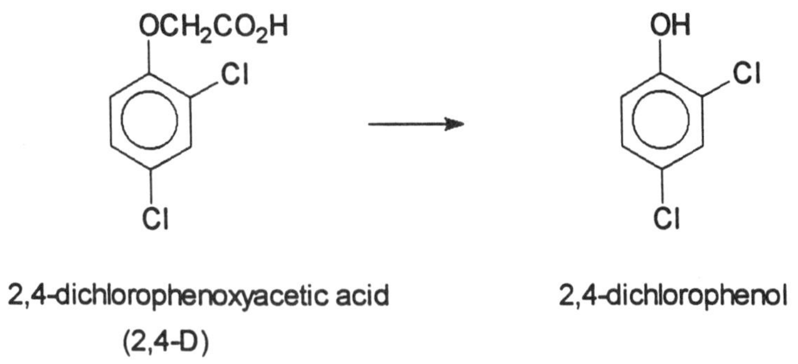

Figure 6.9 Detoxification of 2,4-D by the mono-oxygenase from *A. eutrophus*.

of the gene-specific transcript (Streber and Willmitzer, 1989). Better resistance was conferred using the 35S promoter rather than a light-specific promoter (Streber and Willmitzer, 1989) and also by the inclusion of an alfalfa mosaic virus leader sequence in the construct (Lyon *et al.*, 1989), putatively enhancing translation efficiency.

Cotton has been engineered for resistance to 2,4-D by transfer of a similar *tfd* A gene from *A. eutrophus* DSM 4059 (Bayley *et al.*, 1992). The gene was isolated by probing with oligonucleotides corresponding to sequences of the pJP4 *tfd* A gene. Cotton hypocotyls were transformed conventionally with *Agrobacterium*. The 35S promoter enabled gene expression to give at least 10-fold tolerance to 2,4-D, i.e. to a rate of 1.5 kg/ha. Similar tobacco transformants had resistances up to 20 kg/ha. Transgenic tobacco showed an enhanced capacity to evolve $^{14}CO_2$ from [^{14}C]2,4-D, further indicating functionality of the new gene.

Interestingly, the *tfd A* gene does not enable satisfactory cross-resistance to the closely related phenoxyacetic acid herbicides MCPA or 2,4,5-T, indicating a very restricted substrate specificity. In none of these crops was *tfd A* able to function as a selectable marker. There are as yet no data evaluating the resistance of these plants to 2,4-D in the field.

6.3.5 Other herbicides

Two other examples currently exist where microbial herbicide detoxification genes have been utilized to generate herbicide resistance in plants. Isolation of a gene from the soil bacterium *Pseudomonas putida* specifying dechlorination of the herbicide dalapon (2,2-dichloropropionic acid) has enabled the engineering of dalapon-resistant tobacco. The gene functioned promisingly as a selectable marker and imparted up to 5-fold tolerance to the herbicide in greenhouse plants (Buchanan-Wollaston *et al.*, 1992). Tobacco resistant to cyanamide (H_2NCN) has been achieved by the introduction of a fungal enzyme enabling detoxification of the herbicide. An inducible nitrile hydratase found in the soil fungus *Myrothecium verrucaria* is encoded by the 732 bp *cah* gene (Maier-Greiner *et al.*, 1991b). Transformation of tobacco with *cah* enabled recovery of regenerants which were able to tolerate foliar treatments of cyanamide equivalent to field rates (Maier-Greiner *et al.*, 1991a). Resistant regenerants were shown to contain cyanamide hydratase activity. An attractive feature of this system of resistance is the generation of the natural metabolite urea as the breakdown product of the herbicide. This can subsequently be utilized by the plant's endogenous urease. However, cyanamide is obsolete as a herbicide and therefore resistant plants are not likely to find useful application. Integration of the novel genes for dalapon and cyanamide resistance into the plant genome has not been confirmed.

6.4 CONCLUSIONS AND FUTURE PROSPECTS

6.4.1 What has been achieved?

Remarkable progress has been made in harnessing recombinant DNA techniques for the purpose of herbicide resistance. Field level resistance to several herbicides has been achieved in less than a decade since artificial transfer of the first 'reporter' genes to plant tissues. It is clear that introduction of a single gene can impart complete herbicide resistance. Nevertheless high level expression of the gene and correct intracellular targeting are essential for satisfactory efficacy. Resistance is more completely created by bringing about enhanced herbicide detoxification than manipulating the herbicide target site. A detoxification gene need only be expressed at a very low percentage of the total protein content to enable satisfactory field performance. Therefore it is unnecessary to know anything about a herbicide's mode of action in order to create resistance. On the other hand, target site manipulation provides a greater chance of obtaining cross resistance between different herbicides with a common mode of action.

The range of crop plants transformed for herbicide resistance is limited at the moment. However, given the fast progress in development of reproducible transformation techniques for major crops, it is only a matter of time before this range can be expanded. Among the prospects we can look forward to are the engineering of multiple herbicide resistance and enhancing the tissue-specificity of expression. Progress so far has served to lay the ground rules for successful introduction of herbicide resistance. It should therefore be emphasized that creation of commercially viable resistance has not necessarily been the intention in every case discussed here.

So far more has been learnt about the construction of resistant plants than their performance in the field. Extensive trials still need to be completed before we can be assured about the agronomic performance of these new varieties, and the stability of the new gene(s) over many generations. For further coverage of this area see Chapter 8.

6.4.2 Marketing herbicide-resistant plants – prospects and issues

Access to transgenic herbicide-resistant crops will not revolutionize arable agriculture. Rather, their availability together with the appropriate herbicides presents the farmer with an additional tool, complementing his existing management techniques. Because the market is in a constant state of flux it is difficult to predict the impact such altered crops will have. However, it is certain that farmers will only take up the new crops if there is a cost-benefit. This could entail the ability to control a hitherto difficult weed or to spray on a case-by-case basis when weeds become a threat, rather than preventatively. What is certain is that plant breeders will be keenest to use the traits of tolerance to the most environmentally benign herbicides compatible with the specific weed problems to be addressed. This will be driven by regulatory constraints which can be expected to become ever more exacting. The net result will be a trend to the use of fewer, more acceptable herbicides and the obsolescence of some others. Because more modern herbicides are used at low dose rates, the herbicide burden in the environment will be reduced.

The prospect of cultivating these altered crops has given rise to a number of concerns. These centre on the risk of increased weed resistance to herbicides by enhanced selection pressure, by transmission of resistance genes from the crop to related weeds and the appearance of resistant 'volunteer' crop plants. Crops possessing a gene directing enhanced metabolism may accumulate greater quantities of herbicide metabolites. However it must be emphasized that none of these problems are new. 'Natural' tolerance exists in crops to conventional selective herbicides and weed resistance is a known risk where persistent herbicides are used repetitively on the same ground. Wild plants acquiring herbicide resistance following pollination by a resistant crop plant will only out-compete the wild-type weeds if they have enhanced 'fitness' in the environment. This will be unlikely in the absence of the herbicide. Nevertheless, transmissibility is a legitimate concern and is properly receiving the attention of ecologists (Tynan *et al.*, 1990; Dale, 1992). As now, these problems will not develop where crop and herbicide treatments are sensibly rotated. This will ideally entail the use of different herbicides of varying modes of action and metabolic breakdown routes.

The acceptance of artificially resistant crops has additional implications for herbicide manufacturers. Engineering multi-crop resistance to one herbicide will reduce the costs of research and development and of registration where previously differing analogues within a chemical series have had to be developed for differing crops (Brown *et al.*, 1991). One may speculate that if the need to screen for crop selectivity in a synthesis programme is removed, chemists will be able to focus on potency and ecotoxicological safety at an earlier stage. This would be carried out in parallel with the identification of appropriate detoxification genes for the promising candidates. Optimization of use during the patent life of a herbicide will be a major goal for manufacturers.

Finally, the behaviour of herbicide resistance genes as dominant selectable markers suggests these as new tools in crop development. Some of these genes function well as selectable markers in the recovery of plants from transformed tissue cultures. They may well replace the present use of antibiotic resistance genes as selection aids. Indeed it is desirable to avoid the insertion of antibiotic resistance genes into crops which are going to be disseminated into the environment.

The possibilities of linking herbicide resistance genes directly with agronomically desirable traits will provide obvious benefits in conventional plant breeding programmes. It is often difficult to readily identify useful traits by their phenotype. These can be tracked more readily in a breeding programme by an easily assayable herbicide resistance phenotype which enables co-selection of the desired genes.

REFERENCES

Acaster, M.A. and Weitzman, P.D.J. (1985) Kinetic analysis of glutamine synthetases from various plants. *FEBS Letters*, **189**, 241–4.

Amrhein, N., Deus, B., Gehrke, P. and Steinrücken, H.C. (1980) The site of the inhibition of the shikimate pathway by glyphosate II. Interference of glyphosate with chorismate formation *in vivo* and *in vitro*. *Plant Physiology*, **66**, 830–4.

Barry, G., Kishore, G., Padgette, S., Taylor, M., Kolacz, K., Weldon, M., Re, D., Eichholtz, D., Fincher, K. and Hallas, L. (1992) Inhibition of amino acid biosynthesis: strategies for imparting glyphosate tolerance to crop plants, in *Biosynthesis and Molecular Regulation of Amino Acids in Plants*, (eds B.K. Singh, H.E. Flores and J.C. Shannon), American Society of Plant Physiologists, Rockville, MD, pp. 139–45.

Bayley, C., Trolinder, N., Ray, C., Morgan, M., Quisenberry, J.E. and Ow, D.W. (1992) Engineering 2,4-D resistance into cotton. *Theoretical and Applied Genetics*, **83**, 645–9.

Bevan, M.W., Flavell, R.B. and Chilton, M.-D. (1983) A chimaeric antibiotic resistance gene as a selectable marker for plant cell transformation. *Nature*, **304**, 184–7.

Bozak, K.R., O'Keefe, D.P. and Christoffersen, R.E. (1992) Expression of a ripening-related Avocado (*Persea americana*) cytochrome P450 in yeast. *Plant Physiology*, **100**, 1976–81.

Brasileiro, A.C.M., Tourneur, C., Leplé, J.-C., Combes, V. and Jouanin, L. (1992) Expression of the mutant *Arabidopsis thaliana* acetolactate synthase gene confers chlorsulfuron resistance to transgenic poplar plants. *Transgenic Research*, **1**, 133–41.

Brown, G.M. (1962) The biosynthesis of folic acid II. Inhibition by sulfonamides. *Journal of Biological Chemistry*, **237**, 536–40.

Brown, H.M., Dietrich, R.F., Kenyon, W.F. and Lichtner, F.T. (1991) Prospects for the biorational design of crop selective herbicides, in *Proceedings Brighton Crop Protection Conference*, The British Crop Protection Council, Farnham, UK, pp. 847–56.

Buchan-Wollaston, V., Snape, A. and Cannon, F. (1992). A plant selectable marker gene based on the detoxification of the herbicide dalapon. *Plant Cell Reports*, **11**, 627–31.

Chaleff, R.S. and Mauvais, C.J. (1984) Acetolactate synthase is the site of action of two sulfonylureas in higher plants. *Science*, **224**, 1443–5.

Charest, P.J., Hattori, J., De Moor, J., Iyer, V.N. and Miki, B.L. (1990) *In vitro* study of transgenic tobacco expressing *Arabidopsis* wild type and mutant acetohydroxyacid synthase genes. *Plant Cell Reports*, **8**, 643–6.

Cheung, A.Y., Bogorad, L., Van Montagu, M. and Schell, J. (1988) Relocating a gene for herbicide tolerance: a chloroplast gene is converted into a nuclear gene. *Proceedings of the National Academy of Sciences USA*, **85**, 391–5.

Christou, P., Ford, T.L. and Kofron, M. (1992) The development of a variety-independent gene-transfer method for rice. *Trends in Biotechnology*, **10**, 239–46.

Cole, D.J., Edwards, R. and Owen, W.J. (1987) The role of metabolism in herbicide selectivity, in *Herbicides. Progress in Pesticide Biochemistry and Toxicology*, Vol. 6, (eds D.H. Hutson and T.R Roberts), Wiley, Chichester, pp. 57–104.

Comai, L., Facciotti, D., Hiatt, W.R., Thompson, G., Rose, R. and Stalker, D. (1985) Expression in plants of a mutant *aroA* gene from *Salmonella typhimurium* confers tolerance to glyphosate. *Nature*, **317**, 741–4.

Comai, L., Sen, L.C. and Stalker, D.M. (1983) An altered *aroA* gene product confers resistance to the herbicide glyphosate. *Science*, **221**, 370–1.

Cossins, E.A. (1987) One-carbon metabolism, in *The Biochemistry of Plants – A Comprehensive Treatise*, Vol. II, (ed. D.D. Davis), Academic Press, New York, pp. 318–418.

Dale, P.J. (1992) Spread of engineered genes to wild relatives. *Plant Physiology*, **100**, 13–5.

Das Sarma, S., Tischer, E. and Goodman, H.M. (1986) Plant glutamine synthetase complements a *glnA* mutation in *Escherichia coli*. *Science*, **232**, 1242–4.

Datta, S.K., Datta, K., Soltanifar, N., Donn, G. and Potrykus, I. (1992) Herbicide resistance in Indica rice plants from IRRI breeding line IR 72 after PEG-mediated transformation of protoplasts. *Plant Molecular Biology*, **20**, 619–29.

De Block, M. (1988) Genotype-independent leaf disc transformation of potato (*Solanum tuberosum*) using *Agrobacterium tumefaciens*. *Theoretical and Applied Genetics*, **76**, 767–74.

De Block, M., Botterman, J., Vandewiele, M., Dockx, J., Thoen, C., Gosselé, V., Movva, N.R., Thompson, C., Van Montagu, M. and Leemans, J. (1987). Engineering herbicide resistance in plants by expression of a detoxifying enzyme. *EMBO Journal*, **6**, 2513–8.

De Block, M., De Brouwer, D. and Tenning, P. (1989) Transformation of *Brassica napus* and *Brassica oleracea* using *Agrobacterium tumefaciens* and the expression of the *bar* and *neo* genes in the transgenic plants. *Plant Physiology*, **91**, 694–701.

De Greef, W., Delon, R., De Block, M., Leemans, J. and Bottermans, J. (1989) Evaluation of herbicide resistance in transgenic crops under field conditions. *Bio/Technology*, **7**, 61–4.

della-Cioppa, G., Bauer, S.C., Taylor, M.L., Rochester, D.E., Klein, B.K., Shah, D.M., Fraley, R.T. and Kishore, G.M. (1987) Targeting a herbicide-resistant enzyme from *Escherichia coli* to chloroplasts of higher plants. *Bio/Technology*, **5**, 579–84.

Devillard, C. (1992) Transformation *in vitro* of aspen (*Populus tremula* × *Populus alba*) by *Agrobacterium rhizogenes* and regeneration of plants tolerant to Basta. *Compte Rendu Seances Académie des Sciences Series 3*, **314**, 291–8.

Devine, M., Duke, S.O. and Fedtke, C. (1992) *Physiology of Herbicide Action*, Prentice-Hall, Englewood Cliffs, NJ.

D'Halluin, K., Bossut, M., Bonne, E., Mazur, B., Leemans, J. and Botterman, J. (1992) Transformation of sugarbeet (*Beta vulgaris* L.) and evaluation of herbicide resistance in transgenic plants. *Bio/Technology*, **10**, 309–14.

D'Halluin, K., Botterman, J. and De Greef, W. (1990) Engineering of herbicide-resistant alfalfa and evaluation under field conditions. *Crop Science*, **30**, 866–71.

Dodge, A.D. (1991) Photosynthesis, in *Target Sites for Herbicide Action* (ed. R.C. Kirkwood), Plenum, New York, pp. 1–27.

Don, R.H., Weightman, A.J., Knackmuss, H.-J. and Timmis, K.N. (1985) Transposon mutagenesis and cloning analysis of pathways for degradation of 2,4-dichlorophenoxyacetic acid and 3-chlorobenzoate in *Alcaligenes eutrophus* JMP 134 (pJP4). *Journal of Bacteriology*, **161**, 85–90.

Dröge, W., Broer, I. and Pühler, A. (1992) Transgenic plants containing the phosphinothricin *N*-acetyltransferase gene metabolize the herbicide L-phosphinothricin (glufosinate) differently from untransformed plants. *Planta*, **187**, 142–51.

Eckes, P., Schmitt, P., Daub, W. and Wengenmayer, F. (1989) Over production of alfalfa glutamine synthetase in transgenic tobacco plants. *Molecular and General Genetics*, **217**, 263–68.

Edwards R. and Owen, W.J. (1986) Comparison of glutathione *S*-transferases of *Zea mays* responsible for herbicide detoxification in plants and suspension cultured cells. *Planta*, **169**, 208–15.

Falco, S.C. and Dumas, K.D. (1985) Genetic analysis of mutants of *Saccharomyces cerevisiae* resistant to the herbicide sulfometuron methyl. *Genetics*, **109**, 21–35.

Fillatti, J.J., Kiser, J., Rose, R. and Comai, L. (1987) Efficient transfer of a glyphosate tolerance gene into tomato using a binary *Agrobacterium tumefaciens* vector. *Bio/Technology*, **5**, 726–30.

Fraley, R.T. (1988) Field tested genetically engineered plants, in *Genetic Improvements of Agriculturally Important Crops: Progress and Issues*, (eds R.T. Fraley, N.M. Frey and J. Schell), Cold Spring Harbor Laboratory Press, Cold Spring Harbor, pp. 83–6.

Freyssinet, G., Leroux, B., Lebrun, M., Pelissier, B., Sailland, A. and Pallett, K.E. (1989) Transfer of bromoxynil resistance into crops, in *Proceedings of the Brighton Crop Protection Conference*, The British Crop Protection Council, Farnham, UK, 1225–34.

Gordon-Kamm, W.J., Spencer, T.M., Mangano, M.L., Adams, T.R., Daines, R.J., Start, W.G., O'Brien, J.V., Chambers, S.A., Adams, W.R., Willetts, N.G., Rice, T.B., Mackey, C.J., Krueger, R.W., Kausch, A.P. and Lemaux, P.G. (1990) Transformation of maize cells and regeneration of fertile transgenic plants. *Plant Cell*, **2**, 603–18.

Guerineau, F., Brooks, L., Meadows, J., Lucy, A., Robinson, C. and Mullineaux, P. (1990) Sulfonamide resistance gene for plant transformation. *Plant Molecular Biology*, **15**, 127–36.

Guerineau, F. and Mullineaux, P. (1989) Nucleotide sequence of the sulfonamide resistance gene from plasmid R46. *Nucleic Acids Research*, **17**, 4370.

Hall, L.M. and Devine, M.D. (1990) Cross resistance of a chlorsulfuron resistant biotype of *Stellaria media* to a triazolopyrimidine herbicide. *Plant Physiology*, **93**, 962–6.

Hartnett, M.E., Chui, C.-F., Mauvais, C.J., McDevitt, R.E., Knowlton, S., Smith, J.K., Falco, S.C. and Mazur, B.J. (1990) Herbicide resistant plants carrying mutated acetolactate synthase genes, in *Managing Resistance to Agrochemicals*, (eds M.B. Green, H.M. LeBaron and W.K. Moberg), ACS Symposium Series **421**, ACS, Washington, DC, pp. 459–75.

Hattori, J., Rutledge, R., Labbé, H., Brown, D., Sunohara, G. and Miki, B. (1992) Multiple resistance to sulfonylureas and imidazolinones conferred by an acetohydroxyacid synthase gene with separate mutations for selective resistance. *Molecular and General Genetics*, **232**, 167–73.

Haughn, G.W., Smith, J., Mazur, B. and Somerville, C. (1988) Transformation with a mutant *Arabidopsis* acetolactate synthase gene renders tobacco resistant to sulfonylurea herbicides. *Molecular and General Genetics*, **211**, 266–71.

Herrera-Estrella, L., Depicker, A., Van Montagu, M. and Schell, J. (1983) Expression of chimaeric genes transferred into plant cells using a Ti-plasmid-derived vector. *Nature*, **303**, 209–13.

Hinchee, M.A.W., Connor-Ward, D.V., Newell, C.A., McDonnell, R.E., Sato, S.J., Gasser, C.S., Fischhoff, D.A., Re, D.B., Fraley, R.T. and Horsch, R.B. (1988) Production of transgenic soybean plants using *Agrobacterium* mediated DNA transfer. *Bio/Technology*, **6**, 915–22.

Hirschberg, J. and McIntosh, L. (1983) Molecular basis of herbicide resistance in *Amaranthus hybridus*. *Science*, **222**, 1346–49.

Hirschberg, J., Yehuda, A.B., Pecker, I. and Ohad, N. (1987) Mutations resistant to photosystem II herbicides in *Plant Molecular Biology*, (eds D. von Wettstein and N.-H. Chua), Plenum, New York, pp. 357–66.

Hoagland, R.E., Duke, S.O. and Elmore, C.D. (1978) Effects of glyphosate on metabolism of phenolic compounds II. Influence on soluble hydroxyphenolic compound, free amino acid and soluble protein levels in dark grown maize roots. *Plant Science Letters*, **13**, 291–9.

Jordan, M.C. and McHughen, A. (1988) Glyphosate tolerant flax plants from *Agrobacterium* mediated gene transfer. *Plant Cell Reports*, **7**, 281–4.

Kidd, B.R., Stephen, N.H. and Duncan, H.J. (1982) The effect of asulam on purine biosynthesis. *Plant Science Letters*, **26**, 211–7.

Kishore, G.M., Padgette, S.R. and Fraley, R.T. (1992) History of herbicide-tolerant crops, methods of development and current state of the art – emphasis on glyphosate tolerance. *Weed Technology*, **6**, 626–34.

La Rossa, R.A. and Schloss, J.V. (1984) The sulfonylurea herbicide sulfometuron methyl is an extremely potent and selective inhibitor of acetolactate synthase in *Salmonella typhimurium*. *Journal of Biological Chemistry*, **259**, 8753–7.

Larson-Kelly, N., Comai, L., Kiser, J., Mau, C., Pokalsky, A.R., McBride, K., Jones, A., Shewmaker, C. and Stalker, D.M. (1988) Chloroplast delivery of a bacterial EPSP synthase in transgenic plants and tolerance to glyphosate. *SAAS Bulletin of Biochemistry and Biotechnology*, **1**, 37–40.

Lawther, R.P., Nichols, B., Zurawaski, G. and Hatfield, G.W. (1979) The nucleotide sequence preceding and including the beginning of the *ilv E* gene of the *ilv GEDA* operon of *Escherichia coli* K12. *Nucleic Acids Research*, **7**, 2289-301.

Leason, M., Dunliffe, D., Parkin, D., Lea, P.J. and Miflin, B.J. (1982) Inhibition of pea leaf glutamine synthetase by methionine sulfoximine, phosphinothricin and other glutamate analogues. *Phytochemistry*, **21**, 855–7.

Le Baron, H.M. (1991) Distribution and seriousness of herbicide resistant weed infestations worldwide, in *Herbicide Resistance in Weeds and Crops, 11th Long Ashton Symposium*, (eds J.C. Caseley, G.W. Cussons and R.K. Atkin), Butterworth-Heinmann, Oxford, pp. 27–43.

Lee, K.Y., Townsend, J., Tepperman, J., Black, M., Chui, C.-F, Mazur, B., Dunsmuir, P. and Bedbrook, J. (1988) The molecular basis of sulfonylurea herbicide resistance in tobacco. *EMBO Journal*, **7**, 1241–8.

Li, Z., Hayashimoto, A. and Murai, N. (1992) A sulfonylurea herbicide resistance gene from *Arabidopsis thaliana* as a new selectable marker for production of fertile transgenic rice plants. *Plant Physiology*, **100**, 662–8.

Liu, B., Yue, S., Hu, N., Li, X., Zhai, W., Li, N., Zhu, R., Zhu, L., Mao, D. and Zhou, P. (1990) Transfer of the atrazine resistance gene of black nightshade to the soybean chloroplast genome and its expression in transgenic plants. *Science in China, Series B*, **33**, 444–52.

Lyon, B.R., Llewellyn, D.J., Huppatz, J.L., Dennis, E.S. and Peacock, W.J. (1989) Expression of a bacterial gene in transgenic tobacco confers resistance to the herbicide 2,4-dichlorophenoxyacetic acid. *Plant Molecular Biology*, **13**, 533–40.

Maier-Greiner, U.H., Klaus, C.B.A., Estermaier, L.M. and Hartmann, G.R. (1991a) Herbicide resistance in transgenic plants through degradation of the phytotoxin to urea. *Angewandte Chemie International Edition*, **30**, 1314–5.

Maier-Greiner, U.H., Obermaier-Skobranek, B.M.M., Estermaier, L.M., Kammerloher, W., Freund, C., Wülfing, C., Burkert, U.I., Matern, D.H., Breuer, M., Eulitz, M., Küfrevioglu, O.I. and Hartmann, G.R. (1991b) Isolation and properties of a nitrile hydratase from the soil fungus *Myrothecium verrucaria* that is highly specific for the fertilizer cyanamide and cloning of its gene. *Proceedings of the National Academy of Sciences USA*, **88**, 4260–4.

Mazur, B.J., Chui, C.-F. and Smith, J.K. (1987) Isolation and characterisation of plant genes coding for acetolactate synthase, the target enzyme for two classes of herbicides. *Plant Physiology*, **85**, 1110–7.

McBride, K.E., Kenny, J.W. and Stalker, D.M. (1986) Metabolism of the herbicide bromoxynil by *Klebsiella pneumoniae* subsp. *ozaenae*. *Applied and Environmental Microbiology*, **52**, 325–30.

McHughen, A. (1989) *Agrobacterium* mediated transfer of chlorsulfuron resistance to commercial flax cultivars. *Plant Cell Reports*, **8**, 445–9.

McHughen, A. and Holm, F. (1991) Herbicide resistant transgenic flax field test: agronomic performance in normal and sulfonylurea-containing soils. *Euphytica*, **55**, 49–56.

Miki, B.L., Labbé, H., Hattori, J., Ouellet, T., Gabard, J., Sunohara, G., Charest, P.J. and Iyer, V.N. (1990) Transformation of *Brassica napus* canola cultivars with *Arabidopsis thaliana* acetohydroxyacid synthase genes and analysis of herbicide resistance. *Theoretical and Applied Genetics*, **80**, 449–58.

Mousedale, D.M. and Coggins, J.R. (1984) Purification and properties of 5-enolpyruvylshikimate-3-phosphate synthase from seedlings of *Pisum sativum* L. *Planta*, **160**, 78–83.

Mousedale, D.M. and Coggins, J.R. (1985) Subcellular localization of the common shikimate pathway enzymes in *Pisum sativum* L. *Planta*, **163**, 241–9.

Murakami, T., Anzai, H., Imai, S., Satoh, A., Nagaoka, K. and Thompson, C.J. (1986) The bialaphos biosynthetic genes of *Streptomyces hygroscopicus*: molecular cloning and characterisation of the gene cluster. *Molecular and General Genetics*, **205**, 42–50.

Newman, T., Friden, P., Sutton, A. and Freundlich, M. (1982) Cloning and expression of the *ilv B* gene of *Escherichia coli* K12. *Molecular and General Genetics*, **186**, 378–84.

Odell, J.T., Caimi, P.G., Yadav, N.S. and Mauvais, C.J. (1990) Comparison of increased expression of wild-type and herbicide resistant acetolactate synthase genes in transgenic plants, and indication of post transcriptional limit on enzyme activity. *Plant Physiology*, **94**, 1647–54.

Odell, J.T., Nagy, F. and Chua, N.-H. (1985) Identification of DNA sequences required for activity of the cauliflower mosaic virus 35 S promoter. *Nature*, **313**, 810–2.

Ray, T.B. (1984) Site of action of chlorsulfuron. Inhibition of valine and isoleucine biosynthesis in plants. *Plant Physiology*, **75**, 827–31.

Riemenschneider, D.E., Haissig, B.E., Sellmer, J. and Fillatti, J.J. (1988) Expression of an herbicide tolerance gene in young plants of a transgenic hybrid poplar clone, in *Somatic Cell Genetics of Woody Plants*, (ed. M.R. Ahuja), Kluwer, Dordrecht, pp. 73–80.

Roland, S., Ferone, R., Harvey, R.J., Styles, V.L. and Morrison, R.W. (1979) The characteristics and significance of sulfonamides as substrates for *Escherichia coli* dihydropteroate synthase. *Journal of Biological Chemistry*, **254**, 10337–45.

Sanders, G.E. and Pallett, K.E. (1985) *In vitro* activity and binding characteristics of the hydroxybenzonitriles in chloroplasts isolated from *Matricaria indora* and *Viola arvensis*. *Pesticide Biochemistry and Physiology*, **24**, 317–25.

Sathasivan, K., Haughn, G.W. and Murai, N. (1991) Molecular basis of imidazolinone herbicide resistance in *Arabidopsis thaliana* var. Columbia. *Plant Physiology*, **97**, 1044–50.

Saxena, P.M. and King, J. (1988) Herbicide resistance in *Datura innoxia*. Cross-resistance of sulfonylurea resistant cell lines to imidazolinones. *Plant Physiology*, **86**, 863–7.

Shah, D.M., Horsch, R.B., Klee, H.J., Kishore, G.M., Winter, J.A., Turner, N.E., Hironaka, C.M., Sanders, P.R., Gasser, C.S., Aykent, S., Siegel, N.R., Rogers, S.G. and Fraley, R.T. (1986) Engineering herbicide tolerance in transgenic plants. *Science*, **233**, 478–81.

Shaner, D.L., Anderson, P.C. and Stidham, M.A. (1984) Imidazolinones. Potent inhibitors of acetohydroxyacid synthase. *Plant Physiology*, **76**, 545–6.

Squires, C.H., DeFelice, M., Wessler, S.R. and Calvo, J.M. (1981) Physical characterization of the *ilv Hl* operon of *Escherichia coli* K12. *Journal of Bacteriology*, **147**, 797–804.

Stalker, D.M. (1991) Developing herbicide resistance in crops by gene transfer technology, in *Plant Genetic Engineering*, (ed. D Grierson), Blackie, Glasgow, pp. 82–104.

Stalker, D.M., Hiatt, W.R. and Comai, L. (1985) A single amino acid substitution in the enzyme 5-enolpyruvylshikimate-3-phosphate synthase confers resistance to the herbicide glyphosate. *Journal of Biological Chemistry*, **260**, 4724–8.

Stalker, D.M., and McBride, K.E. (1987) Cloning and expression in *Escherichia coli* of a *Klebsiella ozaenae* plasmid born gene encoding a nitrilase specific for the herbicide bromoxynil. *Journal of Bacteriology*, **169**, 955–60.

Stalker, D.M., McBride, K.E. and Malyj, L.D. (1988a) Herbicide resistance in transgenic plants expressing a bacterial detoxification gene. *Science*, **242**, 419–43.

Stalker, D.M., Malyj, L.D. and McBride, K.E. (1988b) Purification and properties of a nitrilase specific for the herbicide bromoxynil and corresponding nucleotide sequence analysis of the *bxn* gene. *Journal of Biological Chemistry*, **263**, 6310–4.

Steinrücken, H.C., and Amrhein, N. (1980) The herbicide glyphosate is a potent inhibitor of 5-enolpyruvylshikimic acid-3-phosphate synthase. *Biochemical and Biophysical Research Communications*, **94**, 1207–12.

Strauch, E., Wohlleben, W. and Pühler, A. (1988) Cloning of a phosphinothricin *N*-acetyltransferase gene from *Streptomyces viridochromogenes* Tü 494 and its expression in *Streptomyces lividans* and *Escherichia coli*. *Gene*, **63**, 65–74.

Streber, W.R., Timmis, K.N. and Zenk, M.H. (1987) Analysis, cloning and high-level expression of 2,4-dichlorophenoxyacetate monooxygenase gene *tfd A* of *Alcaligenes eutrophus* JMP 134. *Journal of Bacteriology*, **169**, 2950–5.

Streber, W.R. and Willmitzer, L. (1989) Transgenic tobacco plants expressing a bacterial detoxifying enzyme are resistant to 2,4-D. *Bio/Technology*, **7**, 811–6.

Subraminian, M.V., Hung, H.-Y., Dias, J.M., Miner, V.W., Butler, J.H. and Jachetta, J.J. (1990) Properties of mutant acetolactate synthases resistant to triazolopyrimidine sulfonamide. *Plant Physiology*, **94**, 239–44.

Tachibana, K., Watanabe, T., Sekizawa, Y. and Takematsu, T. (1986) Accumulation of ammonia in plants treated with bialaphos. *Journal of Pesticide Science*, **11**, 33–7.

Thompson, C.J., Movva, N.R., Tizard, R., Crameri, R., Davies, J.E., Lauwereys, M. and Botterman, J. (1987) Characterisation of the herbicide resistance gene *bar* from *Streptomyces hygroscopicus*. *EMBO Journal*, **6**, 2519–23.

Tynan, J.L., Williams, M.K. and Conner, A.J. (1990) Low frequency of pollen dispersal from a field trial of transgenic potatoes. *Journal of Genetics and Breeding*, **44**, 303–6.

Vasil, V., Castillo, A.M., Fromm, M.E. and Vasil, I.K. (1992) Herbicide resistant fertile transgenic wheat plants obtained by microprojectile bombardment of regenerable embryogenic callus. *Bio/Technology*, **10**, 667–74.

Veerasekaran, P., Kirkwood, R.C. and Parnell, E.W. (1981) Studies on the mechanism of action of asulam in plants. Part II: effect of asulam on the biosynthesis of folic acid. *Pesticide Science*, **12**, 330–8.

Vermeulen, A., Vaucheret, H., Pautot, V. and Chupeau, Y. (1992) *Agrobacterium* mediated transfer of a mutant *Arabidopsis* acetolactate synthase gene confers resistance to chlorsulfuron in chicory (*Cichorium intybus*). *Plant Cell Reports*, **11**, 243–7.

Wang, Z., Takamizo, T., Iglesias, V.A., Osusky, M., Nagel, J., Potrykus, I. and

Spangenberg, G. (1992) Transgenic plants of tall fescue (*Festuca arundinacea*) obtained by direct gene transfer to protoplasts. *Bio/Technology*, **10**, 691–6.
Wohlleben, W., Arnold, W., Broer, I., Hillemann, D., Strauch, E. and Pühler, A. (1988) Nucleotide sequence of the phosphinothricin *N*-acetyltransferase gene from *Streptomyces viridochromogenes* Tü 494 and its expression in *Nicotiana tabacum*. *Gene*, **70**, 25–37.
Yadav, N., McDevitt, R.E., Benard, S. and Falco, S.C. (1986) Single amino acid substitutions in the enzyme acetolactate synthase confer resistance to the herbicide sulfometuron methyl. *Proceedings of the National Academy of Sciences USA*, **83**, 4418–22.

7

Genetic manipulation of crops for insect resistance

A.M.R. Gatehouse and V.A. Hilder

7.1 INTRODUCTION

Crop protection plays a vital and integral role in modern-day agricultural production. The ever increasing demands on yield and the intensification of farming practice, particularly in the Developed world, has increased the problem of pest damage and hence control. It has long been recognized that extensive cultivation of certain crops, to the exclusion of other plants, i.e. monoculture, may favour drastic increases in the populations of insects which feed upon these crops; the number of herbivores exploiting a given host species increasing with the area occupied by that host (Strong, 1979). Not only does this situation occur in the field, but also during storage of food crops, where accumulation of large quantities of seed means that any insect which can utilize them as a food source is almost certain to undergo a population explosion, thus resulting in significant damage and loss. In West Africa up to 100% of the stored cowpea crop may be damaged within 5 months due to attack by the bruchid beetle *Callosobruchus maculatus*. Although the build-up of pests to a specific host plant is generally reduced in diverse plant communities (Cromartie, 1981), agricultural practices such as combining crops where each acts as host for a particular pest (e.g. cotton and maize, both of which are attacked by the corn earworm *Heliothis zea*), can have the opposite effect, thereby exacerbating the situation (Pimental *et al.*, 1977).

At present crop pest control is primarily dependent upon the use of agrochemicals and to a lesser extent upon inherent varietal resistance, with biological control methods being successfully employed only in a few specific cases. However, it is only since the second world war that an almost

Molecular Biology in Crop Protection. Edited by George Marshall and Dale Walters. Published in 1994 by Chapman & Hall, London. ISBN 0 412 54390 7(HB) 0 412 54400 8(PB).

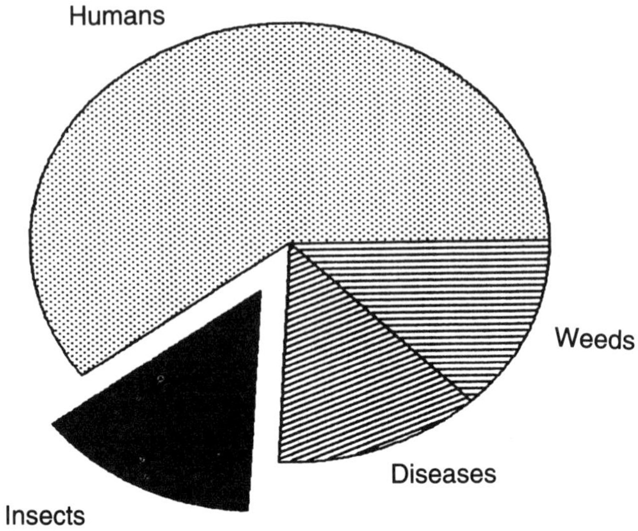

Figure 7.1 Relative proportions of total world crop production lost to insects, diseases and weeds. Only 63% of the total potential crop production is available for human consumption.

exclusive reliance has been placed upon insecticide use. This, in many cases, has resulted in the rapid build-up of resistance by insect pests to such compounds, as is illustrated by the rapidly developed resistance to the organochloride insecticides by the cotton bollworm, *Heliothis virescens*. Indeed, there are many examples of resistance in a major pest being observed within the first year of field use (Metcalf, 1986). In some cases the indiscriminate application of pesticides has exacerbated the problem of insect herbivory, where elimination of a wide range of predatory species along with the primary pests, has resulted in secondary pests becoming primary pests themselves, with even more devastating consequences.

Despite an annual insecticide expenditure of about US $7 billion, it is estimated that 37% of all crop production is lost worldwide to pests and diseases, with at least 13% lost directly to insects (Figure 7.1). There is a need therefore, to provide a greater level of protection to our crops. An integrated pest control programme comprising a combination of practices including the judicious use of pesticides, crop rotation, field sanitation, but above all exploiting inherently resistant plant varieties, would appear to provide the best option for improving pest control. Within this last category the use of genetically engineered insect-resistant crops may be included.

7.2 NATURAL DEFENCE MECHANISMS OF PLANTS – INHERENT RESISTANCE

Crop plants have been selected primarily for consistent high yields, nutritional value and, where necessary, adaptation to certain environmental conditions, together with low mammalian toxicity. This selection pressure over the centuries has severely disrupted the co-evolutionary relationship between plants and insects, with the result that very few cultivated species have retained the degree of resistance exhibited by their wild relatives (Feeny, 1976). For example, since the complex phenolic, gossypol, is toxic to mammals thus adversely affecting the utilization of cotton seed as an animal feed, lines devoid of this compound were selected for. However, these new lines exhibiting reduced mammalian toxicity were extremely susceptible to attack by the cotton budworm *H. virescens*, towards which gossypol is toxic (Berardi and Goldblatt, 1980).

To some extent the plant breeder has sought to redress the balance so as to exploit some of the natural defence mechanisms which exist in nature. These plant-based defence mechanisms depend on three factors:

1. *Temporal avoidance.* In some plants the life cycle is such that it avoids producing sensitive tissues at a time when pest activity is maximal. Such a strategy has been mimicked in some integrated pest management agricultural schemes where the use of 'early' or 'late' varieties may allow the crop to escape serious predation.
2. *Physical defences.* Many mechanisms have evolved which render sensitive plant tissues moderately inaccessible. Prime examples are the hard, lignified coats which enclose many seeds. Vegetative tissues may be made inaccessible to many types of insect by surface hairs, spines or thickened, waxy cuticles. There has been an interesting demonstration of how the overall morphology of the plant can affect the relative performance of herbivores and their predators: ladybirds (*Coccinella septempunctata* and *Hippodamia variegata*) are less likely to fall off leafless peas and are, therefore, more effective at predating the pea aphid (*Acyrthosiphon pisum*) on leafless varieties (Kareiva and Sahakian, 1990). This example is cited because it highlights the complexity of real biological systems. It is seldom likely to be sufficient to consider just the short-term interaction of a particular plant phenotype with a single species of herbivore. Most crops are subject to a whole complex of pests, which in turn are susceptible to a complex of predators and pathogens.
3. *Chemical defences.* One of the most important classes of natural defence mechanisms is chemical defence. Plants have a vast metabolic capability and produce many secondary chemicals, which are toxic, antimetabolic or aversive to species which might otherwise be potential predators (Norris and Kogan, 1980). For the most part these secondary compounds are relatively small organic molecules which are produced by complex

metabolic pathways involving the controlled interaction of many different gene products. Examples include the pyrethrins from chrysanthemums and alkaloids like nicotine from tobacco. Other classes of plant secondary compounds which have been implicated in protection from insect attack include the terpenoids, steroids, flavanoids, phenolics, glucosinolates, cyanogenic glycosides, rotenoids, saponins and non-protein amino acids (Gatehouse *et al.*, 1991a). Production of some of these compounds imposes a demonstrable metabolic cost on the plant, indicated by a reduced fitness in the absence of predation, which suggests that their production in the plant is a selective response to insect feeding (Baldwin *et al.*, 1990). As secondary compounds are the products of multienzyme pathways which involve the interaction of many gene products, such defence systems are in most cases too complex for the state of the art of plant genetic engineering, although work is in hand towards exploiting them (e.g. Dawson *et al.*, 1989; Hallahan *et al.*, 1992). A few plant defence mechanisms, however, are based on proteins themselves, the product of a single gene. Because proteins are non-volatile, they must be ingested to have an effect, and the target site of many is the insect digestive system. For example, resistance of certain wild accessions of the common bean (*Phaseolus vulgaris*) to the Mexican bean beetle (*Zabrotes subfasciatus*) is due to a protein, arcelin-4, which largely replaces the conventional bean storage protein, phaseolin. Arcelin-4 is not digestible by this insect, and the larvae therefore starve to death on these resistant beans (Minney *et al.*, 1990). Proteinaceous inhibitors of insect digestive enzymes have a similar effect (Gatehouse *et al.*, 1979) and certain lectins bind to, and disrupt, cells of the insect gut epithelium (Gatehouse *et al.*, 1984). Ribosome-inactivating proteins (RIPs) such as ricin from castor beans have also been shown to be highly toxic to certain insects (Gatehouse *et al.*, 1990).

Most of these types of single gene characters are eminently suitable for gene transfer using current technology. Being of plant origin, they have the added advantage that they are likely to have a high degree of compatibility with the metabolic system of the transgenic host plants.

7.3 ADVANTAGES OF INSECT-RESISTANT CROPS

The advantages of using crop varieties which are inherently resistant to, or tolerant of, insect pests, as opposed to using exogeneously applied chemicals, are many. Perhaps the most obvious ones include:

1. Season long protection.
2. Insects are always treated at the most sensitive stage.
3. Protection is independent of the weather.
4. It affords protection of plant tissues which are difficult to treat using

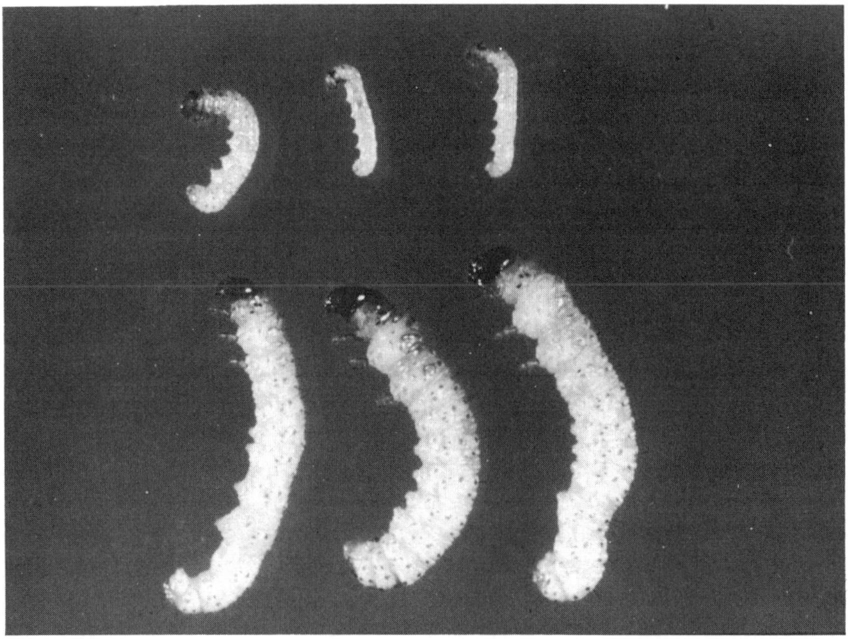

Figure 7.2 Effects of BPTI when fed to 3 month old porina (*Wiseana copularis*) larvae. Top, larvae fed BPTI; bottom, control larvae (courtesy of E. P. J. Burgess, The Horticulture and Food Research Institute of New Zealand, Auckland.)

in insects. That they affect the nutritional biochemistry of *C. maculatus* has been clearly demonstrated, since methionine supplementation has been shown to overcome the antinutritional effects of the cowpea trypsin inhibitor (CpTI) (Gatehouse and Boulter, 1983). Broadway suggests that they are only one part of the complex interaction between plant nutritional value and the insects' digestive physiology (Broadway *et al.*, 1986); this may explain why there are some cowpea varieties with high levels of CpTI which bruchids are able to infest.

7.4.3 Thiol protease inhibitors as insecticidal agents

Many insects, particularly Coleoptera, such as *Callosobruchus maculatus* (Gatehouse *et al.*, 1985), rely on thiol proteases for their major digestive proteolysis. It would therefore be logical to determine whether inhibitors of these particular enzymes would be effective in controlling this group of insect pests. To this end, *in vitro* proteolysis studies have been carried out and the inhibitor from rice, oryzacystatin, has been shown to inhibit the digestive proteases of stored grain Coleoptera (Liang *et al.*, 1991). Not only has this group of inhibitors be shown to be effective *in vitro*, but their insecticidal properties have also been demonstrated in artificial diet studies.

7.4.4 α-Amylase inhibitors

Protein inhibitors of mammalian α-amylases are abundant in cereal grains and are also present in legume and other seeds (Richardson, 1991). There are several distinct sequence families, with some being homologous to the Bowman–Birk type protease inhibitors. Generally these inhibitors are of relatively low molecular weight and although some may inhibit both mammalian and insect α-amylase activity, others are far more specific, being effective against either one or the other.

7.4.5 α-Amylase inhibitors as insecticidal agents

Unlike protease inhibitors, induced synthesis of amylase inhibitors by insect attack has not been observed and therefore the precise physiological role of these proteins remains open to speculation. However, those purified from wheat and *Phaseolus vulgaris* have been shown to be insecticidal to non-pest species when tested in artificial diet (Gatehouse *et al.*, 1986; Ishimoto and Kitamura, 1988). A very specific α-amylase inhibitor has been isolated from a particular wild line of *P. vulgaris* (G12953) which is resistant to attack by a major storage pest of *P. vulgaris, Zabrotes subfasciatus* (Minney *et al.*, 1990). Gatehouse *et al.* (1987) demonstrated that its presence was correlated with seed resistance to this particular pest species and that *in vitro* it was a very potent inhibitor. Very recently, it has been shown to be toxic to *Z. subfasciatus* when tested in artificial seeds.

Recent studies have shown that the different types of α-amylase inhibitor present in wheat endosperm are differentially active against different lepidopteran pest species (Gutierrez *et al.*, 1993).

7.4.6 Lectins

Lectins are carbohydrate-binding proteins. The first plant lectin was described over a century ago when Stillmark demonstrated that extracts from castor bean (*Ricinus communis*) agglutinated red blood cells. Many belong to a homologous family of proteins based on an amino acid chain of approximately 220 residues, although totally different sequence types have been shown to have similar functional properties. (For a detailed review on plant lectins the reader is referred to Pusztai, 1991.)

7.4.7 Lectins as insecticidal agents

Much interest in these particular proteins has arisen as a result of some of them being very toxic to mammals and birds, and detailed studies on their mechanisms of toxicity have been carried out (de Oliveira, 1986). The possibility that they play a protective role in plants was first demonstrated by Janzen *et al.* (1976). Subsequently, Gatehouse *et al.* (1984) confirmed that the purified lectins from seeds of *P. vulgaris* were toxic to the developing larvae of the bruchid beetle *C. maculatus*. On the basis of indirect immunofluorescence investigations using monospecific antisera for globulin lectins, they showed that the molecules, when ingested, bound to the midgut epithelial cells; a similar mechanism is known to occur in rat. In pest species of *P. vulgaris*, such as *Acanthoscelides obtectus* (Say), the lectin molecules do not bind to these epithelial cells, and presumably, are unable to cause the harmful effects demonstrated previously (Gatehouse *et al.*, 1989). More recent results have suggested that the toxicity of *P. vulgaris* lectin preparations towards bruchids is due to a contaminating α-amylase inhibitor which, while it is without lectin functionality, has sequence homology to the lectins. However, this observation appears to be specific to *P. vulgaris* and is not the case with lectin preparations from other species. The lectins present in the mature seeds of the winged bean *Psophocarpus tetragonolobus*, have been shown to be involved in seed resistance to non-pest species (Gatehouse *et al.*, 1991b).

Since the demonstration of lectins as protective agents in plant tissues, much interest has arisen in their potential in crop protection. To this end several studies have been carried out involving the screening of purified lectins against insect pests in an attempt to identify insecticidal proteins, and hence isolate the genes encoding them for subsequent plant transformation. Shukle and Murdoch (1983) demonstrated that the lectin from soybean was toxic to the developing larvae of the lepidopteran *Manduca sexta* when tested in bio-assay and, recently, wheat germ agglutinin has been shown to have an inhibitory effect on development of the European corn borer (*Ostrinia nubilalis*) and the Southern corn root worm (*Diabrotica undecimpunctata*), two important pests of maize (Czapla and Lang, 1992). The disadvantage, however, of using lectins from *P. vulgaris*, winged bean, soybean or wheat germ, is that they have also been shown to exhibit mammalian toxicity. This toxicity will limit their potential use in the production of transgenic crop plants. The lectin isolated from pea, *Pisum sativum*, on the other hand, has been shown to be innocuous to mammals, as it is readily broken down in the gut. However, it was shown to be insecticidal when the purified protein was incorporated into artificial seeds (Boulter and Gatehouse, 1986). A lectin purified from snowdrop has also been shown to be insecticidal, but does not exhibit mammalian toxicity.

Not only have certain lectins been shown to be insecticidal to members of the orders Coleoptera and Lepidoptera, but recent results have demonstrated their toxicity to some homopteran pests. The snowdrop lectin (GNA) and wheatgerm agglutinin (WGA) were shown to be highly insecticidal to the brown plant hopper (*Nilaparvata lugens*), a major economic pest of rice, when incorporated into an artificial liquid diet medium at 0.1% (w/v) (Powell *et al.*, 1993; Figure 7.3); similar results were also obtained with the green leaf hopper (*Nephotettix*). As with members from other insect orders, not all lectins tested against these sap-sucking insects were found

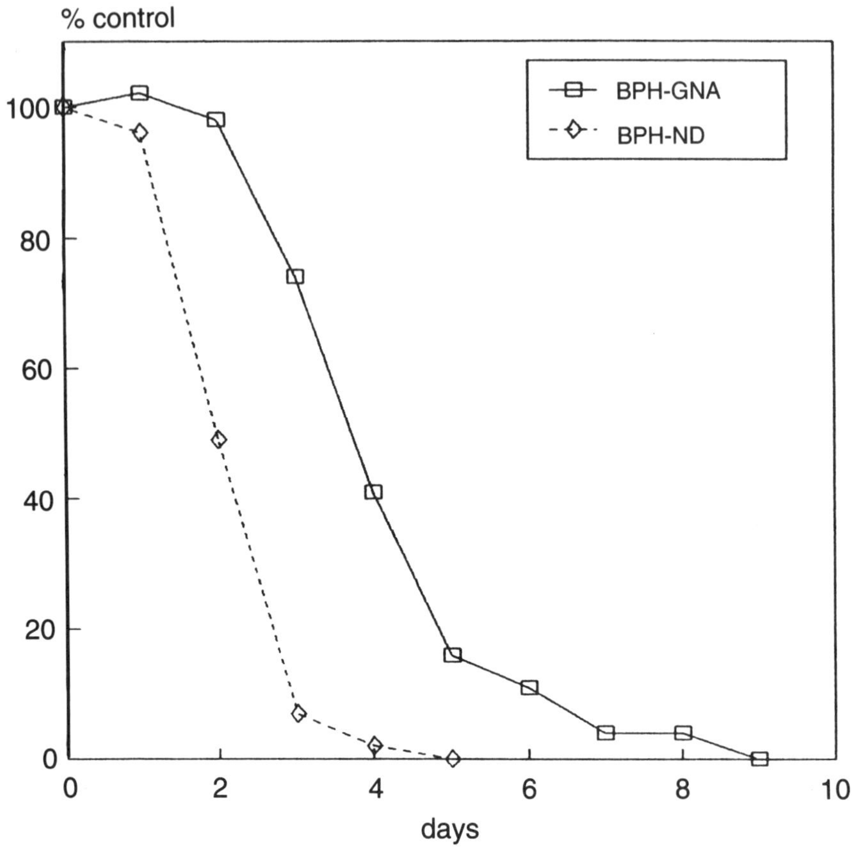

Figure 7.3 Effects of GNA on survival of rice brown plant hopper (*N. lugens*). GNA was incorporated into an artificial diet at 0.1% (w/v) (GNA). Insects receiving GNA were compared to others receiving no diet (ND).

to be entomocidal. This is the first direct evidence of proteins being effective against Homoptera.

7.5 INSECT-RESISTANT TRANSGENIC PLANTS EXPRESSING PLANT DERIVED GENES

7.5.1 Cowpea protease inhibitors (CpTI)

The first gene of plant origin to be successfully transferred to another plant species resulting in enhanced insect resistance was that isolated from

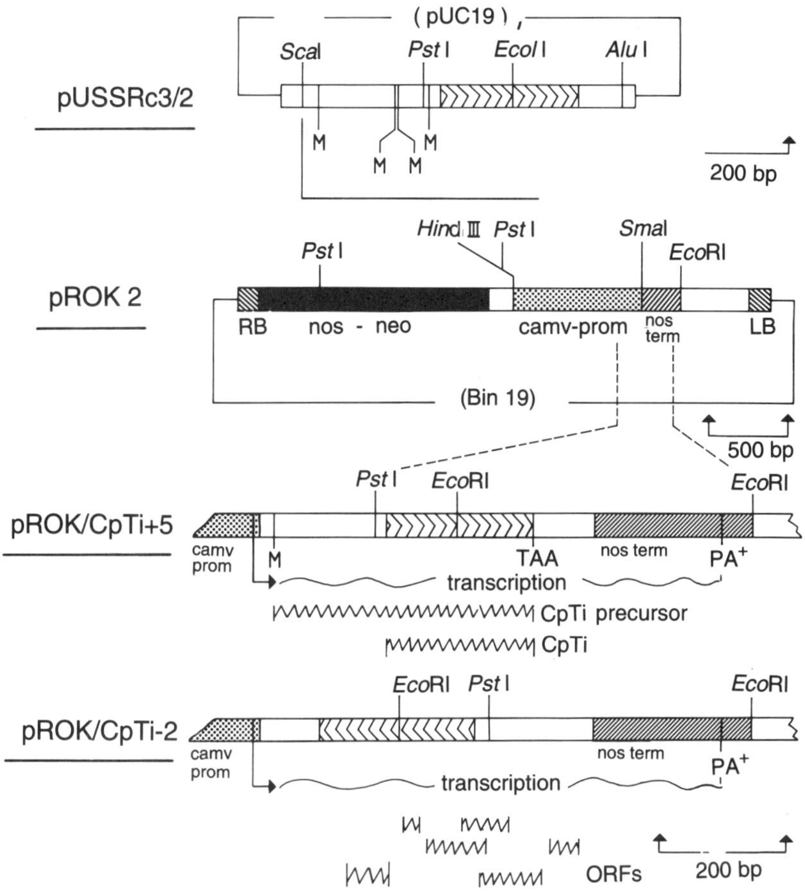

Figure 7.4 Construction of a CpTI expression vector for plant transformation (Hilder *et al.*, 1987). The CpTI cDNA pUSSRc3/2, containing a complete mature CpTI-coding sequence (indicated as >>>; in frame, initiator codons are marked M), was restricted with *Alu*I and *Sca*I and ligated into the *Sma*I site of the expression vector pROK 2. Clones with the coding sequence in the correct orientation relative to the promoter (pROK/CpTI + 5) and in the incorrect orientation (pROK/CpTI-2) were generated. Transcripts generated by the clone with the CpTI-coding sequence in the correct orientation will be translated to produce a CpTI precursor polypeptide; transcripts from the clone with the CpTI-coding sequence in the incorrect orientation contain six short open reading frames.

cowpea, encoding a trypsin/trypsin inhibitor (Hilder *et al.*, 1987). This protein was considered to be a particularly suitable candidate for transfer to other plant species via genetic engineering for a number of reasons. It had been shown to be an effective antimetabolite against a range of field

Table 7.1 Insect pests against which cowpea trypsin inhibitors (CpTI) are effective

Order	*Insect pest*	*Primary crops attacked*
FIELD PESTS		
Lepidoptera	*Heliothis virescens**	Tobacco, cotton
	*Heliothis zea**	Maize, cotton, beans, tobacco
	Helicoverpa armigera	Cotton, beans, maize, sorghum
	*Spodoptera littoralis**	Maize, rice, cotton, tobacco
	Chilo partellus	Maize, sorghum, sugarcane, rice
	*Autographa gamma**	Sugarbeet, lettuce, cabbage, beans, potato
	*Manduca sexta**	Tomato, tobacco, potato
Orthoptera	*Locusta migratoria*	Polyphagous but preference for wild and cultivated grasses
Coleoptera	*Diabrotica undecimpunctata*	Maize
	Costelytra zealandica	Grasses, clover
	Anthonomus grandis	Cotton
STORAGE PESTS		
Coleoptera	*Callosobruchus maculatus*	Cowpea, soyabean
	Tribolium confusum	Most flours

*Insects to which CpTI transgeneic tobacco plants exhibit significantly enhanced levels of resistance.

and storage pests, including members of the Lepidoptera, Coleoptera and Orthoptera (Table 7.1). There was however, no evidence that it had any deleterious effects upon mammals. It is a small polypeptide of about 80 amino acids, belonging to the Bowman–Birk inhibitor family (Gatehouse *et al.*, 1980). Homologous sequences are encoded by a moderately-repetitive gene family in the cowpea genome (Hilder *et al.*, 1989).

A full-length cDNA clone encoding a trypsin/trypsin inhibitor from cowpea was produced and the coding sequence was placed under the control of a CaMV 35S promoter (Hilder *et al.*, 1989) in the final construct produced for transfer to plants (Figure 7.4). The construct employed the *Agrobacterium tumefaciens* Ti plasmid binary vector pROK2. A terminator from the nopaline synthetase gene was placed 3′ to the coding sequence, and the construct also contained a *nos-aph (3′)II* (usually referred to as *nos-neo*) gene to allow transformants to be selected on the basis of kanamycin resistance. The vector was mobilized into *Agrobacterium* and the bacteria were used to infect tobacco leaf discs by standard protocols. Subsequent production of rooted plants, after selection of regenerating shoots on kanamycin-containing media, also followed normal procedures (Horsch *et al.*, 1985). By taking cuttings from the original transformants and rooting them, numbers of clonal plants sufficient for insect bioassay could be produced from each of the original transformants. The transformed plants were shown to express CpTI in the leaves at levels varying from undetectable to nearly 1% of total soluble protein, using a dot blot immunoassay. This range of values has subsequently been found to be fairly typical for plant genes driven from the CaMV promoter. In control plants transformed with a construct where the coding sequence of CpTI had been inserted in the incorrect (i.e. 3′–5′) orientation relative to the CaMV promoter, no expression of the protein was detected. The expression of CpTI was confirmed in the 'correct' transformants by Western blotting, and by a direct *in vitro* assay for inhibition of bovine trypsin. The former technique showed that tobacco was capable of processing the precursor CpTI polypeptide, encoded by the inserted coding sequence to a polypeptide resembling native CpTI on sodium docecyl sulphate–polyacrylamide gel electrophoresis (SDS–PAGE). Other plant proteins have been shown to be correctly processed in transgenic tobacco plants (Ellis *et al.*, 1988). The latter technique showed that the CpTI synthesized in transgenic tobacco possessed its normal functional integrity. The lack of complication in obtaining relatively high levels of expression of functional CpTI in tobacco illustrates the advantage of expressing plant proteins in transgenic plants. No problems with codon usage, mRNA stability and protein processing appeared to occur, as have been observed when proteins derived from non-plant sources are expressed in plants.

The critical test was to ascertain whether the CpTI producing tobacco plants exhibited enhanced levels of resistance/tolerance to insect infestation compared to the control plants. In the first instance, bioassay of clones of selected transformants was carried out using first instar larvae of the tobacco budworm (*H. virescens*). This insect was chosen as it is a serious pest of tobacco, cotton and maize and thus represents a pest of major

Figure 7.5 Bioassay of control and CpTI-expressing transgenic tobacco plants against larvae of *H. virescens* (tobacco budworm). Left, a control plant showing almost complete destruction; right, a transgenic CpTI expressor, showing minimal damage.

economic importance. With these clonal plants, and subsequent generations derived from their self-set seed, the CpTI expressing plants showed only minor damage compared to the control plants (Figure 7.5), which in some instances were reduced to stalks. Although the larvae begin to feed on the CpTI-expressing plants, causing some limited damage, they either die or fail to develop as they would on control plants. These observations are consistent with a mechanism of CpTI toxicity initially proposed by Gatehouse and Boulter (1983). This protection afforded by CpTI has subsequently been demonstrated for other lepidopteran pests including *H. zea*, *Spodoptera littoralis* and *Manduca sexta* (Gatehouse *et al.*, 1991a). Statistical analysis of the bioassay in terms of plant damage by leaf area, and insect survival and biomass, confirmed the highly significant protection afforded by CpTI. Recent trials carried out in California showed that expression of CpTI in tobacco afforded significant protection in the field against *H. zea*. Results from these trials closely resembled those obtained previously in trials carried out under controlled environmental conditions in growth chambers (Hoffman *et al.*, 1992). Unfortunately, it is not possible to test the efficacy of CpTI against coleopteran pests in transgenic tobacco plants, as most coleopterans of economic interest do not appear to attack tobacco. However, other species of CpTI expressing plants which are susceptible to coleopteran attack, including potato, oilseed rape and lettuce, have now become available.

Despite CpTI being an effective antimetabolite against a wide spectrum of insect pests (Table 7.1), recent mammalian feeding trials incorporating the purified protein at levels of 10% of the total protein, have failed to demonstrate toxicity, at least in the short term (Pusztai *et al.*, 1992). CpTI has been demonstrated to directly inhibit insect gut proteases, although this may not be its only site of action. It is not effective against the acidic proteases of the mammalian stomach. The differences in the organization of the insect and mammalian digestive systems and the vast range of secondary compounds available from plants means that it should not be that difficult to find compounds within plants which are toxic to herbivorous insects but not to mammals (Hilder and Gatehouse, 1989; Hilder *et al.*, 1993).

7.5.2 Protease inhibitor II (PI-II)

The tomato inhibitor II gene, when expressed in tobacco, has also been shown to confer insect resistance (Johnson *et al.*, 1989). The tomato and potato inhibitor II gene encodes a trypsin inhibitor (with some chymotrypsin inhibitory activity) and expression of this gene in tobacco, on the constitutive promoter CaMV 35S, resulted in increased levels of protection against the larvae of the lepidopteran *M. sexta*. These workers showed that the decrease in larval weight was roughly proportional to the level of protease inhibitor II being expressed: at levels over 100 μg of the foreign protein/g of tissue, larval growth was severely retarded, whereas at lower levels (about 50 μg/g tissue) growth was retarded to a lesser degree. Several of the transgenic plants were shown to contain inhibitor levels over 200 μg/g tissue; these levels are within the range that is routinely induced by wounding leaves of either tomato or potato plants (Graham *et al.*, 1986). However, tobacco plants expressing tomato inhibitor I (specific for chymotrypsin) at levels of 130 μg/g had no deleterious effects upon larval development. Interestingly, when PI-II was expressed off the wound-inducible promoter, the resulting transgenic plants did not show significantly enhanced levels of resistance to insect attack.

7.5.3 Thiol (cysteine) protease inhibitors

To date, genes encoding various thiol protease inhibitors have been expressed in several different plant species. The cysteine protease inhibitor gene from rice has been expressed in poplar trees and preliminary trials against the two insect pests *Chrysomela populi* and *C. tremula* have yielded encouraging results. Furthermore, this gene has also been expressed in oilseed rape with the intention of conferring resistance to the seed weevil *Ceuthorhyncus* (Jouanin, personal communication).

7.5.4 α-Amylase Inhibitors

Strong evidence has been presented to suggest that an α-amylase inhibitor present in the seeds of *P. vulgaris* is encoded by an already identified lectin gene, whose product is referred to as LLP (Moreno and Chrispeels, 1989). A chimeric gene, consisting of the coding sequence of the lectin gene that encodes LLP and the 5′ and 3′ flanking sequences of the lectin gene that encode PHA-2, has been constructed and expressed in tobacco (Altabella and Chrispeels, 1990). Not only did seeds from these transgenic plants show antigenicity to bean α-amylase inhibitor, but they also expressed a series of polypeptides of molecular weight corresponding to these particular inhibitors. Furthermore, seed extracts were found to be active against both porcine pancreatic α-amylase and the α-amylase present in the midgut of meal worm, *Tenebrio molitor*. This last observation led to the suggestion that introduction of this lectin gene (*αai*) into other leguminous plants may be a strategy to protect the seeds from seed-eating larvae of Coleoptera. Indeed, it has recently been reported that transgenic pea plants expressing the *αai* are resistant to attacks by the coleopteran *Bruchus tisorum* (Higgins and Spencer, pers. comm.). Although transgenic tobacco plants expressing this gene are available, no insect bioassays on these plants appear to have been reported. Other workers have claimed to have expressed a gene encoding an α-amylase inhibitor from wheat in tobacco, resulting in increased protection against the army worm (*Spodoptera* spp.) and greasy cut worm (*Agrotis* spp.). However, no data appear to be available as yet to indicate that expression of the foreign gene corresponds with enhanced levels of insect resistance.

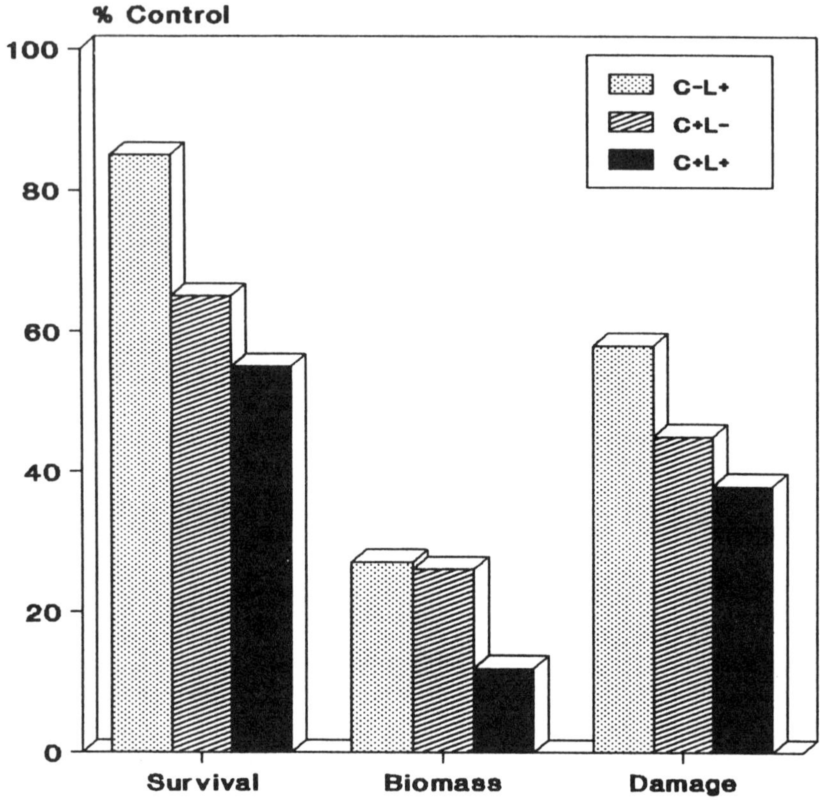

Figure 7.6 Insect survival and biomass, and leaf damage to plants, from assays of CpTI and pea lectin expressing transgenic tobacco plants against *H. virescens*.

7.5.5 Pea lectin (P-Lec)

Transgenic tobacco plants expressing pea lectin (P-Lec) at high levels off the CaMV 35S promoter were found to show enhanced levels of resistance/ tolerance to the budworm *Heliothis virescens*, both in terms of significantly reduced larval biomass and in reduced leaf damage, as determined by computer aided image analysis (Boulter *et al.*, 1990) (Figure 7.6).

7.6 BACTERIAL TOXINS AS INSECTICIDAL AGENTS

The earliest record of *Bacillus thuringiensis* (*B.t.*) was in 1901 when Ishiwato isolated this bacterium from diseased silkworm larvae. He observed that whereas young actively growing cultures of the bacterium were innocuous to the larvae, older cultures undergoing sporulation were toxic. Subsequently, Berliner isolated this same bacterium from diseased larvae of the Mediterranean flour moth *Ephestia kuhniella* and, whilst toxic to this species, it was not toxic to larvae of the meal worm *Tenebrio molitor*, thus indicating that *B.t.* was specific in its insecticidal activity.

B.t. has been used for over 30 years as an insecticidal crop spray, usually *Bacillus thuringiensis* as formulations of whole bacteria (Dulmage, 1981; Chapter 3). The bacterium produces a crystalline protein on sporulation, which, when consumed by the insect, is hydrolysed in the gut to release toxic protein fragments (Sacchi *et al.*, 1986). The basis of toxicity is thought to be a disruption of gut membrane function mediated through specific receptor sites in the brush border of the midgut epithelial cells. There have been many experiments carried out on a range of different lepidopteran larvae demonstrating that the subsequent toxicity of the crystal protein correlates with its binding-affinity (Van Rie *et al.*, 1990). However, Wolfersberger (1990) found an inverse correlation between receptor affinity and toxicity in gypsy moth larvae. Thus, although specific binding of the crystal proteins to the midgut epithelial receptors may be essential for toxicity, the level of toxicity may be determined by the efficiency of pore formation in this membrane. What happens after the crystal proteins have bound to the receptor sites is not clear and whilst some workers suggest that these proteins interfere with potassium transport (Sacchi *et al.*, 1986), others suggest the formation of non-specific pores as a result of colloidosmotic lysis (Knowles and Ellar, 1986; Hendrickx *et al.*, 1990).

Different strains of *B.t.* produce toxins effective against different insect species, the basis of this specificity being due, at least in part, to the conditions in the insect gut where proteolysis of the protoxin occurs. Lepidoptera in general have highly alkaline gut conditions for proteolysis of ingested proteins, so that *B.t.* toxins effective against lepidopteran pests are theoretically of very low toxicity towards mammals, which do not produce the gut conditions necessary to generate active toxins.

Apart from the lepidopteran specific *B.t.* toxins, another class with potentially significant agricultural importance, are those exhibiting coleopteran toxicity, i.e. those encoded by *CryIII* genes (Hofte and Whiteley, 1989). These toxins form rhomboidal crystal inclusions and so far have been found in three morphologically different *B.t.* isolates. The DNA sequences of the protoxin genes from each are identical, although the precise character of the crystallized protoxin or the final insecticidal fragments may differ, due to the potential for alternative proteolytic processing occurring within the several hosts.

7.7 INSECT-RESISTANT TRANSGENIC PLANTS EXPRESSING *B.t.* TOXINS

Since the field use of *B.t.* as an insecticide is limited by its poor stability under environmental conditions and high cost of production, genetic

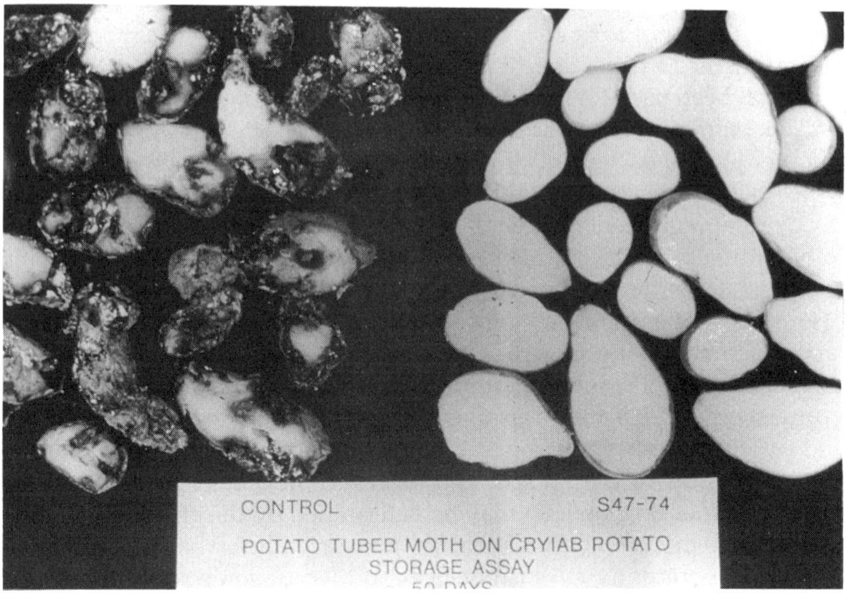

Figure 7.7 Damage to Kennebec (Control) and Cry 1 A(b) engineered Kennebec potato tubers (S47-74) by potato tuber moth larvae (*P. operculella*) (courtesy of M. Peferoen, Plant Genetic Systems, Gent, Belgium).

engineering of plants to produce this toxin endogenously has clear attractions and is being pursued actively. However, much attention has also been paid to methods for producing the *B.t.* toxin in genetically engineered bacteria for use as a spray insecticide. Three different research groups reported the production of insect-resistant transgenic plants expressing *B.t.* toxins within months of each other, and although details of the methods used differ, the general strategies were basically similar (Barton *et al.*, 1987; Fischhoff *et al.*, 1987; Vaeck *et al.*, 1987).

While the isolation of genes encoding *B.t.* toxins is a straightforward exercise, especially as they are usually carried on plasmids in *B.t.*, achieving expression of an active toxin in plants gave many difficulties. The complete protoxin molecule is non-toxic until it is proteolytically cleaved, and attempts to express the complete toxin in plants were unsuccessful. Only very low levels of expression were achieved, and the complete *B.t.* protoxin appeared to have deleterious effects on the transgenic plants. Instead, truncated toxin genes had to be constructed, containing only the N-terminal part of the coding sequence of the protoxin, which contains the active toxin fragment. This required *in vitro* mutagenesis of the coding sequence and the addition of sequences from synthetic oligonucleotides. The original bacterial promoter and terminator sequences on the genes would not in any case function in plants, and were thus replaced by a constitutive promoter and a suitable terminator. The gene constructs were once again engineered into transgenic plants using *Agrobacterium*-based vector systems. Both transgenic tobacco and tomato plants were produced initially.

Levels of expression of the truncated *B.t.* toxins in transgenic plants were low, at best approximately 0.02% of total soluble leaf protein (detected by immunoassay), even when using a highly expressed (CaMV 35S) promoter. This reflects the difficulty in expressing a bacterial protein in plants, although recently, expression levels of 0.1% of the cell protein have been obtained using synthetic genes encoding two different *B.t.* toxins; these levels of expression are sufficient to provide complete mortality to susceptible insects. Despite the initial low levels of expression, bioassay of transformed tobacco plants under growth-cabinet conditions against tobacco hornworm (*Manduca sexta*) showed a very significant level of protection compared to controls. The protection observed at low levels of expression of the foreign protein is an advantage of this strategy for plant protection. On many of these transgenic tobacco plants 100% mortality of applied larvae was observed within a few days, and feeding damage was minimal compared with control plants which were substantially defoliated. This is a reflection of the (known) high toxicity of the strain of *B.t.* toxin used against the test insect. However, transgenic tomato plants expressing truncated *B.t.* toxins also exhibited enhanced resistance towards two *Heliothis* species (tobacco budworm and corn earworm) which are not as sensitive to the strain of *B.t.* used. Tests were made on the fruit of these transgenic tomato plants, which showed that expression of the *B.t.* toxin caused decreases of about 50% in the weight gain of *H. virescens* larvae fed on them, compared to controls. Potatoes engineered with the *B.t.* toxin gene encoding the *CryIIA* toxin were found to show significant levels of resistance to feeding damage by the larvae of the Colorado potato beetle (*Leptinotarsa decemlineata*) (McPherson *et al.*, 1989), and more recently, the leaves and tubers from different potato cultivars, engineered with the gene encoding *CryIA(b)*, proved highly resistant to feeding and tunnelling damage by tuber moth larvae (*Phtorimaea operculella*) (Peferoen *et al.*, 1990) (Figure 7.7).

The first field trials on insect resistant transgenic plants was carried out in 1986 in North Carolina, using *B.t.* expressing tobacco plants engineered by Plant Genetic Systems. Subsequently, field trials were carried out on engineered tomato plants and although all the fruits were damaged by the tomato fruitworm (*Helicoverpa armigera*) and tomato pinworm (*Keiferia lycopersicella*), damage on the transgenic plants was significantly less. However, the levels of protection afforded by the *B.t.* expressing fruits is, as yet, commercially unacceptable (Delannay *et al.*, 1989). Transformed cotton plants, expressing the *B.t.* toxins, have also been tested in field trials and shown to exhibit significant levels of resistance (see Chapter 10). Initially, it was thought that *B.t.* toxins would confer resistance at very low levels, but in cotton, for example, levels of expression of about 0.5% are required for protection. It is thought that these *B.t.* expressing cotton plants will be the first of the insect-resistant transgenic plants to be commercialized. For more information on engineering insect resistant plants with *B.t.* crystal protein genes, the reader is referred to very recent reviews by Peferoen (1992) and Barton and Miller (1993).

Unfortunately, a major problem which is at present being encountered with this strategy, is the development of resistance to *B.t.* crystal proteins in the field (Tabashnik *et al.*, 1990) and, particularly, the possibility of broad spectrum resistance to *B.t.* toxins (Gould *et al.*, 1992). This is of great concern, and it is hoped that a study of these resistant insects will not only provide valuable information on the mechanism of action of *B.t.* proteins, but also lead to strategies to retard the build-up of resistance to them in insect populations.

7.8 PYRAMIDING GENES

One of the goals of the plant breeder is to 'pyramid' genes expressing agriculturally desirable characteristics. This strategy has also been adopted by the biotechnologist. In order to increase the protective efficacy, spectrum of activity and durability of resistance, it is envisaged that 'packages' of different genes will be introduced into crops. The components of such packages should each act on different targets within the insect, thus mimicking the multimechanistic resistance which occurs in nature. Protease inhibitors should be particularly valuable in this respect since, apart from their inherent insecticidal effects, they would protect other introduced gene products from premature digestion in the insect gut.

The first demonstration of such an approach has been the introduction of both CpTI and pea lectin into tobacco. These plants were obtained by cross-breeding plants derived from the two primary transformed lines (Boulter *et al.*, 1990). Although the insecticidal effects of the two genes were not synergistic, they were additive (Figure 7.6), with insect biomass on the double expressors being only 11% compared to those from control plants and 50% of those from plants expressing either CpTI or P-Lec alone. Leaf damage was also least on the double expressing plants. Not only is this the first example of a lectin gene being successfully transferred to another plant species and resulting in enhanced insect resistance, but it is also the first demonstration of additive protective effects of different plant-derived insect resistance genes. It is also interesting to note that trypsin inhibitors were reported to have a marked potentiating effect on *B.t.* toxins (MacIntosh *et al.*, 1990).

7.9 CONCLUSION

The use of insect resistant transgenic plants is a viable means of producing crops with significantly enhanced levels of resistance. This technology should complement, rather than replace, other forms of crop protection and it is envisaged that it will have a major impact on agricultural systems both in the Developed and Developing parts of the world. The limitations to its uptake are not so much focused on the availability of suitable candidate genes, or even to the range of crops amenable to transformation and regeneration, but to regulatory barriers (see Connett and Barfoot, 1992 and Chapter 11) and consumer acceptability.

REFERENCES

Altabella, T. and Chrispeels, M.J. (1990) Tobacco plants transformed with the bean αai gene express an inhibitor of insect α-amylase in their seeds. *Plant Physiology*, **93**, 805–10.

Baldwin, I.T., Sims, C.L. and Kean, S.E. (1990) The reproductive consequences associated with inducible alkaloidal responses in wild tobacco. *Ecology*, **71**, 252–62.

Barton, K., Whiteley, H. and Yang, N.-S. (1987) *Bacillus thuringiensis* δ-endotoxin in transgenic *Nicotiana tabacum* provides resistance to lepidopteran insects. *Plant Physiology*, **85**, 1103–9.

Barton, K.A. and Miller, M.J. (1993) Production of *Bacillus thuringiensis* insecticidal proteins in plants, in *Transgenic Plants*, Vol. 1: Engineering and Utilization, (eds S.-D. Kung and R. Wu), Academic Press, New York, pp. 297–315.

Berardi, L.C. and Goldblatt, L.A. (1980) Gossypol, in *Toxic Constituents of Plant Foodstuffs*, 2nd edn, (ed. I.E. Liener), Academic Press, New York, pp. 183–237.

Boulter, D. and Gatehouse, A.M.R. (1986) Isolation of genes involved in pest and disease resistance, in *Biomolecular Engineering in the European Community*, (ed. E. Magnien), Martinus Nijhoff, Dordrecht, pp. 715–25.

Boulter, D., Edwards, G.A., Gatehouse, A.M.R., Gatehouse, J.A. and Hilder, V.A. (1990) Additive protective effects of incorporating two different higher plant derived insect resistance genes in transgenic tobacco plants. *Crop Protection*, **9**, 351–4.

Broadway, R.M. and Duffy, S.S. (1986) Plant proteinase inhibitors: mechanism of action and effect on the growth and digestive physiology of larval *Heliothis zea* and *Spodoptera exigua. Journal of Insect Physiology*, **32**, 827–33.

Broadway, R.M., Duffy, S.S., Pearce, G. and Ryan, C.A. (1986) Plant proteinase inhibitors: a defence against herbivorous insects? *Entomologia experimentalis et applicata*, **41**, 33–8.

Brown, W.E., Takio, K., Titani, K. and Ryan, C.A. (1985) Wound-induced trypsin inhibitor in alfalfa leaves: identity as a member of the Bowman–Birk inhibitor family. *Biochemistry*, **24**, 2105–8.

Burgess, E.P.J., Stevens, P.S., Keen, G.K., Laing, W.A. and Christeller, J.T. (1991) Effects of protease inhibitors and dietary protein level on the black field cricket *Teleogryllus commodus. Entomololgia experimentalis et applicata*, **61**, 123–30.

Campos, F.A.P. and Richardson, M. (1983) The complete amino acid sequence of the bifunctional α-amylase/trypsin inhibitor from seeds of ragi (Indian finger millet, *Eleusine coracana* Geartn.). *FEBS Letters*, **152**, 300–4.

Christeller, J.T. and Shaw, B.D. (1989) The interaction of a range of serine proteinase inhibitors with bovine trypsin and *Costelytra zealandica* trypsin. *Insect Biochemistry*, **19**, 233–41.

Connett, J.A. and Barfoot, P.D. (1992) The development of genetically modified varieties of agricultural crops by the seed industry, in *Plant Genetic Manipulation for Crop Protection*, (eds A.M.R. Gatehouse, V.A. Hilder and D. Boulter), CAB, Wallingford.

Cromartie, W.J. (1981) The environmental control of insects using crop diversity, in *Handbook of Pest Management in Agriculture* (ed. D. Pimental), CRC Press, Boca Raton, FL, pp. 223.

Czapla, T.H. and Lang, B.A. (1990) Effect of plant lectins on the larval development of European corn borer (Lepidoptera: Pyralidae) and Southern corn rootworm (Coleoptera: Chrysomelidae). *Journal of Economic Entomology*, **83**, 2480–5.

Dawson, G.W., Hallahan, D.L., Mudd, A., Patel, M.M., Pickett, J.A., Wadhams, L.J. and Wallsgrove, R.M. (1989) Secondary plant metabolites as targets for genetic modification of crop plants for pest resistance. *Pesticide Science*, **27**, 191–201.

Delannay, X., La Vallee, B.J., Proksch, R.K., Fuchs, R.L., Sims, S.R., Greenplate, J.T., Marrone, P.G., Dodson, R.B., Augustine, J.J., Layton, J.G. and Fischhoff, D.A. (1989) Field performance of transgenic tomato plants expressing the *Bacillus thuringiensis* var. *kurstaki* insect control protein. *Bio/Technology*, **7**, 1265–9.

de Oliveira, J.T.A. (1986) Seed lectins: the effects of dietary *Phaseolus vulgaris* lectin on the general metabolism of monogastric animals. PhD Thesis, Aberdeen University.

Dulmage, H.T. (1981) Insecticidal activity of isolates of *Bacillus thuringiensis* and their potential for pest control, in *Microbial Control of Pests and Plant Diseases 1970–1980*, (ed. H.D. Burgess), New York, Academic Press, pp. 193–222.

Dymock, J.J., Laing, W.A., Shaw, B.D., Gatehouse, A.M.R. and Christeller, J.T. (1992) Behavioural and physiological responses of grass grub larvae (*Costelytra zealandica*) feeding on protease inhibitors. *New Zealand Journal of Zoology* **19**, 123–31.

Ellis, J.R., Shirsat, A.H., Hepher, A., Yarwood, J.N., Gatehouse, J.A., Croy, R.R.D. and Boulter, D. (1988) Tissue specific expression of a pea legumin gene in seeds of *Nicotiana plumbaginifolia. Plant Molecular Biology*, **10**, 203–14.

Farmer, E.E. and Ryan, C.A. (1992) Octadecanoid precursors of jasmonic acid activate the synthesis of wound-inducible proteinase inhibitors. *Plant Cell*, **4**, 129–34.

Feeny, P.P. (1976) Plant apparency and chemical defence. *Recent Advances in Phytochemistry*, **10**, 1-40.

Fischhoff, D.A., Bowdish, K.S., Perlak, F.J., Marrone, P.G., McCormikck, S.M., Niedermeyer, J.G., Dean, D.A., Kusano-Kretzmer, K., Mayer, E.J., Rochester, D.E., Rogers, S.G. and Fraley, R.T. (1987) Insect tolerant transgenic tomato plants. *Bio/Technology*, **5**, 807–13.

Gatehouse, A.M.R. and Boulter, D. (1983) Assessment of the anti-metabolic effects of trypsin inhibitors from cowpea (*Vigna unguiculata*) and other legumes on development of the bruchid beetle *Callosobruchus maculatus. Journal of the Science of Food and Agriculture*, **34**, 345–50.

Gatehouse, A.M.R., Gatehouse, J.A., Dobie, P., Kilminster, A.M. and Boulter, D. (1979) Biochemical basis of insect resistance in *Vigna unguiculata. Journal of the Science of Food and Agriculture*, **30**, 948–58.

Gatehouse, A.M.R., Gatehouse, J.A. and Boulter, D. (1980) Isolation and characterisation of trypsin inhibitors from cowpea. *Phytochemistry*, **19**, 751–56.

Gatehouse, A.M.R., Dewey, F.M., Dove, J., Fenton, K.A. and Pusztai, A. (1984) Effect of seed lectin from *Phaseolus vulgaris* on the development of larvae of *Callosobruchus maculatus*; mechanism of toxicity. *Journal of the Science of Food and Agriculture*, **35**, 373–80.

Gatehouse, A.M.R., Butler, K.J., Fenton, K.A. and Gatehouse, J.A. (1985) Presence and partial characterisation of a major proteolytic enzyme in the larval gut of *Callosobruchus maculatus. Entomologia experimentalis et applicata*, **39**, 279–86.

Gatehouse, A.M.R., Fenton, K.A., Jepson, I. and Pavey, D.J. (1986) The effects of α-amylase inhibitors on insect storage pests: inhibition of α-amylase *in vitro* and effects on development *in vivo*. *Journal of the Science of Food and Agriculture*, **37**, 727–34.

Gatehouse, A.M.R., Dobie, P., Hodges, R.J., Meik, J., Pusztai, A. and Boulter, D. (1987) Role of carbohydrates in insect resistance in *Phaseolus vulgaris*. *Journal of Insect Physiology*, **33**, 843–50.

Gatehouse, A.M.R., Shackley, S.J., Fenton, K.A., Bryden, J. and Pusztai, A. (1989) Mechanism of seed lectin tolerance by a major insect storage pest of *Phaseolus vulgaris, Acanthoscelides obtectus*. *Journal of the Science of Food and Agriculture*, **47**, 269–80.

Gatehouse, A.M.R., Barbieri, L., Stirpe, F. and Croy, R.R.D. (1990) Effects of ribosome inactivating proteins on insect development – differences between Lepidoptera and Coleoptera. *Entomologia experimental et applicata*, **54**, 43–51.

Gatehouse, A.M.R., Howe, D.S., Flemming, J.E., Hilder, V.A. and Gatehouse, J.A. (1991b) Biochemical basis of insect resistance in winged bean seeds (*Psophocarpus tetragonolobus*) seeds. *Journal of the Science of Food and Agriculture*, **55**, 63–74.

Gatehouse, J.A., Hilder, V.A. and Gatehouse, A.M.R. (1991a) Genetic engineering of plants for insect resistance, in *Plant Genetic Engineering*, (ed. D. Grierson), Plant Biotechnology Series **1**, Blackie & Son, London/Chapman and Hall, New York, pp. 105–35.

Gould, F., Martinez-Ramirez, A., Anderson, A., Ferre, D. Silva, F.J. and Moar, W.J. (1992) Broad spectrum resistance to *Bacillus thuringensis* toxins in *Heliothis virescens*. *Proceedings National Academy Sciences USA*, **89**, 7986–90.

Graham, J.S., Hall, G., Pearce, G. and Ryan, C.A. (1986) Regulation of synthesis of proteinase inhibitors I and II mRNAs in leaves of wounded tomato plants. *Planta*, **169**, 399–405.

Green, T.R. and Ryan, C.A. (1972) Wound-induced proteinase inhibitor in plant leaves: a possible defence mechanism against insects. *Science*, **175**, 776–77.

Gutierrez, C., Garcia-Casado, G., Sanchez-Monge, R., Gomez, L., Castanera, P. and Salcedo, G. (1993) Three inhibitor types from wheat endosperm are differentially active against α-amylases of Lepidopteran pests. *Entomologia experimentalis et applicata*, **66**, 47–52.

Hallahan, D.L., Pickett, J.A., Wadhams, L.J., Wallsgrove, R.M. and Woodcock, C.M. (1992) Potential of secondary metabolites in genetic engineering of crops for resistance, in *Plant Genetic Manipulation for Crop Protection – Biotechnology in Agriculture*, Vol. **7** (eds A.M.R. Gatehouse, V.A. Hilder and D. Boulter), CAB International, Wallingford, pp. 215–248.

Hendrickx, K., De Loof, A. and Van Mellaert, H. (1990) Effects of *Bacillus thuringiensis* delta-endotoxin on the permeability of brush border membrane vesicles from tobacco hornworm (*Manduca sexta*) midgut. *Comparative Biochemistry and Physiology*, **95C**, 241–45.

Hilder, V.A., Gatehouse, A.M.R., Sheerman, S.E., Barker, R.F. and Boulter, D. (1987) A novel mechanism of insect resistance engineered into tobacco. *Nature*, **330**, 160–3.

Hilder, V.A., Barker, R.F., Samour, R.A., Gatehouse, A.M.R., Gatehouse, J.A. and Boulter, D. (1989) Protein and cDNA sequences of Bowman–Birk protease inhibitors from the cowpea (*Vigna unguiculata* Walp). *Plant Molecular Biology*, **13**, 701–10.

Hilder, V.A. and Gatehouse, A.M.R. (1989) Genetic engineering for insect resistance using genes of plant origin, in *Genetic Engineering of Crop Plants*, (eds G.W. Lycett and D. Grierson), Butterworths, London, pp. 51–66.

Hilder, V.A., Gatehouse, A.M.R. and Boulter, D. (1993) Transgenic plants for conferring insect tolerance: protease inhibitor approach, in *Transgenic Plants*, Vol. 1, (eds S. Kung and R. Wu), Academic Press, London, pp. 317–38.

Hoffman, M.P., Zalom, F.G., Wilson, L.T., Smilanick, J.M., Malvy, L.D. Kiser, J., Hilder, V.A. and Barnes, W.M. (1992) Field evaluation of transgenic tobacco containing genes encoding *Bacillus thuringiensis* δ-endotoxin or Cowpea trypsin inhibitor: efficacy against *Helicoverpa zea*. *Journal of Economic Entomology*, **85**, 2516–22.

Hofte, H. and Whitley, H.R. (1989) Insecticidal crystal proteins of *Bacillus thuringiensis*. *Microbiological Reviews*, **53**, 242–55.

Horsch, R.B., Fry, J.E., Hoffman, N.L., Wallroth, M., Eichholtz, D., Rogers, S.G. and Fraley, R.T. (1985) A simple and general method for transferring genes into plants. *Science*, **227**, 1229–31.

Ishimoto, M., and Kitamura, K. (1988) Identification of the growth inhibitor on azuki bean weevil in kidney bean (*Phaseolus vulgaris* L.) *Japanese Journal of Breeding*, **38**, 367–70.

Janzen, D.H., Juster, H.B. and Liener, I.E. (1976) Insecticidal action of the phytohemagglutinin in black bean on a bruchid beetle. *Science*, **192**, 795–6.

Johnson, R., Narvaez, J., An, G. and Ryan, C.A. (1989) Expression of proteinase inhibitors I and II in transgenic tobacco plants: effects on natural defence against *Manduca sexta* larvae. *Proceedings of the National Academy of Sciences USA*, **86**, 9871–75.

Kareiva, P. and Sahakian, R. (1990) Tritrophic effects of a simple architectural mutation in pea plants. *Nature*, **345**, 433–4.

Knowles, B.H. and Ellar, D.J. (1986) Colloid-osmotic lysis is a general feature of the mechanism of action of *Bacillus thuringiensis* δ-endotoxins with different insect specificities. *Biochimica et Biophysica Acta*, **924**, 509–18.

Laskowski, M. and Sealock, R.W. (1977) Protein proteinase inhibitors – molecular aspects, in *The Enzymes*, Vol. 3, 3rd edn, (ed. P. Boyer), Academic Press, New York, pp. 375–473.

Liang, C., Brookhart, G., Feng, G.H., Reeck, G.R. and Kramer, K.J. (1991) Inhibition of digestive proteinases of stored grain Coleoptera by oryzacystatin, a CPI from rice seed. *FEBS Letters*, **278**, 139–42.

Lipke, H., Fraenkel, G.S. and Liener, I.E. (1954) Effect of soybean inhibitors on growth of *Tribolium confusum*. *Journal of Agricultural Food Chemistry*, **2**, 410–5.

MacIntosh, S.C., Kishore, G.M., Perlak, F.J., Marrone, P.G., Stone, T.B., Sims, S.R. and Fuchs, R.L. (1990) Potentiation of *Bacillus thuringiensis* insecticidal activity by serine protease inhibitors. *Journal of Agricultural and Food Chemistry*, **38**, 1145–52.

McPherson, S., Perlak, F., Fuchs, R., MacIntosh, S., Dean, D. and Fischhoff, D. (1989) Expression and analysis of the insect control protein from *Bacillus thuringiensis* var. *tenebrionis*, in *Abstracts of the First International Symposium on the Molecular Biology of the Potato*, Bar Harbor, ME, p. 51.

Metcalf, R.L. (1986) The ecology of insecticides and the chemical control of insects, in *Ecological Theory and Integrated Pest Management*, (ed. M. Kogan), John Wiley, New York, pp. 251–97.

Minney, B.H., Gatehouse, A.M.R., Dobie, P., Dendy, J., Cardona, C. and Gatehouse, J.A. (1990) Biochemical bases of seed resistance to *Zabrotes subfasciatus* (bean weevil) in *Phaseolus vulgaris* (common bean); a mechanism for arcelin toxicity. *Journal of Insect Physiology*, **36**, 757–67.

Moreno, J. and Chrispeels, M.J. (1989) A lectin gene encodes the α-amylase inhibitor of the common bean. *Proceedings of the National Academy of Sciences USA*, **86**, 7885–9.

Norris, D.M. and Kogan, M. (1980) Biochemical and morphological bases of resistance, in *Breeding Plants Resistant to Insects*, (eds F.G. Maxwell and P.R. Jennings), Wiley, London, pp. 23–60.

Pearce, G., Strydom, D., Johnson, S. and Ryan, C.A. (1991) A polypeptide from tomato induces wound-inducible proteinase inhibitor proteins. *Science*, **253**, 895–898.

Peferoen, M., Jansens, S., Reynaerts, A. and Leemans, J. (1990) Potato plants with engineered resistance against insect attack, in *Molecular and Cellular Biology of the Potato*, (ed. M.E. Vayda and W.C. Park), CAB International, Wallingford, pp. 193–204.

Peferoen, M. (1992) Engineering of insect-resistant plants with *Bacillus thuringiensis* crystal protein genes. In *Plant Genetic Manipulation for Crop Protection – Biotechnology in Agriculture*, No. 7 (eds A.M.R. Gatehouse, V.A. Hilder and D. Boulter), CAB International, Wallingford, pp. 135–53.

Pimental, D., Shoemaker, C., La Due, E.L., Rovinsky, R.B. and Russel, N.P. (1977) Alternatives for reducing insecticides on cotton and corn, in Economic and *Environmental Impact*, Environmental Research Laboratory, Office of Research and Development, EPA, Athens, GA, pp. 145.

Powell, K.S., Gatehouse, A.M.R., Hilder, V.A. and Gatehouse, J.A. (1993) Antimentabolic effects of plant lectins and plant and fungal enzymes on the nymphal stages of two important rice pests, *Nilaparvata lugens* and *Nephotettix cinciteps*. *Entomologia experimentalis et applicata*, **66**, 119–26.

Pusztai, A. (1991) *Plant Lectins*. Chemistry and Pharmacology of Natural Products Series. Cambridge University Press, Cambridge.

Pusztai, A., Grant, G., Bardocz, S., Brown, D.J., Stewart, J.C., Ewen, S.W.B., Gatehouse, A.M.R. and Hilder, V.A. (1992) Nutritional evaluation of the trypsin inhibitor from cowpea. *British Journal of Nutrition*, **68**, 783–91.

Richardson, M.J. (1991) Seed storage proteins: the enzyme inhibitors, in *Methods in Plant Biochemistry*, Vol. **5**, (ed. L.J. Rogers), Academic Press, New York, pp. 259–305.

Sacchi, V.F., Parenti, P., Hanozet, G.M., Giordana, B., Luthy, P. and Wolfersberger, M.G. (1986) *Bacillus thuringiensis* toxin inhibits K^+-gradient-dependent amino acid transport across the brush-border membrane of *Pieris brassicae* midgut cells. *FEBS Letters*, **204**, 213–8.

Shukle, R.H. and Murdock, L.L. (1983) Lipoxygenase, trypsin inhibitor, and lectin from soybeans: effects on larval growth of *Manduca sexta* (Lepidoptera: Sphingidae). *Environmental Entomology*, **12**, 787–91.

Steffens, R., Fox, F.R. and Kassel, B. (1978) Effect of trypsin inhibitors on growth and metamorphosis of corn borer larvae *Ostrinia nubilalis* (Hubner). *Journal of Agricultural and Food Chemistry*, **26**, 170–74.

Strong, D.R. (1979) Biogeographic dynamics of insect–host plant communities. *Annual Review of Entomology*, **24**, 89.

Tabashnik, B.E., Cushing, N.L., Finson, N. and Johnson, M.W. (1990) Field development of resistance to *Bacillus thuringiensis* in diamondback moth (Lepidoptera: Plutellidae). *Journal of Economic Entomology*, **83**, 1671–6.

Vaeck, M., Reynaerts, A., Hofte, H., Jansens, S., De Beuckeleer, M., Dean, C., Zabeau, M., Van Montagu, M. and Leemans, J. (1987) Transgenic plants protected from insect attack. *Nature*, **327**, 33–7.

Van Rei, J., Jansens, S., Hofte, H., Degheele, D. and Van Mellaert, H. (1990) Receptors on the brush border of the insect midgut as determinants of the specificity of *Bacillus thuringiensis* delta-endotoxins. *Applied and Environmental Microbiology*, **56**, 1378–85.

Wolfersberger, W. (1990) The toxicity of *Bacillus thuringiensis* δ-endotoxin to gypsy moth larvae is inversely related to the affinity of binding sites on midget brush border membranes for the toxins. *Experentia*, **46**, 475–7.

Assessment of the impact and performance of transgenic plants

A. McHughen

8.1 INTRODUCTION

Every organism is defined by its genome. The genome is the sum total of genetic information carried by the organisms' nucleic acid, which, for all organisms except some viruses, is DNA. Traits are genetic characteristics in every organism; they are determined by the expression (or non-expression) of specific genes. In turn, a gene is a specific base sequence of DNA. The base sequence of DNA, the actual order of the A/T/C/G residues along the DNA molecule, provides the recipe to the cell protein synthetic machinery to arrange amino acid residues in a particular order. The arrangement of amino acid residues results in a polypeptide, and the arrangements and modifications of the polypeptides result in a particular protein. Ultimately, it is the presence (or absence) of a protein or group of proteins that provides the trait we observe – the phenotype. In its simplest form, it is helpful to think of the genome as a collection of recipes: each gene is a recipe that the cell follows to make a specific protein. We can extend this analogy and suggest that each chromosome is a volume of recipes. Just as a recipe book usually contains other information, such as menu planning, recommendations on which dishes complement each other and nutritional information, the genome is more than a series of protein recipes, in that it also carries regulatory and other information. However, for the purpose of this discussion, we will focus on the admittedly simplistic set of recipes.

The basic purpose of genetic engineering, or genetic enhancement as some prefer to call it, is to provide an organism with additional recipes,

Molecular Biology in Crop Protection. Edited by George Marshall and Dale Walters. Published in 1994 by Chapman & Hall, London. ISBN 0 412 54390 7(HB) 0 412 54400 8(PB).

either for desired traits it ordinarily lacks or to counteract existing detrimental traits. Genome size varies greatly between organisms. While it is possible to measure the physical amount of DNA in a cell with reasonable accuracy, it is not so easy to determine the number of structural gene recipes it carries, as the recipe component of the genome will vary according to organism, and also each gene recipe will vary in length. On average, however, a typical protein of 100 amino acids will require 300 base pairs of actual coding DNA; additional base pairs for the recipe will consist of regulatory sequences (promotor, terminator, enhancer, etc.) and, in eukaryotes, sequences called introns, located inside a coding sequence. It has been estimated (Kamalay and Goldberg, 1980) that a tobacco plant comprises some 60 000 diverse structural genes, different recipes in a genome of about 4500 million base pairs. The 'recipes' account for only 4.6% of the total DNA in the tobacco genome. In higher eukaryotes, only a relatively small portion of the total genome is used for structural gene recipes.

If we define a transgenic plant as one having genes transferred from a different species, then transgenic plants are nothing new. Plant breeders for years have been manipulating the transfer of genes across species' borders, with varying degrees of success. Triticale is a notable example; Triticale plants contain genes originally donated from the wheat genus (*Triticum*) and other genes from the rye genus (*Secale*). Triticale was produced by human manipulation, producing a fertile and genetically stable recombination of DNA from the two different parent genera.

The triticale (*Triticale*) plant contains many genes exclusively from wheat, many genes exclusively from rye, and many genes common to both genera. Transgenic plants are similar in that they contain genes originating from different parental sources, but the vast majority of genes comes from one source, while as little as one gene might come from another. If we imagine an average structural gene to be approximately 1000 bases of DNA and we transfer this average gene into the genome of a flax plant (*Linum usitatissimum*), which possesses the smallest genome of any major field crop at 350 million base pairs, we see that the proportion of foreign genetic information is exceedingly small. Depending on the expression and function of that DNA, however, the resulting transgenic plant might behave quite differently from its primary parent. The triticale plant differs from the transgenic flax plant in that the triticale has a large recombinant DNA proportion in its genome, while the transgenic flax has a small proportion.

Among the genes so far isolated and transfered into plants, those affecting crop protection have had the greatest impact, both on the transgenic plant and on public opinion. While disease, herbicide and pest resistance traits can be dramatic to observe in a population of plants, they can be conferred by relatively simple genetic mechanisms. It is astounding to think that just one gene, comprising less than 1/350 000 of the plant genome, can protect a plant from a lethal dose of a herbicide or from a predaceous insect.

With the development of transformation technology to deliver isolated genes into viable cells of many different crop species, and with the technology to regenerate whole, fertile plants from those genetically transformed cells, we have an ability to produce plants with novel traits. While the number of genes available for transfer (i.e. isolated, purified and cloned) is increasing dramatically, there are still relatively few such genes conferring traits of agronomic value. The majority of those that do exist are in the crop protection area.

8.1.1 Disease resistance

Disease resistance is one of the ever-present goals of plant breeding. Traditional plant breeders have been extremely successful in using conventional breeding techniques to produce cultivars with various types of disease resistance, but it is an ongoing battle as the pathogens evolve to overcome the resistance mechanisms.

One of the problems faced by scientists involved in genetic engineering for disease resistance is that there is a multitude of mechanisms employed by plants to confer resistance. Few of these have been fully analysed at a molecular genetic level, or even at a biochemical or physiological level. However, if a mechanism is not well understood at the molecular genetic level, then its use in genetic engineering is not practicable.

Research into conferring disease resistance via genetic engineering has been most successful with viral diseases. Several transgenic plants transformed to express viral coat proteins have been grown in field trials with promising results. Viruses have an extremely small genome, typically consisting of only a few genes. One common gene codes for a coat protein to envelop the nucleic acid genetic material of the virus. Several researchers have found that, if the plant produces coat protein, the plant is resistant if not completely immune to the virus from which the coat protein gene is derived (Powell Abel *et al.*, 1986; Beachy *et al.*, 1990; further details are contained in Chapter 5).

Field test results with transgenic lines of viral coat protein-producing plants have been successful, in that the transgenic lines show minimal damage from viral infection, while the control, non-transformed lines are damaged by the viral infection (Kaniewski *et al.*, 1990; Jongedijk *et al.*, 1992 and others in Table 8.1).

8.1.2 Herbicide resistance

Resistance to chemical herbicides is probably the most dramatic example of an application of genetic engineering to plants, as the visual impact of living and thriving transgenic plants next to dead, non-transgenic plants is striking and memorable. The development of herbicide-resistant plants has been driven both by scientific and commercial ambitions. The opportunities, techniques applicable and the results from a wide range of herbicide examples have been previously described in Chapter 6.

8.1.3 Insect resistance

Several different genes conferring resistance to insect pests through transgenic research are being developed. It is fortuitious that an effective, if rather narrow spectrum, insect control agent is a simple bacterial product which has been used on farms since the 1950s. The bacterium, *Bacillus thuringiensis* (*B.t.*), carries a gene coding for a protein which, when ingested by a susceptible insect, is fatal. Different genotypes of the bacteria produce different forms of the toxic protein, affecting different types of insect. The most common strains are effective against lepidopteran insects (caterpillars), while some are effective against dipteran or coleopteran pests (Feitelson *et al.*, 1992: further details in Chapter 7).

8.2 EVALUATING THE TRANSFORMATION STRATEGY

In order for agriculture to benefit from the advances in gene transfer technology, the potentially useful traits (such as those described above) need to be incorporated into commercial cultivars and proven to be effective and non-deleterious to other agronomic features of the cultivar. That is, the transgenic line has to prove its worth over a battery of tests, starting in the laboratory and continuing through to large-scale field trials. The ultimate test, of course, will be conducted in the arena of commercial production.

Ordinarily, the research team follows the transformation experiment with various tests and assays. As soon as plant cells are exposed to the foreign DNA vector, whether it is *Agrobacterium* or a microprojectile, they undergo analysis to determine success at transfer. This involves selection to eliminate (or, in practice, to reduce) the non-transformed cells, investigate integration of the foreign gene into the new host genome, stability of integration of DNA into the host genome, and degree of expression of the transferred genes. In addition, the researchers must stimulate the identified transformed cell to grow and regenerate a whole plant. For many species, the actual transfer of DNA to host plant cells is relatively straightforward; it is the regeneration of plants from the transformed cells that has been difficult or not yet achieved. Cereals and grain legumes fall into this category, where DNA can be successfully transferred either by microprojectile (for cereal species, such as maize, *Zea mays* and wheat, *Triticium aestivum*) or *Agrobacterium* (for grain legumes, such as lentil, *Lens culinaris*). However, examples of successful regeneration from transformed cells of these two important groups of crop species are few.

Most transformation strategies make use of regenerable tissues *in vitro*, partly to overcome this limitation. All major crop species have a tissue culture regeneration method reported, although some are more efficient than others. The putatively regenerable tissue is exposed to the transformation vector, and then individually transformed cells within that tissue are stimulated to regenerate, i.e. to grow into whole plants. In any given tissue, only a certain number of cells will actually become transformed. Again, depending on the efficiency of the transformation system and the genotype, the number of transformed cells might be high (e.g. using *Agrobacterium* on tobacco tissue) or relatively low (e.g. *Agrobacterium* on lentil, *Lens culinaris*). Practical difficulties encountered in generating transgenic plants are in identifying and partitioning transformed cells from non-transformed cells in the tissue mass, and also in coaxing the transformed cells to regenerate into whole plants.

8.2.1 Laboratory tests

In initial experiments to produce transgenic plants, researchers usually try to transfer a group of several genes, consisting of a scorable marker gene, a selectable marker gene and perhaps a gene of potential agronomic value. The most common scorable marker gene is β-glucuronidase, commonly called 'GUS' (for details see Chapter 1). When expressed in plant tissue, GUS reacts with its substrate in a histochemical assay to produce an indigoblue dye in the transformed cell. Investigators use GUS to track the efficiency of cellular transformation and expression. In practice, a portion of the explant tissue is sacrificed to estimate the relative efficiency of the method used to transfer the genes. The assays are not definitive, however, in that some plant tissues naturally produce a similar blue pigmentation, leading to 'false positive' results. Also, false positive results can occur if *Agrobacterium* is used as the vector and the bacterial protein synthetic machinery expresses the gene while it is in the bacteria. In order to overcome this problem, researchers have introduced a plant intron into the GUS coding sequence. The plant protein synthetic machinery can deal with the presence of the intron, but the bacterial system cannot (Vancanneyt *et al.*, 1990).

The most common selectable marker gene is neomycin phosphotransferase, type II (*npt*-II), which codes for an enzyme that detoxifies aminoglycosidic antibiotics in the neomycin class such as kanamycin (see Chapter 1). When kanamycin is added to the culture medium on which putatively transformed tissue is growing, the kanamycin is supposed to kill normal, non-transformed cells, while the transformed ones can detoxify the antibiotic and so continue growth. Thus kanamycin is a selection agent because it can be used to select for the transformed cells, as opposed to the scorable markers (such as GUS), which give no growth advantage to transformed cells. In practice, however, selectable markers are inefficient. Usually some cells are able to survive or even thrive on the selection agent, but then turn out to be non-transformed. Thus putatively transformed cells, cell lines and regenerant plants derived from them must be carefully scrutinized throughout the process, at least until seed progeny can be tested.

The gene of interest can sometimes be used as the selection marker, if the gene product (the protein coded for by the gene) confers some sort of selectable advantage to the transformed cells. An example here is the PAT gene, intended to provide resistance to the herbicide phosphinothricin. This gene seems to work as effectively in cells as in whole plants, so phosphinothricin can be added to the culture medium to give selective advantage to those cells successfully transformed with the gene. Using the gene of potential agronomic value as the selection agent is advantageous in at least two ways. First, as mentioned above the transferred DNA is inserted as a cassette, so the presence in a transformed cell of one gene on the cassette (say the *npt*-ll) indicate the presence of another gene on the same cassette (say the gene of interest). However, in practice this does not always happen. Sometimes there are errors in the integration of the foreign DNA, such that the marker gene might be successfully integrated, while the gene of interest is not. Having the gene of interest also serve as the marker eliminates this problem. Secondly, elimination of an agronomically unnecessary gene, such as *npt*-ll, from a PAT-containing cassette facilitates processing the resulting transgenic plant and its progeny-derived cultivar through the regulatory process, as there have been concerns raised that *npt*-ll in the environment might pose some risk, although at present data to support that assertion are lacking.

Other tests designed to verify the integration of transferred DNA include the Southern blot, named after the scientist who developed it (see Chapter 1). In the standard Southern, a DNA segment (probe) corresponding to a known sequence of the transferred cassette DNA is made radioactive, then permitted to interact with the genomic DNA from the putative transformed plant or cell line. If the transferred DNA is incorporated into the plant genome, then the radioactive probe, being complementary, will bind to the transferred sequence and can be detected by its radioactivity. Unbound probe DNA is washed off. The Southern is a very powerful tool to determine the presence of the given sequence, but it too is not definitive, in that false positive results have been reported. Although these are often due to contaminating DNA from the vector, and this can be overcome by a slight modification of the technique, the Southern is still limited to determining only the presence of the gene. It tells nothing of the gene's activity or expression. The Northern blot is similar to the Southern, but provides a measure of gene expression in that it tests for the presence of mRNA. The Western, (see Chapter 1), which tests for the gene product protein itself, provides a better assay of the activity of the transferred gene.

In recent years, an easier method to test for the presence of transferred DNA uses the polymerase chain reaction (PCR), which amplifies segments of DNA specifically and rapidly (see Chapter 1). If a small sample of tissue contains the foreign DNA, then a simple overnight PCR reaction will generate enough copies of the sequence to detect easily and score positively.

None of these tests, however, provides information on efficacy of the transferred genes. In the early days of plant genetic engineering, there was a concern about the stability of transferred DNA. Reports indicated the new DNA would undergo rearrangements before, during or after integration into the new host genome, and that this instability could be embarrassing if a transgenic plant suddenly stopped expressing its novel characteristics. Because we still do not understand exactly where in the host genome transferred DNA is inserted (other than that the location appears to be at random), it has been difficult to predict and address such instabilities. In order to overcome this, most new transgenic plants are permitted to mature, set seed, and the appropriate analyses (e.g. exposure of putative herbicide resistant plants to herbicide) are conducted on the progeny. This strategy is based on the assumption that if the transferred DNA remains stable through meiosis, it will probably be because it is stably integrated into the host genome. Empirical results seem to bear this out, in that transferred DNA instability in progeny generations does not appear to be a problem.

8.2.2 Controlled environment tests

Analyses on transgenic plant lines in glasshouse or growth chambers is where most of our data on the performance of transgenic lines have been generated. It is at this stage that segregation patterns are determined, to compare the transferred genes with normal Mendelian laws. The plants are usually first tested for efficacy of the new, desired trait at this time. Primary regenerants tend not to be sprayed with a herbicide to which they are supposed to be resistant, for example, because a spraying error might kill the one plant from which a valuable new cultivar might have been derived. Also, the initial regenerant will be hemizygous for the gene (i.e. not having comparable loci on homologous chromosomes: one chromosome would have the transferred DNA, while the homologue would not), so the response of the initial regenerated plant will probably differ from the homozygous progeny of a self cross. Finally, the results of one test on one plant provides little useful information. Continuing with the example of the herbicide resistant plant, an application of herbicide to the transgenic plant will only tell us if it survives or not, or at most perhaps some indication of the degree of damage.

With progeny, replicated tests can be conducted with homozygous lines (which, in the case of self-pollinated plants, should be clones), using, for example, several rates of the herbicide to determine just how much more resistant the transgenic line is compared with the non-transgenic parent cultivar. Similarly, with a putative disease resistant transgenic line, inoculations can be performed under controlled conditions to measure the quantitative response of the plants compared with the control parent.

In addition to being able to estimate the phenotypic activity of the transferred genes, the controlled environment tests can give an indication of general health of the transgenic line. Some transgenic plants acquire deleterious traits in the transformation process. If, for example, the transferred DNA is inserted into an active endogenous gene, that gene is inactivated and the homozygous progeny will have a recessive phenotype for the native gene at that locus. Depending on the nature of the gene, the plant might be less fit than its non-transgenic siblings, and if it is a critical gene, the plant might be infertile or even non-viable. This problem does not appear to be common (McHughen and Rowland, 1991).

Another reason for a transgenic plant to be less fit is if it had undergone somaclonal variation *in vitro*. This phenomenon, which is not uncommon in tissue culture derived plants and their progeny, is unpredictable and probably uncontrollable. Somaclonal variants are predominantly deleterious, although some beneficial types have been generated in this way. Somaclonal variation can be minimized by restricting the period of disorganized cell growth *in vitro* as much as possible, but is difficult, if not impossible, to eliminate. Conventional breeding can also be used if an otherwise good transgenic line is adulterated by somaclonal variation; here, the transgenic line is simply backcrossed to a good recurrent parent, with progeny selection based on the transgenic trait and against the deleterious trait.

Controlled environment testing can be used to select the most promising lines to take to the field. If the generation of transgenic plants is relatively easy, as for most Solanaceous species [tobacco, (*Nicotiana tabacum*), petunia, (*Petunia hybrida*), potato, (*Solanum tuberosum*)], it is important to cull inferior lines prior to going to the field. For any transgenic material it is critical not to waste time, effort and field space on lines unlikely to yield worthwhile data, so any mechanism to identify poor performers prior to going to the field is worthwhile. However, some transgenic traits are not easily measured in the laboratory or growth chamber/glasshouse, e.g. plants transformed to be resistant to sulphonylurea herbicides. Because these powerful herbicides (e.g. metsulfuron methyl) are applied in extremely low rates (4.5 g/ha) it is almost impossible to apply herbicide uniformly to individual pots or beds in a controlled environment. While various laboratory tests can be used to determine the presence and relative level of expression of the transferred gene, the actual degree of resistance conferred requires large-scale field testing, in order to 'even-out' the non-uniformity of application of the herbicide.

Some traits are perhaps better characterized in controlled environments, as opposed to field tests. An example of this might be transgenic insect resistant lines. The population of predaceous insects and the damage they cause to a plant is more easily measured in a controlled environment than in a field, where different species might compete, either for the plant or the pest, thus adding a measure of variability easily controlled indoors.

When the goal is to produce a genetically improved field crop cultivar, nothing short of full-scale field tests are adequate to determine commercial viability of a transgenic genotype. While various preliminary tests, conducted in the laboratory or growth chamber/glasshouse are useful in eliminating the obviously deficient lines and perhaps to give an indication of promising ones, eventually the lines must be tested in the open field under production conditions.

8.2.3 Field tests

Regardless of the trait or the manner of preliminary evaluation, ultimately the transgenic lines must perform under conditions that prevail in commercial production. It is only under production-type conditions that a true and accurate evaluation can be made of the commercial potential of a given line.

Only in the past few years have there been field tests of transgenic plants and, while they now number in the hundreds each year, there are still few reports in the literature of the data derived from the tests. Table 8.1 provides information on the tests reported in the scientific literature to date.

The reasons so few field trials have been conducted and reported are several and varied. First, before it is feasible to conduct a field trial, one needs to have obtained sufficient data on the nature of the transgenic line from laboratory and confined environment (glasshouse or growth chamber) tests since there is no point conducting a field trial if one is not absolutely sure the transferred gene is being expressed. Second, a proper field trial requires a supply of seed. The technology to transform and regenerate crop species has developed only recently, and it usually takes several generations of seed increases from the initial transgenic regenerant to provide sufficient seed to conduct a field trial. Third, many researchers feel no need to go to a field trial until they have generated a transgenic line from a commercial cultivar of their crop of interest. Many of the transformation techniques have been developed for model cultivars within a species and it then takes time to either modify the technique to apply to commercial cultivars, or else to use conventional breeding to transfer the novel trait from the model cultivar to the commercial one. Similarly, most of the early transformation experiments used marker genes to follow and develop the transformation process for a given species and most researchers feel no need to go to a field trial with a transgenic line containing a novel gene with no agronomic value. However, transgenic plants carrying marker genes have been put to good advantage in field tests, e.g. to determine incidence of escape through outcrossing (McHughen *et al.*, 1990). Transgenic plants carrying genes of potential agronomic value have also been put to similar good use, e.g. the use of cotton expressing *B.t.* toxin to track pollen dispersal via insects (Umbeck *et al.*, 1991).

Table 8.1 Field tests of transgenic crop species as relevant to crop protection and reported in the refered literature

Crop type	*Primary transgene*	*Primary trait(s) measured in test*	*Sample units*[1]	*Design Ra*	*Re*	*SY*[1]	*Reference*
Disease resistance:							
tomato	TMV cp	resistance to ToMV	rows	Y	4	1	Nelson *et al.* (1988)
tomato	TMV cp	resistance to ToMV	rows	Y	4	4	Sanders *et al.* (1992)
potato	PVX, PVY cp	resistance to PVX/Y	rows	Y	4	1	Kaniewski *et al.* (1990)
potato	PVX cp	resistance to PVX	rows	Y	4	3	Jongedijk *et al.* (1992)
cucumber	CMV cp	resistance to CMV	rows	Y	1	3	Gonsalves *et al.* (1992)
Herbicide resistance:							
alfalfa	PAT	resistance to ppt	rows	?	1	1	D'Halluin *et al.* (1990)
sugarbeet	PAT	resistance to ppt	rows	?	9	1	D'Halluin *et al.* (1992)
tobacco	PAT	resistance to ppt	plots	?	2	1	DeGreef *et al.* (1989)
potato	PAT	resistance to ppt	plots	?	V	1	DeGreef *et al.* (1989)
linseed	mutant ALS	resistance to SUs	plots	Y	4	1	McHughen and Holm (1991)
Insect resistance:							
tomato	*B.t.* toxin	insect resistance	rows	?	?	3	Delannay *et al.* (1989)
Other relevant tests:							
linseed	*npt*-II	pollen dispersal	rows	N	1	2	McHughen *et al.* (1990)
cotton	*B.t.* toxin	pollen dispersal	rows	N	1	1	Umbeck *et al.* (1991)
potato	GUS	agronomic traits	plants	Y	3	1	Dale and McPartlan (1992)
canola	*npt*-II	agronomic traits	plants	N	1	1	Arnoldo *et al.* (1992)
linseed	mutant ALS	agronomic traits	plots	Y	3	3	McHughen and Rowland (1991)

Abbreviations: Ra, randomization; Re, replications; SY, station years (total of sites and times); TMV, tobacco mosaic virus; cp, coat protein; ToMV, tomato mosaic virus; PVX, potato virus X; PVY, potato virus Y; CMV, cucumber mosaic virus; PAT, phosphinothricin acetyl transferase; ppt, phosphinothricin; ALS, acetolactate synthase; SUs, sulphonylureas; *npt*-II, neomycin phosphotransferase; GUS, β-glucuronidase; Y, yes; N, no; V, variable depending on genotype, treatment, location or year; ?, unknown or not stated.

[1]The most important aspects of the field test design include randomization (Ra), number of sites and years of an experiment, and the sample units, i.e. whether data were collected from single plants, rows of plants or complete plots.

In addition, there are more practical reasons why so few field trials have been conducted. Many researchers conducting work on plant transformation, especially those in university or government research laboratories, lack the field facilities to grow a test of transgenic material, and/or lack the expertise to design and analyse the data from a field test. A research farm, even a small one, is an expensive and valuable facility.

Finally, regulatory restrictions have severely limited the number of field tests conducted. The governments of most countries with the research capability to generate transgenic plants have regulations controlling the environmental release of transgenic material and many of these include or have included an outright ban on the environmental exposure of transgenic organisms. In recent years, however, the regulations have become less onerous in most countries. Further details of the regulatory and environmental issues of genetically-modified plants are contained in Chapter 10.

8.3 FIELD TEST DESIGN

There are usually two questions to be answered in analysing a novel transgenic line destined for commercial production. First, and most obvious, is whether or not the new line expresses the novel trait to the extent required to provide a benefit to the grower. That is, a transgenic line might actually be transformed and express the novel gene to a certain extent, but not enough to provide a real benefit. For example, flax transformed to be resistant to glyphosate herbicide was able to withstand an application of the herbicide in field trials while the non-transformed controls died, but the herbicide did enough damage to delay maturity for several days, thus making the novel transgenic lines commercially useless (McHughen and Mitchell, 1990). So in spite of being able to survive a normally lethal dose of the herbicide, these transgenic lines were not commercially viable. The second question, often overlooked, concerns the normal agronomic function of the transgenic line. While it is obvious to ensure that a novel transgenic line does express the novel trait, it is also critical to ensure that it maintains the other agronomic characteristics expected of a commercial cultivar.

8.3.1 Measuring the activity of the novel trait

Replication and randomization are critically important in establishing credibility of the data derived from the test, as they help to 'smooth out' environmental variables that might influence the results. For example, no field is absolutely uniform in its soil characteristics. If one side of a field is more fertile, then the plants growing there will perform better, not due to any genetic superiority, but rather due to the soil conditions. Such an effect might go unrecognized in a non-replicated, non-randomized trial leading to, at best, ambiguous results and, at worst, an incorrect interpretation. Replication involves having each block of control versus treatment comparison repeated at more than one location within the trial area. Randomization is simply the placement of the test variables in random order within the treatment block. For example, it is important that the control genotype is not always placed next to test genotype number 1 or that herbicide treatment number 3 is not always adjacent to herbicide treatment number 1. Randomization helps to minimize any unforeseen interaction between adjacent treatments and to sample the environmental variation within the test site. The example of soil fertility is just one of many potentially unforeseen variables to confound, confuse or invalidate data from non-replicated, non-randomized trials. In general, data from such non-randomized, non-replicated tests are discounted, if not totally disregarded, by agronomists and plant breeders.

Almost all of the reported field tests of transgenic plants report efficacy of the transgene. Most of these tests have been very simple assays; one type, for example, is the 'spray and pray' test of herbicide resistance, where the herbicide is sprayed on the plot and the researcher 'prays' the transgenic plants survive. Most of the early tests were small, rarely incorporating proper design features of randomization or replication, thus relegating the derived data to the status of qualitative observation. One example was a test from 1989, where single rows of glyphosate resistant transgenic flax plants were alternated with non-transgenic rows of flax, then sprayed with the herbicide at one standard rate of application (recommended field rate at the time).

In another early field test, transgenic *B.t.* toxin-producing tomatoes (*Lycopersicum esculentum*) were grown in single rows, with no indication of replication or randomization (Delannay *et al.*, 1989). The obvious visible effects of insect damage (in this case, *Manduca sexta*) provided sufficient impetus to continue the work on a larger, more sophisticated scale. Even with this primitive experiment, it was obvious that the *B.t.* toxin-producing plants were more resistant to the insects than the non-transgenic ones.

Tests have become more sophisticated. One report (Gonsalves *et al.*, 1992) clearly illustrates the evolution of field test design, as it covers 3 years of field tests. In this report, the authors describe their studies of a transgenic cucumber line carrying a coat protein gene to confer resistance to cucumber mosaic virus (CMV). The first year test was a fairly simple one using individual plants as the unit of replication; observations were made on individual plants, but no fruit was set due to time limitations. The following 2 years were marked with progressively more elaborate (and statistically sophisticated) field test designs to provide better information on the performance of the transgenic cucumber lines compared to the control, non-transgenic lines. The final trial was designed as a randomized complete block, providing a much greater level of confidence in the results than their earlier tests.

Figure 8.1 A properly-designed field test of transgenic sulphonylurea herbicide resistant linseed flax. This test consisted of transgenic and non-transgenic (control) genotypes sown in area plots in random placement, replicated four times and repeated (with modifications) over several years. The herbicide treated plots are evident where a non-transgenic genotype is located. This test provided reliable quantitative data on agronomic performance as well as on the degree of herbicide resistance expressed by the various genotypes. From McHughen and Holm (1991).

Several types of field design have been employed to provide statistically meaningful data. An early test of herbicide resistance in tobacco (DeGreef *et al.*, 1989) used a design called a 3×4 factorial. In this, three plant genotypes (one non-transformed parental cultivar and two transgenic lines) were tested in four treatments of the herbicide (zero plus three increasing concentrations of the chemical). An important feature was the use of area plots as the unit of measurement, as opposed to single plants or rows.

Area plots were also used in a test of transgenic herbicide resistant flax (McHughen and Holm, 1991). In this case, the field design was called a split-plot, which allowed the comparison of more than one type of variable. Here, the major variable was the type of herbicide (no herbicide treatment, as the control, plus two different chemicals with the same mode of action), while the minor variable was the genotype (a non-transgenic parental cultivar, serving a control, and two transgenic lines derived from the control cultivar). In addition to having area plots, this test was randomized and replicated four times (Figure 8.1).

8.3.2 Measuring agronomic characteristics

Commercial success of a new cultivar, transgenic or conventional, depends on the overall performance of the genotype. Farmers will not be interested in a novel cultivar if the yield is significantly lower, for example, even if it does have an additional and useful attribute. As indicated earlier, laboratory and controlled environment tests can be used to determine if the transformation event had caused dramatic deleterious effects (such as infertility or non-viability), larger scale field tests are the only way to discover whether less obvious deleterious effects exist.

Again, it is invalid to assume that a transgenic line will perform the same as its parent cultivar for traits other than the one introduced. We have already noted that the process of insertion of a foreign gene itself can interrupt and inactivate an endogenous gene. If that endogenous gene controlled a minor trait or was a minor but significant contributor to a major trait, the earlier tests might not detect the phenotypic lesion. Similarly, a minor somaclonal variation event in the culture process might not be detected with smaller scale tests. Probably the greatest reason for a transgenic line not performing the same as its parent, however, has nothing to do with the transformation process at all.

If one wheat seed from a registered cultivar is grown to maturity, self-pollinated and the resultant seed sown to produce a second generation in a similar manner, one might expect little variation because of minimal inter-generation variability in self-pollinated species. In theory, the select line should be almost identical to the original population seed. In practice, however, there is considerable genetic variability even in registered cultivars. So, the performance of the select line will depend on the chance of the initial seed being a representative from the higher performing genotypes in the population of seed in the bag or from the poorer performers. This is exactly the same with transgenic lines, as any transgenic plant can be traced back to a single individual genotype, usually from a randomly chosen seed or seedling. If that seed happened to be genetically inferior to the majority of the seeds of that cultivar, then a transgenic derivative cultivar will also exhibit inferior performance compared to the parent cultivar.

Most qualitative traits (e.g. flower colour, growth habit) can be scored in early tests using individual plants or rows. However, in order to determine the performance of quantitative traits, especially 'yield' (i.e. seed yield or biomass depending on crop type), larger scale trials need be conducted with area plots serving as the unit of measure. Yield is a function of area and of populations. Yield performance of a single plant is not a reliable indicator of the performance of a population of plants derived from that individual. Only area plots provide accurate and reliable yield estimates. This is why area plots are so critical in measuring yield and why yield data are lacking from almost all reported field trials published to date where plants or linear rows have been used as the unit of measurement.

Another critical aspect often overlooked is that of seed source for field trials, especially in early generation tests of transgenic lines. Because the test lines are considered quite valuable and productive seed increases are a prerequisite for field trials, the seed plants of the transgenic lines are nurtured to ensure good health (and seed set). Meanwhile, the seed of the control (usually the non-transformed parental cultivar) is readily available from either the original stock of seed, now several years older, from which the transgenic parent was taken or seed is available from a seed store, in which case the growing conditions cannot be easily ascertained. It is important to ensure differences in performance between the transgenic and the control lines are not due to differences in quality of the planted seed.

Data from field tests designed exclusively to determine agronomic performance of transgenic lines (as opposed to those designed primarily to measure efficacy of the novel trait) are just beginning to appear in the literature. The first of these compared the agronomic performance of five transgenic flax lines against four major commercial cultivars and also against 24 conventionally-bred candidate cultivars (i.e. those undergoing analysis to determine suitability for registration) (McHughen and Rowland, 1991). The test design was a randomized complete block, with three replicates. In addition, this test was conducted simultaneously at three separate sites, thus assisting in determining possible effects of location. Parameters measured in the test included days to anthesis, relative maturity, height, lodging, seed yield, seed weight, oil content and oil quality, being the major agronomic (and quality) characteristics of commercial importance. In this particular test, four of the five transgenic lines performed as well as or better than the standard cultivars, while one line was inferior (and therefore not tested further). The result indicated that transformation itself does not necessarily interfere with normal plant productivity, but also that large-scale field evaluation is required to cull the occasional poor performing transgenic line.

A different type of field test was conducted using transgenic canola (*Brassica napus*). Eleven lines of canola were transformed to contain the marker gene *npt*II, conferring resistance to kanamycin (Arnoldo *et al.*, 1992). The gene has no agronomic value, so the objective of the test was to determine the effect of transformation on agronomic characteristics. Maturity, seed yield, oil content and protein content were scored and determined to be similar between each transgenic line and its parent. While these results indicate the benign nature of the transformation process, and the *npt*II gene product, it should be noted that the researchers did not record whether or not the test was randomized or replicated.

It is important to remember that the scientific goals involved in genetic engineering technology often have to conform to the evolving commercial environment of cultivar development. Sometimes the technical objectives of a test differ from commercial objectives. While we are interested in applying genetic engineering technology to crop improvement, the questions asked will reflect the personal interests of the observer. For example, scientists involved in the transformation process are more interested in how the transgenic line performs relative to the parental cultivar. The farmer, however, is more interested in how the transgenic line performs relative to his/her current favourite cultivar. Conventional plant breeding is extrememly successful at providing incremental improvements to the cultivars, thus the commercial half-life of a cultivar can be very short. Therefore, even if the current top cultivar is improved through transformation, by the time a derived line is ready for commercial release, the parent cultivar might be obsolete. The novel transgenic cultivar has to compete against a different, probably improved, standard for a market share. While the new commercial favourite probably lacks the transgenic traits, the other agronomic traits are presumably superior to those of the transgenic line. Depending on the importance of the transgenic trait, farmers might not be willing to sacrifice other performance standards. One way around this problem is for breeding institutions to incorporate transgenic technology into the breeding programs, to ensure that new candidate cultivars will be composed of both conventional and transgenic improvements. Fortunately, this practice is gaining in popularity.

8.3.3 Production and evaluation of a transgenic inbreeding crop cultivar: a summary

Figure 8.2 follows the basic procedure required to generate a transgenic flax cultivar. Other crops, especially those which are not self pollinated may require some modification of this protocol. Essentially there are 10 steps as follows:

1. A putative transgenic shoot regenerates from transformed tissue. The shoot is excised from the source and a stem segment removed to be used for biochemical/molecular tests of transformation, e.g. GUS or PCR. The shoot is rooted and further shoot growth is encouraged.
2. Assuming the stem segment assays are positive, the shoot is considered transgenic. Following effective rooting, the plantlet is transferred to soil in a controlled environment. This primary regenerant is permitted to grow, mature and set selfed seed under ideal conditions to maximize seed production. If the stem segment assays are negative for transformation, the shoot is discarded.
3. Seed progeny of the transgenic regenerant is divided into three lots. Preferably 20, but at least 10, progeny plants are grown separately for seed increase. A second set of seed is germinated and the seedlings used for biochemical and molecular tests to confirm the stable, heritable transgenic status of the new line. The third seed lot is set aside as reserve seed.
4. If the tests on the second seed lot confirm their expected Mendelian

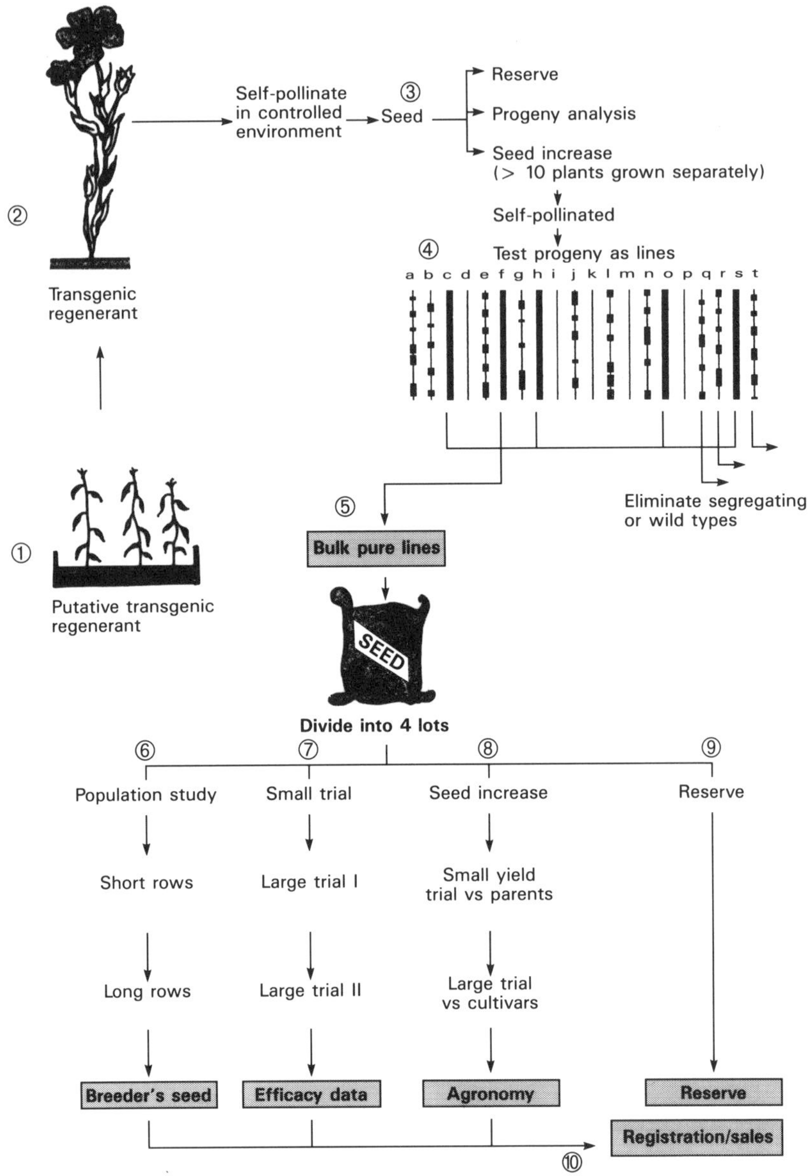

Figure 8.2 Production and evaluation of transgenic inbreeding crop cultivar: a summary.

segregation pattern for a single locus insert, seed produced by the first lot is grown and each plant is tested for transgene expression. Preferably 20, but at least 10 plants are grown in a row from a minimum of 10 progeny. If the transgenic event was a single locus insert, the expected result from a 20-row test would be four to six rows with no positive plants (indicating the homozygous recessive, wild-type segregants), four to six rows all positive plants, indicating homozygous dominant (transgenic) segregants, and the remaining eight to 12 rows would have some positive and some negative plants, from the heterozygous segregants.

5. Seed harvested from the rows with all transgenic plants (the homozygous transgenic plants) can be bulked. This will be the seed stock for a new transgenic cultivar, and is divided into four lots.
6. Seed from the first seed lot is used to produce at least 150 plants. The 150 plants are grown under ideal conditions to produce 150 separate populations of selfed seed. The populations are grown side-by-side in short rows in the field. The rows are inspected for homogeneity and uniformity. Any rows containing off-type plants are eliminated. The remaining, uniform rows are harvested separately and grown again the following season in long rows. Again the rows are inspected and any containing non-uniform plants are eliminated. If necessary, plants can be sampled from the rows and assayed for transgenic expression.
7. The second seed lot is used for efficacy trials. Here the plants are tested to determine if the degree of gene expression or activity is commercially viable. For a herbicide resistant line, this will begin as a small trial, involving single row s of plants or even single plants as plots, with perhaps three treatments (zero, $\times$ 1 and $\times$ 2 recommended rates of herbicide application) and two genotypes (the transgenic line and its parent cultivar). The seed produced in this trial is saved and used in the following years' test, which because of increased seed availability, can be considerably larger, using area plots and perhaps with more replications and locations.
8. The third seed lot is used as a seed increase source to provide sufficient seed stock to conduct agronomic and quality evaluations. After the seed increase, a small agronomic trial is grown in which the transgenic line is compared to its parent cultivar and perhaps other current commercial cultivars. The seed increase is required because agronomic studies require replication of area plots to provide reliable data on quantitative traits such as yield. Similar but more elaborate trials are conducted in the following season, at multiple locations to determine how the transgenic cultivar performs in the absence of the herbicide (or relevant agent). The results of these trials are used to provide a comprehensive evaluation of agronomic performance (as described in the previous section).
9. The fourth seed lot is maintained in a safe place as a reserve.
10. Each country has its own requirements of registration of a new cultivar, but common requirements are met with the following. (i) The efficacy of the novel trait must be determined and documented for each transgenic line. (ii) Assuming that the degree of expression of the novel trait is deemed sufficient for commercial use, the agronomic performance

must be at least as good as current commercial cultivars. (iii) Assuming the efficacy (seed lot 2) and agronomic data (seed lot 3) are acceptable to the relevant regulatory bodies for new cultivar registration, the breeders seed (seed lot 1) is ready for marketing.

8.3.4 Legislative/regulatory issues

The advent of transgenic organisms evoked a great deal of public controversy. The spectre of genetically engineered plants and public concern related to their release into the environment resulted in widespread bans, moratoria or other statutory restrictions. For an account of environmental and regulatory aspects of using genetically-modified plants in the field the reader is directed to Chapter 10.

8.4 FUTURE PROSPECTS

The first transgenic cultivars are now coming to commercial fruition. The initial releases carry genetically simple yet, in terms of their impact on crop protection practices, dramatic new traits. These include, for example, cultivars of cotton and flax resistant to such herbicides as glyphosate, phosphinothricin or sulphonylurea (eg. metsulfuron methyl). Other examples include canola and cotton transformed to produce *B.t.* toxin (for insect resistance).

Even more dramatic changes are in store outside of the crop protection arena. Tomatoes with enhanced shelf life, potatoes with enhanced starch content, canola and flax with novel, more valuable fatty acid profiles. As the technology evolves, we can expect to see wholesale changes to the nature of the crop species. A locally well adapted crop species might be used to produce a whole array of different products, including pharmaceuticals (e.g. γ-linolenic acid, antibodies, vaccines), novel oils (e.g. palm oil or olive oil from flax) and speciality proteins. With these different products, the biological plant will become a manufacturing plant. Regardless of what genetic modifications are made, however, we will still need plant breeders and agronomists to evaluate the performance of the new genotypes under stringent field conditions mimicking as nearly as possible commercial production and environmental conditions.

REFERENCES

Arnoldo, M., Baszczynski, C., Bellemare, G., Brown, G., Carlson, J., Gillespie, B., Huang, B., MacLean, N., MacRae, W., Rayner, G., Rozakis, S., Westecott, M. and Kemble, R. (1992) Evaluation of transgenic canola plants under field conditions. *Genome*, **35**, 58–63.

Beachy, R., Loesch-Fries, S., and Tumer, N. (1990) Coat protein mediated resistance against virus infection. *Annual Review of Phytopathology*, **28**, 451–74.

Dale, P. and McPartlan, H. (1992) Field performance of transgenic potato plants compared with controls regenerated from tuber discs and shoot cuttings. *Theoretical and Applied Genetics*, **84**, 585–91.

DeGreef, W., Delon, R., DeBlock, M., Leemans, J. and Botterman, J. (1989) Evaluation of herbicide resistance in transgenic crops under field conditions. *Bio/Technology*, **7**, 61–4.

Delannay, X., LaVallee, B., Proksch, R., Fuchs, R., Sims, S., Greenplate, J., Marrone, P., Dodson, R., Augustine, J., Layton, J. and Fischhoff, D. (1989) Field performance of transgenic tomato plants expressing the *Bacillus thuringiensis* var. *Kurstaki* insect control protein. *Bio/Technology*, **7**, 1265–9.

D'Halluin, K., Botterman, J. and DeGreef, W. (1990) Engineering of herbicide resistant alfalfa and evaluation under field conditions. *Crop Science*, **30**, 866–71.

D'Halluin, K., Bossut, M., Bonne, E., Mazur, B., Leemans, J. and Botterman, J. (1992) Transformation of sugarbeet (*Beta vulgaris* L.) and evaluation of herbicide resistance in transgenic plants. *Bio/Technology*, **10**, 309–14.

Feitelson, J., Payne, J. and Kim, L. (1992) *Bacillus thuringiensis*: Insects and beyond. *Bio/Technology*, **10**, 271–5.

Gonsalves, D., Chee, P., Provvidenti, R., Seem, R. and Slightom, J. (1992) Comparison of coat protein-mediated and genetically-derived resistance in cucumbers to infection by cucumber mosaic virus under field conditions with natural challenge inoculations by vectors. *Bio/Technology*, **10**, 1562–70.

Haughn, G., Smith, J., Mazur, B., and Somerville, C. (1988) Transformation with a mutant *Arabidopsis* acetolactate synthase gene renders tobacco resistant to sulfonylurea herbicides. *Molecular and General Genetics*, **211**, 266–71.

Jongedijk, E., deSchutter, A., Stolte, T., van den Elzen, P. and Cornelissen, B. (1992) Increased resistance to potato virus X and preservation of cultivar properties in transgenic potato under field conditions. *Bio/Technology*, **10**, 422–9.

Kamalay, J., and Goldberg, R. (1980) Regulation of structural gene expression in tobacco. *Cell*, **19**, 935–46.

Kaniewski, W., Lawson, C., Sammons, B., Haley, L., Hart, J., Delannay, X. and Tumer, N. (1990) Field resistance of transgenic Russet Burbank potato to effects of infection by potato virus X and potato virus Y. *Bio/Technology*, **8**, 750–4.

McHughen, A. and Holm, F.A. (1991) Herbicide resistant transgenic flax field test: agronomic performance in normal and sulfonylurea-containing soils. *Euphytica*, **55**, 49–56.

McHughen, A., Jordan, M., and McSheffrey, S. (1990) Two years of transgenic flax field tests: what do they tell us?, in *Progress in Plant Cellular and Molecular Biology*, (eds H.J.J. Nijkamp, L.H.W. Van Der Plas and J. Van Aartrijk), Kluwer, Dordrecht, pp. 207–12.

McHughen, A., and Mitchell, A. (1990) Preliminary field test results of Roundup resistant transgenic flax, in *Proceedings of the 53rd Flax Institute Conference*, (ed. J.F. Carter), North Dakota State University, Fargo, ND, pp. 109–11.

McHughen, A. and Rowland, G.G. (1991) The effect of T-DNA on the agronomic performance of transgenic flax plants. *Euphytica*, **55**, 269–75.

Nelson, R., McCormick, S., Delannay, X., Dube, P., Layton, J., Anderson, E., Kaniewska, M., Proksch, R., Horsch, R., Rogers, S., Fraley, R. and Beachy, R. (1988) Virus tolerance, plant growth, and field performance of transgenic tomato plants expressing coat protein from tobacco mosaic virus. *Bio/Technology*, **6**, 403–9.

Powell-Abel, P., Nelson, R., De, B., Hoffman, N., Rogers, S., Fraley, R. and Beachy,

R. (1986) Delay of disease development in transgenic plants that express the tobacco mosaic virus coat protein gene. *Science*, **232**, 738–43.

Sanders, P., Sammons, B., Kaniewski, W., Haley, L., Layton, J., LaVallee, B., Delannay, X. and Tumer, N. (1992) Field resistance of transgenic tomatoes expressing the tobacco mosaic virus or tomato mosaic virus coat protein genes. *Phytopathology*, **82**, 683–90.

Umbeck. P., Barton, K., Nordheim, E., McCarty, J., Parrott, W. and Jenkins, J. (1991) Degree of pollen dispersal by insects from a field test of genetically engineered cotton. *Journal of Economics and Entomology*, **84**, 1943–50.

Vancanneyt, G., Schmidt, R., O'Connor-Sanchez, A., Willmitzer, L. and Rocha-Sosa, M. (1990) Construction of an intron-containing marker gene: splicing of the intron in transgenic plants and its use in monitoring early events in Agrobacterium-mediated plant transformation. *Molecular and General Genetics*, **220**, 245–50.

PART FOUR

Commercial and Legislative Aspects

9

A commercial view of biotechnology in crop protection

M.C. Miller and W. Powell

9.1 INTRODUCTION

In agriculture today the persistent threat of loss of yield and quality from pests and diseases is one of the most important factors which needs to be addressed. At present it is being combated by means of management systems, breeding of resistant varieties, but mostly by the use of agrochemicals (herbicides, insecticides and pesticides). The reliance on chemical control is very high, costly and potentially damaging to the environment, while an alternative, crops genetically engineered with resistance to, for example, viruses and insect pests, is close to the marketplace.

Data collected by the Pesticide Usage Survey Group of the Ministry of Agriculture Fisheries and Food in the UK illustrates the extent of application of pesticides for disease control. In 1991 they reported that herbicides were applied to more than 90% of all crops with a mean of two applications using two products and three active ingredients. Fungicides were applied to more than 95% of potato, wheat and winter barley crops. Insecticides were applied to a mean of 73% of all crops once.

On a worldwide scale, fungicides remain a major input in the costs of crop production in many parts of the World. In 1990, 21% of all agrochemical sales were accounted for by fungicides (US $5.5 million) (Garner *et al.*, 1992). Farmers and growers have a strong motivation to reduce their input costs, particularly at a time of production surpluses, and static or falling demand. There is thus a strong impetus for the seed industry to develop products that will require reduced or no agrochemical inputs.

Molecular Biology in Crop Protection. Edited by George Marshall and Dale Walters. Published in 1994 by Chapman & Hall, London. ISBN 0 412 54390 7(HB) 0 412 54400 8(PB).

Cotton is an intensely managed crop requiring an average pesticide input of US $30.67 per acre, or over US $354 million each year, in the US alone. Nevertheless, arthropod pests still cause US$273 million loss in fibre production annually. Over reliance on chemical insecticides has resulted in the development of chemical resistance in pests and elimination of natural predators. A further concern is the potential damage to the environment due to the application of huge quantities of pesticides. Engineering pest resistant cotton is therefore attractive both economically and environmentally. A biopesticide should be specific to targeted pests, effective at low dosages, develop low or no resistance in the target insects and be harmless to non-target organisms. Insecticidal proteins from bacteria, plants or animals have attracted the greatest interest.

Biotechnology is a rapidly expanding sector with potential benefits, particularly in the area of crop protection, of great economic and social importance. In the waning years of the 20th century people wish to see a reduction in chemical inputs to control pests and diseases in agricultural crops (Coombes, 1991). This is true in the US, Japan and Europe as shown by the interest in genetic engineering for disease resistance and in agricultural biopesticides on the part of the big multinational chemical companies like Zeneca, CIBA Geigy, Monsanto and Rhone-Poulenc. The development of techniques such as genetic engineering, bioprocessing, monoclonal antibodies, protein engineering, tissue culture and protoplast fusion in the 1970s and 1980s promised to open up a new world of opportunities for crop protection and thereby increase profits in agricultural biotechnology. In agriculture a number of areas were identified as having potential for the application of biotechnology. Improvements in plant breeding through genetic engineering were expected to provide new transgenic crops with insect and disease resistance. The crop protection area would see new biological control agents and herbicide resistant crops (Chataway, 1991). These promised to bring good returns on investment and, as a result, money poured into the industry. However, the innovations from agricultural biotechnology would only be of direct use to farmers and generate an income for the producer when they appear in the form of seeds of new varieties.

The extent to which the scientific and technical advances become translated into new varieties would be influenced by many factors, with industry structure, technical progress, the scope for proprietary protection, and the regulatory system of particular importance. Research and development in the area of agricultural biotechnology is costly and time consuming. The protection of this new technology, to ensure that the producer makes a profit out of his investment, is of great importance. At the first International Crop Science Congress in 1992 two of the key areas of interest to plant breeders were those dealing with intellectual property rights and the use of plant biotechnology. The developed countries have evolved legal systems which reward invention of products and processes by the granting of rights of limited monopoly for a defined period of time. These rights take a variety of forms and are generally called intellectual property rights (IPRs). Patents, Trademarks, trade secrets, Copyright and plant variety rights are all forms of IPRs. For the seeds industry, the most important forms of IPRs are plant variety rights (PVRs) [also known as plant breeder rights (PBRs)] and patents. These two systems were designed to be mutually exclusive with each having a particular type of subject matter. Recent developments in biotechnology have, however, blurred the distinction between them.

9.1.1 Public perception

The public perception of the new technologies is also an important factor to be considered in the introduction of plant varieties produced by biotechnology. Fears about the safety of genetically engineered crops may severely disrupt the introduction of genetically engineered varieties, while concerns about the effects of intellectual property protection on the use of new varieties and genes therein in further breeding programmes, is creating disruption within the industry itself.

In this chapter a brief outline of the current methods available for the protection of biotechnology developments will be presented and the main issues facing the introduction of biotechnology in crop protection will be outlined.

9.2 INTELLECTUAL PROPERTY RIGHTS AND PATENTING

Intellectual property rights encompass trade secrets, know how, rights in design and copyright, confidential information, patents and plant breeders rights. All afford protection to varying degrees. The technical developments in agricultural biotechnology have focused on patents and plant breeders rights for intellectual protection. The legislation which protects this intellectual property must stimulate industrial developments and encourage cooperation between plant breeders, the biotechnology industry, farmers and processors. The advent of the new technologies required harmonization between the PBRs and patents, and the system of laws that provide them.

9.2.1 Patents

A patent is a property right granted by state authority, which excludes others from the use or benefit of the protected invention, without the consent of the patentee. Patents are granted to individuals and companies who can lay claim to a new product or manufacturing process, or to an improvement of an existing product or process which was not previously known. The grant of a patent gives the patentee a monopoly to make use of or sell the invention for a fixed period of time, usually 20 years from the date of filing of the application. In return for this right the patentee pays a fee to cover the costs of processing the patent and publicly discloses details of the invention. Eighty per cent of the information which these patents contain is never published elsewhere.

The question of what is patentable under the European Patent Convention (EPC) has to be answered on the basis of articles 52 and 53 EPC which reads as follows:

> "*Article 52*
> Patentable inventions
> European patents shall be granted for any inventions which are susceptible of industrial application, which are new and which involve an inventive step.
>
> *Article 53*
> Exemptions to patentability
>
> a) inventions the publication or exploitation of which would be contrary to 'order public' or morality, provided that the exploitation shall not be deemed to be so contrary merely because it is prohibited by law or regulation in some or all of the contracting states;
> b) plant or animal varieties or essentially biological processes for the production of plants or animals; this provision does not apply to microbiological processes or products thereof."

Thus in order for a patent application to become a valid patent it must meet several criteria (Van Dullen, 1992):

1. It must be 'novel'. This means that it must be original as a patent, and indeed new in any published format. In addition, the application is unlikely to be accepted if the applicant describes it (except in confidence) or, if it is manufactured before the application is submitted, so it is vital that it is kept secret until this time. Otherwise the invention is open to anyone to manufacture it.
2. It must not be obvious. This means that it must not be a predictable improvement of something already in existence or described in the published literature. Theoretically if an inventive person who knows all prior art thinks that an idea is an inventive step, then it is not obvious.
3. It must be useful. It must do or be something of practical benefit, rather than being a scientific observation, or a work of art.
4. It must be capable of being industrially reproduced. This criterion is applied very loosly: many chemical patents, for instance, refer to substances that cannot be reproduced in a factory environment.
5. It must not be illegal or immoral. Examples of illegal patents would be ones for mantraps or counterfeiting machinery. However, some countries

allow 'illegal patents' if the applicant intends to export the product to countries where they are legal.

6. It must be detailed. The patent must be detailed enough so that someone skilled in the art can reconstruct the invention. This is fundamental and failure to give sufficient detail can be cause to refuse a patent.

Some categories of invention are generally unpatentable. These include computer software and higher life forms (such as the genetically altered mouse), although the latter at least may soon be patentable.

Plant patents evolved alongside the plant variety rights. The US allowed plant patents since its Plant Patent Act of 1930 and has issued more than 6500. The Japanese have allowed plant patents since 1970. The European Patent Office (EPO) excludes plant varieties from patenting but the EPO board of appeal ruled in 1991 that plant patents may be awarded for useful and inventive genetic modifications, for example a corn having an additional, useful gene. The first patent issued in this category covered forage crops such as alfalfa (*Medicago sativa*), with increased protein content.

Given the complexity and rapid technical advances in plant biotechnology, there are strong reasons to suppose that the criteria for obtaining patent protection can be met. The types of inventions for which patents have been filed include tissue culture and micropropagation methods, methods for protoplast fusion, techniques for gene insertion, vectors, isolated genes, and gene promoter and terminator sequences. The claims usually extend to plants developed using such methods or containing such genes or parts of genes. Thus several patent applications have claims of the type: *plants of species A having resistance to insect pest B through genetic transfer of gene C*. This claim relates to a plant, not a plant variety.

9.2.2 Plant Breeders' Rights

In the UK, 1964 marked the introduction of plant breeders rights which gave breeders a major incentive to invest in new varieties. This allowed plant breeding companies to expand and prosper. It also benefited farmers and consumers through the introduction of new varieties with better yields and disease resistance. To qualify for PBRs, the new variety must be distinct, uniform and stable (DUS). PBRs are protected under the UPOV convention (International Union for the Protection of New Varieties of Plants 1961) which is implemented in the UK by the Plant Varieties and Seeds Act 1964. Plant breeders, conventional and users of the modern biotechnologies, take the risk that their new variety stands only a slim chance of success. A new variety passes from the seven or so years of a breeding programme to exhaustive statutory evaluation tests before finally being approved by the authorities for sale as certified seed to the farmer. The new plant varieties are protected, allowing the breeder or their agent to sell the seeds. However, there are a number of important exemptions, e.g. other breeders could use the protected material in further breeding programmes and farmers could save seed from the protected crop for resowing. The UK signed a revised version of the UPOV convention in 1991. This maintains the monopoly for the original breeder and provides greater protection of protected varieties. In the patent law style, the 1991 convention put certain acts beyond the breeders. A right in the original convention, that farmers could save and reuse seed has been delegated to the individual governments to decide in the future. The revised treaty no longer prohibits dual protection, the use of both patents and breeders rights to protect intellectual property on plants.

9.2.3 Dependent varieties

In the 1991 convention, the topic of dependent varieties was addressed. This had arisen due to fear that the new technologies would result in plagiarism of new varieties with genetic engineers able to insert genes to confer traits like, for example, insect and virus resistance. Articles 14(1)–(4) cover varieties which are essentially derived from a protected variety or not clearly distinguishable therefrom, or whose production involves the repeated use of the protected variety. In the 1991 text an 'essentially derived variety' is a variety that is predominately derived, but clearly distinguishable from, the 'initial variety' and conforms to the genotype or combination of genotypes of that variety. The derived variety retains almost the totality of the initial varieties' genotype yet is distinguishable from that variety by a limited number of characters (typically one). This could arise via genetic engineering where a gene is inserted into an existing protected variety. The repeated use of a protected variety in the production of a third variety (e.g. an F_1 hybrid) requires the authorization of the holder of the Breeders Rights. Classical breeding methods are not regarded as giving rise to derived varieties.

9.2.4 Farm saved seed

The rate at which farmers saved seed from one harvest, for planting to give the next commercial crop, was estimated in 1990 for small grain crops at 89% in Australia, 85% in Spain, 70% in Canada and the US, 45% in France, and 30% in the UK. In the case of potatoes the estimates were 40% in Germany and 20% in the UK. These constitute very large losses in financial terms for the plant breeders and the seed trade. Under the 1991 convention the person who saves seed is an infringer of the Plant Breeders Rights. In the US, the Plant Varieties Act 1970 amended in 1980 states that a person does not infringe the breeders rights by saving 'seed produced by him from seed obtained or descended from seed obtained by authority of the owner of the variety for seeding purposes'.

Table 9.1 Comparison of main provision of Plant Variety Rights under UPOV 1991 and patent laws in general

Provisions	*UPOV 1991*	*Patent Law*
Protection Coverage	plant varieties of all genera and species	inventions
Requirements	novelty, distinctiveness uniformity	novelty, inventiveness, industrial applicability, utility
Protection term	minimum 20 years (OECD)	17–20 years.
Protection scope	commercial use of all material of the variety	commercial use of protected matter
Breeders' exemption	not for essentially derived varieties	no
Farmers' privilege	up to national laws	no

9.2.5 Patents versus Plant Breeders' Rights

There are many differences between the two main forms of intellectual property protection ranging from cost to legal complexity (Table 9.1). Patents offer greater protection than plant breeders rights since the rights extend (for the lifetime of the patent) to all subsequent generations of the plant. New varieties made by crossing a patented parent with another parent could be eligible for both patents and breeders rights.

Breeders see patent rights as potentially inhibiting the free exchange and use of genetic material which breeders rights have sought to maintain. Companies claim that it is not their intention to restrict access to germplasm but that if a breeder subsequently markets a variety which includes the patented trait, e.g. insect resistance, he should have to obtain the patent holders permission. Enforcing patent rights through successive generations of breeding could be difficult and suggestions have been made that patent rights be limited to the plant varieties initially used. The breeder would pay a reasonable royalty and then be free to use the patented plant in subsequent breeding programmes without repetitive royalties. The need to recoup investment in research and development from a single payment could make the charge prohibitive and provide a disincentive to innovation.

9.2.6 Patent procedure

The procedure for filing a patent is an involved, lengthy process and according to many, drafting good patent claims is one of the most difficult legal tasks. It usually needs collaboration between the inventors and a patent attorney.

An application for patent protection is usually first made in the country of residence or place of business of the applicant. This establishes a priority date which will be recognized in most of the other countries of the world under the provisions of an international convention known as the Paris convention. This postpones the major expense of a foreign patenting programme until towards the end of the first year after the initial filing date in the home country. An application for a European patent is on the same level as national applications in other countries. The one year interim period is very valuable both to industry and other organizations which have the problem of assessing the potential industrial applicability of the new research results. The other advantage is that the inventor can publish details of his invention without detriment to his patent prospects once his priority date has been established. Thus the invention should be clearly defined and well supported by data by the date in the first application and the foreign filing should take place no later than one year from the first application.

9.2.7 The patent application

To obtain a patent, an application must be filed with the relevant national authority (Patent Office) and will be examined for compliance with the legal requirements. The basic procedure of applying for patents is regulated by the Paris Convention for the Protection of Industrial Property of 1883, to which most countries belong. Separate patent applications are usually necessary in each country where protection is sought but a single application in the European Patent Office can cover a number of European countries up to the point at which rights are granted. This occurs under the European Patent Convention (EPC). In contrast, the Patent Cooperation Treaty (PCT) provides, on the basis of a single 'international' application in one language, for an international filing and search which will be effective in any one of the countries which are party to the treaty. This has worldwide membership. The PCT is administered by the World Intellectual Property Organisation (WIPO) in Geneva and the first PCT applications were filed in June 1978.

The invention is defined in the 'claims' which form part of the specifications. Claims are a guide to the scope of the protection conferred by the patent. Patent claims are expressed in the most broad and general language which the patent attorney can devise to avoid loopholes which competitors can exploit (Crespi, 1988).

The patent office will carry out a search of previously published documents including the scientific and patent literature to determine the relevant 'prior art'. Subsequently, the patent application will be examined in the light of the search results. This usually involves argument about the specification, especially the scope of claims, and may take considerable time to settle.

Even after acceptance by the patent office a patent application or granted

First application filed (establishes priority date, public disclosure now acceptable)		
	Reported in *Official Journal (Patents)*	Up to 5 weeks
Preliminary examination and search		
Second application – 1 year later – (consolidating protection in home country) and foreign filing	Publication of specification. *Official Journal (Patents)*	18 months after priority date
Official examination and prosecution		
Grant of patent	Second publication specification. Reported in *Official Journal (Patents)*	Not more than 4.5 years from priority date

Figure 9.1 Progress of a patent application.

patent can, in most countries, be opposed by third parties who may raise objections and prior art similar to, or additional to, those already overcome by the applicant. This is usually termed Opposition and involves argument between the applicant/patentee and Opponent who have equal status as contending parties. The US patent law does not provide for opposition in this sense but allows a third party to request official re-examination of the patent in the light of prior art which has not already been considered. An overview of the timescale for a general patent procedure is presented in Figure 9.1.

9.2.8 Differences between patent law in the UK and in the USA

There are some differences between the patent laws in various countries but one of the most important involves the first to file principle.

(a) The first to file principle in UK law

Priority in patent law is decided not only by the date of filing an application but also by the content of what is filed, i.e. the application must be a proper disclosure of what is claimed. For example, in the case of two Japanese companies Asahi and Dainippon, the question was whether the earlier application by Dainippon in Japan was entitled to its date for priority purposes (Crespi, 1992). The application had disclosed the DNA sequence without providing experimental information on the preparation of the material. Asahi's application had an enabling disclosure and claimed that their patent should take precedence. The law Lords concluded that the Dainippon application was not an enabling disclosure. This would seem to encourage the filing of applications for 'inventions' which may or may not be capable of commercial exploitation, in the hope that in the ensuing 12 months something would turn up to render them patentable. Such applications would have the effect of preventing anyone else who may be better informed from obtaining a patent.

(b) The first to invent principle of US law

The US patent system also recognizes the distinction between bare and enabling disclosure. However, in relation to an inventors own enabling publications, the US law is more generous than the European one. In deciding the question of priority, the actual date of invention is taken into account, proven from laboratory notebook records. An inventor's own prior publications will not ruin the chances of obtaining a US patent provided the application is filed within a 'grace period' of one year from the publication date.

9.2.9 Financial gain from patenting

Once a patent has been granted the next step is to begin to retrieve some of the investment in the biotechnology and cost of protection. This can be achieved in a number of ways but the most common is through licensing agreements. These can be exclusive or non-exclusive and in the 1990s this trading began in earnest. For example, Bio-Rad was given an exclusive license for manufacturing a 'gene gun' developed by DuPont together with associated consumables. Asgrow Seed purchased non-exclusive rights to Mogen International's fungal resistance research in horticultural crops. The two companies will jointly develop fungal resistant plant varieties, with Mogen receiving a royalty from Asgrow on the sales of improved varieties. In Germany, a subsidiary company of Hoechst, located in Paris (Roussel-Uclaf) entered into R&D and marketing agreements worth US $3 million for Ecogen's (Langhorne, PA) bacterial insecticides. Celltech (Slough, UK) and Genetech (San Francisco, CA) agreed to cross license their respective key recombinant monoclonal antibody production patents. In a deal to encourage uptake of technology within the plant biotechnology and seeds industries, Mogen International licensed its binary vector technology to Calgene, the US biotechnology company. Mogen has asked for a fee-based payment instead of the usual royalties on products.

9.2.10 Compulsory license

If a patent holder refuses to permit the use of the invention by another person, when this person offers to pay reasonable compensation and thus to furnish security, that person can be given a compulsory license if permission is indispensable in the public interest.

9.3 OPPORTUNITIES FOR BIOTECHNOLOGY IN CROP PROTECTION

In crop protection, a vast area of new technologies has evolved via biotechnology. Their contribution to crop protection will be covered in this section.

9.3.1 Bacterial donor genes

The use of *Bacillus thuringiensis* (*B.t.*) as a natural means of plant pest control has produced an extensive scientific and patent literature. Crystals produced by the microorganisms are toxic to certain insects. This attracted interest from companies interested in its exploitation as a biological control agent. The first commercial formulation of *B.t.* was marketed in 1957 as 'Thuriade' by Sandoz (Peferoen, 1992). Of the thousands of naturally occurring strains, certain ones have been isolated and patented. Strains have also been manipulated to combine the insecticidal properties of separate strains, providing a broader spectrum of activity. The transfer of the relevant genes involved in the toxin production has produced its own patents which describe the genetic transformation of tomatoes (*Lycopersicon esculentum*), potatoes (*Solanum tuberosum*) and cotton (*Gossypium hirsutum*) to confer resistance against Lepidopteran and Coleopteran insects.

9.3.2 Plant donor genes

The isolation of a gene from one plant and its insertion into another to confer protection against pests, is another area of crop protection which has resulted from biotechnology. The Agricultural Genetics Company isolated and patented the gene responsible for the production of a trypsin inhibitor in cowpea (*Vigna unguiculata*). The trypsin inhibitor produced prevents invading insects from digesting protein so it dies of starvation. The gene has been transferred to other plant genera through genetic engineering and transformation techniques. The protection of cotton against bollworm is just one of the targets in the exploitation of this technology. The advantages of using plants transformed for pest resistance include season long protection independent of the weather and restriction to the pests which are eating the crop.

9.3.3 Biotechniques

A number of important genetic transformation techniques were developed in the commercial environment. This started with the discovery that the soil-borne bacterium *Agrobacterium tumefaciens* could be used for efficient transfer of foreign genes into many other plants (e.g. DeCleene and Deley, 1976; Bevan and Chilton, 1982), and moved on to direct DNA transfer methods such as electroporation (Fromm *et al.*, 1985), polyethylene glycol (PEG)-mediated transformation (Krens *et al.*, 1982) and microinjection into protoplast (Spangenberg *et al.*, 1990), all adding to the number of biotechnology patents. Agracetus (Middleton, WI) received a US patent covering its particle acceleration method of inserting genes into soybean (*Glycine max*). Using this method they have engineered the leading US cotton seed varieties. They have also engineered herbicide resistance in two long grained rice varieties (*Oryza sativa*) and showed stable transmission to subsequent generations. One of the most important of these techniques is the binary vector system of Mogen. Mogen is the owner, by assignment of rights, to a number of key patents and patent applications developed by the University of Leiden in the Netherlands. In 1990 the US patent office issued a patent for the binary vector system, a widely used method for the transformation of broad-leaved crops. The patent was also approved by the European Patent Office in 1991 and it was also filed in Japan.

Mogen has offered licenses on these patents and the licenses will allow the licensees to freely practice the invention for use in research and development programmes, and later to develop and sell products. This was hoped to encourage uptake of licenses of the system without imposing punitive royalties on products developed. The binary vector system offers the simplest and most reliable method for the transformation of plants and many companies used this method to produce transgenic plants. The system offers the advantage of direct cloning in *Escherichia coli*, with vectors carrying multiple cloning sites, markers for direct screening of insertions, and expression cassettes for the transgene. The system is in contrast to the cointegrate-based systems which involves the proper recombination of intermediate vectors into the Ti replicon.

Both Calgene and ICI applied for patents covering antisense technology. The claims made were similar, covering the gene encoding the tomato ripening enzyme polygalacturonase and its use in the antisense form to slow the softening of the tomato cell wall. The patent office had to establish which inventor reduced the specific claims to practice first and should therefore be granted the patent rights. The granting of a broad patent to Calgene covering its antisense technology to regulate plant genes in any plant species may dominate the antisense approach and make the initial Calgene/ICI patent conflict insignificant.

9.3.4 Diagnostics

Patent applications have also been filed in the area of plant disease diagnostics. These include immunoassay tests for the detection of beet necrotic yellows virus, which cause Rhizomania, a test for the detection of tomato black rot bacteria, and work on diagnostic research involving amplification techniques for detecting bacteria in the environment.

9.3.5 Genome mapping

The worldwide programme to map the small genome of *Arabidopsis thaliana* is designed to obtain knowledge about the location and organization of genes in this simple weed. It is hoped that this will allow rapid identification, and hence exploitation and manipulation, of corresponding genes in commercial crops. Genome mapping of the major crop species will provide information about the location of the most important genes and further plant breeding for disease and pest resistance.

9.4 BIOTECHNOLOGY AT WORK IN CROP PROTECTION

A wide range of methods and genes are being used in crop protection today. This can be illustrated by Biosys (CA, USA) who are using nematodes to control insects, while Chinese workers have inserted a plasmid containing modified *B.t.* endotoxin genes into tobacco (*Nicotiana tobacum*) to create plants with increased toxicity to larvae of the tobacco budworm (*Helicoverpa assulta*). The number of different crops in field trials for a particular transgenic trait is shown in Table 9.2.

9.4.1 Herbicide resistance

In 1991, 36% of field trials of genetically-modified plants involved trials of herbicide resistance (Chassaray and Duesing, 1992). Plant Genetic System (PGS) in Gent have carried out work on the introduction of resistance to the herbicide phosphinotricin using a glufosinate acetyltransferase gene whose product degrades (inactivates) the herbicide. Field trials were carried out in Canada, of rapeseed (*Brassica napus*) containing this gene and it has now been introduced into many crops, with effective herbicide tolerance being demonstrated. Monsanto is expected to have herbicide-resistant canola (rapeseed) and soybean (*G. max*) ready for sale in 1996.

9.4.2 Fungal resistance

In 1991 DNA Plant Technology (USA) reported successful results from a field trial of transgenic tobacco plants engineered to make high levels of

Table 9.2 Number (*N*) of different crops involved in field trials for a particular transgenic trait in 1991

Type of the major gene of the construct	*N*
Virus resistance	5
Disease resistance	4
Insect resistance	5
Herbicide resistance	8
Product quality improvement	6
Total	28

chitinase, an antifungal enzyme (Anonymous, 1991). In three tests engineered plants resisted fungal infection and disease control was as good as on plants treated with a chemical fungicide. The gene coding for chitinase was obtained from the bacterium *Serratia marasceris*. The company has also worked on tomato (*L. esculentum*), potato (*Solanum tuberosum*), lettuce (*Lactuca sativa*) and sugarbeet (*Beta vulgaris*). Du Pont (USA) inserted the bean chitinase gene into tobacco plants. The plants showed resistance to *Rhizoctonia solani* (Altman, 1991).

9.4.3 Viral resistance

The genetic engineering of plants for virus resistance (TMV resistant tomatoes) was achieved in the mid 1980s. The method used the expression of the viral coat protein (CP) in the plant. CP-mediated protection may work in a number of ways, possibly decreasing the frequency of infection, or reducing local spread or reducing the virus content. Breeding for virus resistance has been difficult using conventional methods.

9.4.4 Pest resistance

Plant Genetic System (PGS), have several crops engineered with the *B.t.* toxin gene, including potatoes and tomatoes. Scientists at Gent University screened over 5000 strains of *B.t.* for new toxins in order to try and avoid resistance to the toxin becoming a major problem. Resistance to *B.t.* is likely to occur according to Van Montague (1991), but it will be possible to introduce new proteins from new strains. In this respect it is better than dealing with chemical insecticide resistance, because the toxin interacts with specific receptors in the midgut which are required by the insect.

Mycogen is just one of 18 companies researching the application of *B.t.* with AGC, Boehringer-Mannheim, Ecogen, Zeneca and Novo-Enotech all having a stake in non-engineered *B.t.* In 1991 Mycogen received 10 new patents for novel *B.t.* proteins and have signed an international agreement with their long standing Japanese collaborator Kubota Corporation, cover-

organisms from genetic modification programmes and not the ethics of genetic modification *per se*. The advisory group was to consider likely future trends in the use of genetic modification in food production and concerns which might arise from the consumption of food from such programmes, and to recommend how the concerns might be addressed.

Patent Concern, a coalition of 31 consumer, environmental, animal rights and Third World development organizations, is campaigning against the draft directive on patents put forward by the Commission of European Communities. The directive would allow the patenting of animal and plants that had been genetically engineered. The Patent Concern Coalition has called for a moratorium on patents on plants and animals until socio-economic and ethical issues have been properly debated (King, 1991).

9.5.2 Genetic resources

One argument against biotechnology is that the valuable genetic material, often in developing countries, will be destroyed as a result of this search for new genes with unique properties, which will be exploited in developed countries. The issue here is that the genetic material, if patented, is no longer freely available. However, until these genes are isolated they are only available in the original plant, and so little has changed. The resultant new plants will be freely available when the patent expires and a new resource will have been created.

Another argument is that agricultural production will focus on the patented lines leaving the earlier varieties behind and lost for posterity. This ignores the possibility of coexistence and competition on price which will exist between the patented varieties and the existing free varieties. Large gene banks throughout the world have been set up to preserve the plants for generations. This need has been identified in advance of gene technology.

9.6 FINANCIAL ASPECTS

In 1988 few biotechnology companies were making a profit according to a report by Arthur Young, High Technology Group (1988). Also, despite many successes in cloning genes and transforming and regenerating plants, there have been no examples yet of commercial seed products where laboratory techniques have been used to introduce a new trait. The pace and direction of innovation is the outcome of many factors both external and internal, which influence a company. In the case of biotechnology, a number of these influences such as the cost of patents, regulations, the acceptability of the techniques and products, are making themselves felt as constraints to the biotechnology industry.

Of considerable importance to the industry is whether or not they can make any financial return on plant biotechnology. The current economics of many companies is that they cannot. The reasons being the length of time it takes to introduce a product to the market, the expenditure of meeting regulatory requirements and the overall level of public acceptance of the products. A further consideration is the level of return from the farmers and/or consumer that is required to meet the cost of developing the technology.

To ensure a fair return on investment there needs to be a secure structure of intellectual property protection. The industrial argument in favour of patents is an economic one. Many companies claim that they would not invest in research without the protection of a patent guaranteeing financial returns for innovation, as many biotechnology companies have conducted research before it was clear that they would gain patents on their work (Watts, 1991).

9.6.1 Investment

Uncertainties over patent rights and safety regulations are deterring investment by biotechnology companies. In 1991 venture capital organizations invested £62 million in biotechnology, down from £101 million in 1989. In Britain only £15 million was invested, a decrease from the figure of £35 million in 1989. The investment spending by company in 1990 is shown in Table 9.3.

The US agricultural biotechnology companies increased research and development spending 5% from fiscal 1990 to fiscal 1991. On the investment by seed and pesticide companies there was a 2.5% reduction in research and development spending over the same period. A summary of the research and development spending and sales is presented in Table 9.4.

9.6.2 Value of the market

In 1992 Mycogen Corporation received an experimental use permit (EUP) from the US Environmental Protection Agency (EPA) to begin large-scale field testing of its third genetically engineered product, M-Peril bioinsecticides, currently approved by the EPA for commercial sale. The construction of M-Peril involves the transfer and expression of a *B.t.* toxin gene in *Pseudomonas fluorescens* cells which are then killed and the dead cells used. M-Peril controls the European corn borer (*Ostrinia nubilalis*) a major pest of maize in the mid-western USA. The company believes the market for control of the European corn borer exceeds US $20 million annually.

Only a few of the bioinsecticides have been developed into products and today it is estimated that *B.t.* accounts for 90–95% of the insect biocontrol market, the latter estimated at US $105 million (Rigby, 1991).

Table 9.3. Investment in plant biotechnology: research and development spending by company (1990)

No.	*Company*	*Location*	*US $m*
1	Du Pont	USA	20
2	ICI	UK	18
3	Monsanto	USA	18
4	Sandoz	Switzerland	17
5	CIBA-Geigy	Switzerland	15
6	DNA Plant Technology	USA	13
7	Calgene	USA	11
8	Sanofi	France	11
9	Mycogen	USA	10
10	BioTechnica International	USA	8
11	Novo Biokontrol	Denmark	8
12	Bayer	Germany	8
13	Plant Genetic Systems	Belgium	8
14	AGC	UK	7
15	Agracetus	USA	6
16	Abbott	USA	6
17	Ecogen	USA	6
18	BASF	Germany	5
19	Salvay	Belgium	5
20	Calliope	France	5
21	Cyanamid	USA	4
22	Dow Elanco	USA	4
23	Hoechst	Germany	4
24	Upjohn	USA	4
25	Kemira OY	Finland	3

Source: Company Reports and Country NatWest WoodMac.

9.7 CONCLUSIONS AND FUTURE PROSPECTS

Companies which currently have a significant position in the agricultural seeds business are well placed to further dominate the international seeds trade by broadening their crop and country coverage and strengthening their promotion. Their increasing strength will be furthered by intellectual property protection. The small biotechnology companies are finding that without involvement in traditional plant breeding they cannot carry out plant biotechnology successfully. The European Community through its Patent Directive and Plant Variety Rights Legislation is on the way to providing two of the instruments which will encourage the application of biotechnology in crop protection.

The difficulties involved in biotechnology and the timescales and costs have been grossly underestimated but the prizes have not (Roberts, 1993). A recently granted US patent claims all genetically transformed cotton plants, simply claiming all opportunities to obtain an obviously desirable result. The claim may or may not be held valid by the Courts, but the cost of litigation is a powerful deterrent to a competitor to test the position

Table 9.4 Research & Development spending for the fiscal year ending Dec. 91 except as noted

	Research & Development Spending			*Sales*	*Profits*
Company	*Latest FY spending ($m)*	*Change from last FY(%)*	*Versus revenue (%)*	*Latest FY ($m)*	*Latest FY ($m)*
AGBIOTECT COMPANIES					
Biotechnica International	5.142*	-33.5	33.3	15.434	-15.044
Calgene (6/91)	11.151*	7.0	42.7	26.104	-14.379
Crop Genetics International	5.981	13.6	421.8	1.418	-8.386
DNA Plant Technology	6.473	18.7	71.0	9.123	-14.941
Ecogen	6.408*	17.0	102.7	6.240	-10.879
Escagenetics (3/91)	2.851	42.4	112.9	2.525	-4.190
Idexx Labs	3.860	11.0	12.7	30.454	3.152
Mycogen	10.700*	-2.7	58.4	18.312	-3.305
Neogen (5/91)	0.857	33.7	14.4	5.963	-0.774
Syntro (9/91)	2.099*	9.6	52.1	4.026	-0.944
TOTAL	55.522	NA	NA	119.599	-69.690
Average	5.552	5.0	46.4	11.960	-6.969
PESTICIDE AND SEED COMPANIES					
American Cyanamid	492.000	14.3	9.9	4 986.200	358.800
Ciba-Geigy (12/90)	1 483.000	-1.2	10.0	14 274.649	747.034
Dekalb Genetics (8/91)	36.700	10.5	13.3	275.800	15.700
Dow Chemical	1 159.000	2.0	6.2	18 807.000	942.000
Du Pont	1 298.000	-9.1	3.4	38 151.000	1 403.000
FMC	134.644	-14.6	3.5	3 899.449	173.083
ICI	1 115.000	-2.3	4.8	23 353.000	1 014.000
Monsanto	627.000	2.5	7.1	8 864.000	296.000
Pioneer Hi-Bred (8/91)	78.290	8.0	7.0	1 124.902	104.177
Rohm & Haas	183.000	2.7	6.6	2 773.000	163.000
TOTAL	6 606.634	NA	NA	116 509.000	5 216.794
Average	660.663	-2.8	5.6	11 650.900	521.679

[illegible] sponsored expenses. NA, not available

(Roberts, 1993). The value of this patent is enormous but is this the way industry desires to go? The success of such a broad patent will make people reconsider the issues because this will hinder rather than promote technical advances in the area of biotechnology.

REFERENCES

Altman, A. (1991) Frontiers of biotechnology in agriculture. *Trends in Biotechnology*, **9**, 373–4.

Anonymous (1991) Genetic engineering for resistance to fungal and bacterial disease. *Impact AgBioIndustry*, December 1991, 13–20.

Arthur Young, High Technology Group (1988) Biotech 88: Into the Market Place. A Survey of an Industry in Evolution, Arthur Young, San Francisco, CA.

Bevan, M.W. and Chilton, M.D. (1982) T-DNA of *Agrobacterium Ti*, and Ri plasmids. *Annual Review of Genetics*, **16**, 357–84.

Chasseray, E. and Duesing, J. (1992) Field trials of transgenic plants: An overview. *Agro-Food Industry Hi-Tech*, July/August, 5–10.

Chataway, J. (1991) Biotechnology and Business blues. *Ag Biotech News and Information*, **3**, 1003–5.

Coombes, J.A. (1991). Is the crop protection industry meeting the publics need for foods produced without pesticides. *IMPACT AgBioIndustry*, December, 3–4.

Crespi, R.S. (1988) Biotechnology and patent law. *International Industrial Biotechnology*, **8**, 11–4.

Crespi, R.S. (1992) What's 'new' in patent law? *Trends in Biotechnology*, **10**, 108–10.

DeCleene, D. and Deley, J. (1976) The host range of crown gall. *Botanical Review*, **42**, 389–466.

Fromm, M.E., Taylor, L.P. and Walbot, V. (1985) Expression of genes electroported into monocot and dicot plant cells. *Proceedings of the National Academy of Sciences USA*, **82**, 5824–8.

Garner, N., Van Den Elzen, P. and Cornelissen B.J.C. (1992) The potential for the control of fungal disease in crop plants using gene transfer technology. *Biotechnology International*, 111–6.

King, D. (1991) The ultimate claim. *Chemistry and Industry*, June, 404.

Krens, F.A., Molendijk, L., Wullems, G.J. and Schilperoort, R.A. (1982) *In vitro* transformation of plant protoplasts with Ti-plasmid DNA. *Nature*, **296**, 72–4.

Peferoen, M. (1992) *Bacillus thuringiensis* in crop protection. *Agro-Food Industry Hi-Tech*, November/December, 5–9.

Rigby, S. (1991) *B.t.* in crop protection, in Agrow, (ed. J. Sackett), P.J.B. Publications, Richmond, Surrey, UK.

Roberts, T. (1993) Intellectual property for biotechnology, in *Opportunities for Molecular Biology in Crop Protection* (eds D.J. Beadle, D.H.L. Bishop, L.G. Copping, G.K. Dixon and D.W. Hollomon). BCPC Monograph 55. British Crop Protection Council, Farnham, UK, pp. 305–19.

Spangenberg, G., Neuhaus, G. and Potrykus, I. (1990) Micromanipulation in higher plant cells, in *Plant Cell Line Selection: Procedures and Applications*, (ed. P.J. Dix), VCH, Weinheim.

Van Dullen, S. (ed.) (1992) *Introduction of Patents Information*, 2nd edn, The British Library, London.

Van Montague, M. (1991) 'New plants' future in agriculture. *Agro-Industry Hi-Tech*, January/February, 8–15.

Watts, S. (1991) A matter of life and patents. *New Scientist*, **129**, 56–61.

10

Environmental and regulatory aspects of using genetically-modified plants in the field

T.E. Nickson and R.L. Fuchs

10.1 INTRODUCTION

The techniques of genetic engineering have advanced to the point where products from recombinant organisms will soon be available to consumers. Speaking specifically of plants, almost all agronomically important crops have been engineered in an effort to improve upon one or more characteristics (Gasser and Fraley, 1989). Crops that produce a better quality of food are more resistant toward pests and environmental stress, and that are tolerant toward environmentally innocuous herbicides are very close to commercial reality. Such products have the potential to bring significant benefits to food processors, consumers and agriculture. They will be more cost effective to produce and process, of higher nutritional quality, and enable farmers more flexibility in environmentally sound cultural practices. Looking further into the future, plants may be used to produce large quantities of therapeutically useful peptides and otherwise scarce chemotherapy agents like taxol at a lower cost.

The plant biotechnology industry is now at the point where field testing of genetically-modified plants (GMPs) is acknowledged as a critical and realistic step in assessing the safety, commercial potential and environmental impact of each product. Literally hundreds of field tests have occurred throughout the world without incident and the results have been reported (Halvorson *et al.*, 1985; National Research Council, 1989; Ginzburg, 1991;

Molecular Biology in Crop Protection. Edited by George Marshall and Dale Walters. Published in 1994 by Chapman & Hall, London. ISBN 0 412 54390 7(HB) 0 412 54400 8(PB).

Casper and Landsmann, 1992; further details are described in Chapter 8). Getting to this step in the commercialization process requires working with governmental bodies that have the authority to approve field tests of GMPs. An application for a field release permit requires that data and information be provided on the organism, the transformation method, the gene and its origin, expression products and their potential environmental impact. Such data are acquired from studies under controlled conditions such as glass houses and growth chambers.

All studies with GMPs must be undertaken with the proper regulatory oversight by the government presiding over the location in which the studies are done. As GMPs will be used globally, it is critical to work with, and understand the regulations and approval processes applicable to field releases in many countries. This chapter will describe the regulatory processes in place today as they relate to field release of GMPs in Canada, the US, Europe, and Japan. Secondly, a discussion of environmental aspects of GMPs is given. The environmental assessment is presented as two case studies: a herbicide tolerant canola or rapeseed (*Brassica napus*), (*glyphosate tolerant canola*) and insect resistant cotton (*Gossypium hirsutum* and *G. barbadense*).

10.2 REGULATORY ASPECTS

10.2.1 Overview

Plant breeders have been genetically-modifying plants for centuries to improve the agronomic characteristics of crops and ornamentals. In Canada and Europe, there is an approval process for all new plant varieties, while the US has no such requirement. Concern over the uses and potential of genetic engineering for plant improvement has brought about more involvement of governmental bodies in a regulatory role. In Canada, the US, Europe, and Japan, certain public agencies that regulate agricultural and food industries under the jurisdiction of national laws will be responsible to approve the release of GMPs into the environment. With the help of academia, industry and a group of international experts, the Organization for Economic Cooperative Development (OECD), guidelines are being developed to regulate GMPs (Bureau of National Affairs, 1992).

10.2.2 Canada

The Canadian government is strongly supportive of the development of biotechnology in Canada as a means of strengthening economic competitiveness. Canadian regulations and oversight of GMPs are divided among three agencies. Agriculture Canada regulates agricultural products of biotechnology including animal feeds, fertilizers, pesticides, seeds and veterinary biologicals. They also provide food inspection to ensure quality, safety and purity of foods. Field testing of GMPs in Canada requires permits from Agriculture Canada which is also responsible for environmental safety assessments of field trials under the Environmental Assessment and Review Process Guidelines. Health and Welfare Canada regulates foods and food additives, chemicals and other products that may affect the nutritional value of foods. Environment Canada advises Agriculture Canada on environmental issues to preserve and enhance the quality of the Canadian environment.

Agriculture Canada is the lead agency when considering the regulatory aspects of using GMPs in the field (Figure 10.1). They have developed a flexible approach to regulation built on existing laws. Their approach is a product-based, case-by-case assessment with a science-based evaluation of the risk related to each field release. Based on that risk, a four stage system of regulatory oversight is used. Contained glasshouse or lab research (Stage 1) is regarded as minimum risk and involves no regulation. Of incrementally greater perceived risk and regulatory involvement are confined field research trials (Stage 2), unconfined field research trials (Stage 3) and commercial release (Stage 4). Since 1988, there have been over 300 field trials at the Stage 2 level, but only a few at Stage 3. Table 10.1 summarizes the types of field trials approved in 1992 which were conducted at 81 locations across Canada.

Table 10.1 Summary of approved field trials in Canada in 1992

Crop type	*Number of trials*	*Trial type*	*Number of trials*
Alfalfa	2	Herbicide Tolerance	185
Canola	164	Stress Tolerance	2
Flax	24	Virus Resistance	8
Potato	10	Insect Resistance	4
Soybean	1	Increased Oil	1
Tobacco	2	Increased Protein	1

The Plant Products Division of Agriculture Canada has been authorizing field trials of GMPs under the Seeds Act and the Plant Protection Act and Regulations since 1987. In addition, the Plant Industry Directorate, formerly the Pesticide Directorate, of Agriculture Canada may require research permit approvals under the Pest Control Products Act and Regulations. Agriculture Canada approves import movement of seed or plant parts and all field releases of GMPs, issues safety requirements for field tests, and inspects release sites during the course of a test to ensure that the conditions of the permit are being followed. Before an approval is issued, an Environmental Assessment and Review Process (EARP) must be completed. The environmental assessment requires an environmental impact analysis with a detailed site map of the biological ecosystem surrounding the research site and information in four broad categories:

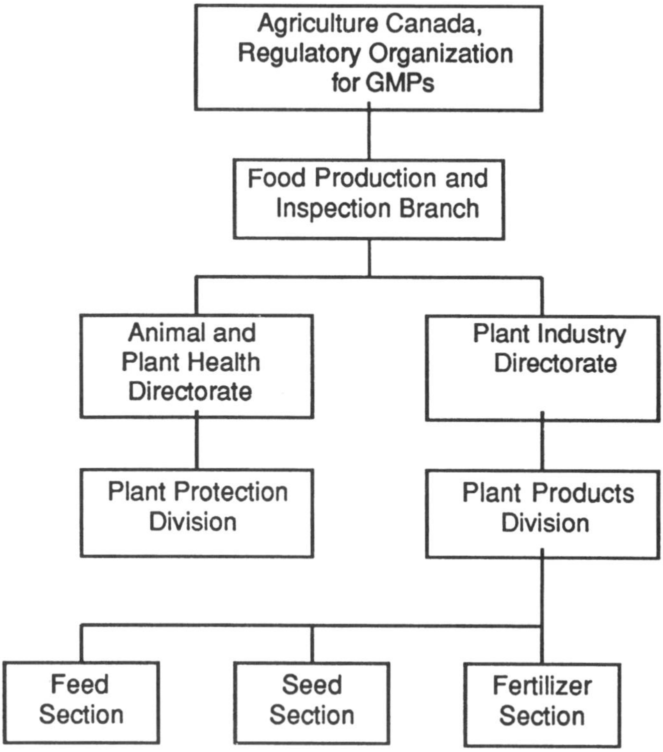

Figure 10.1 The organization of responsibility for GMPs in Agriculture Canada (1993).

1. Ability to predict fate and effects (including organism design and construction, survival, growth and reproduction, gene transfer, dispersal, effects and impact on the biota and ecosystems).
2. Ability to monitor the organism.
3. Ability to contain the organism.
4. Ability to control the experiment, including site monitoring for several years after the experiment.

Also, the relevant provincial governments are notified and asked to comment, as they have legislative authority over the field release site (Kalous and Duke, 1989). Organizations involved in a field release must use a comprehensive approach to environmental safety. This involves planning for movements, containment on the site, waste disposal and contingency emergency procedures for spills or accidental release. No harvested crop from experiments with GMPs is permitted to be used in commercial food or feed without prior approval from Health and Welfare Canada.

The Seed Section of the Plant Products Division of Agriculture Canada has responsibility to approve registration of all new plant varieties prior to sale in Canada. New varieties are approved after they have met the 'merit' requirements in multi-year comparison Cooperative Trials for variety registration. Unlike the US where there is no regulation of variety introduction, Canada requires that any new variety of major crops (excluding horticultural crops) that one intends to grow commercially in Canada must undergo a formal registration through Cooperative Trials. In recent years, genetically-modified canola has made the first appearance in the canola Cooperative Trials with Monsanto's GTCanola (tolerant to glyphosate, the active agent of Roundup® herbicide) and Hoechst's Basta®-tolerant canola (tolerant to glufosinate, the active agent of Basta® herbicide). Two new situations have thus been created. First, there are no formal provisions currently for herbicide tolerance as a merit category in the Cooperative. To respond to this situation, the Western Canada Canola and Rapeseed Recommending Committee (WCC/RRC) has agreed to accept additional information in the form of private data which demonstrates herbicide tolerance, but it is still unclear whether herbicide tolerance will obtain a special merit category. Second, the presence of GMPs has necessitated that Agriculture Canada determine buffer or isolation conditions and crop destruct methods previously not required. Formerly, seed from the trials could be sold into commerce. Since GMPs are not currently regulated in the same way as plants from traditional breeding programs, Cooperative practices were modified.

At this time, Environment Canada (EC) has not been involved directly with field testing of GMPs. Based on the Canadian Environmental Protection Act (CEPA), Environment Canada will not become involved in regulation when another agency is already involved. However, if there were an indication that the Canadian environment was not being protected, they would be obligated to take an active role.

Though not involved with the approval and performance of field tests, Health and Welfare Canada has been closely involved in the development of GMPs from the food and feed safety perspective. Under the Canadian Food and Drug Regulations and CEPA, Health and Welfare Canada assumes authority to regulate the foods and feeds produced through genetic engineering. They have recently issued a proposal as an information letter detailing how they intend to regulate these foods (Health and Welfare, Canada, 1992). It is an assessment of novel foods and processes which would impose a premarket clearance of foods and feeds from genetically-modified organisms. Their proposal is closely aligned with guidelines recently proposed in the UK that calls for process-based regulations. This contrasts with proposed guidelines recently published by the World Health Organization (WHO), OECD and the US Food and Drug Administration (FDA) (*vide infra*). Health and Welfare Canada is currently revising the information letter based upon broad public input from interested parties and industry which requested harmonization with international guidelines and the substantial equivalence concept.

10.2.3 The United States

GMPs and their food products are regulated by two federal agencies in the US. These agencies are the US Department of Agriculture (USDA) and the FDA. A third body, the Environmental Protection Agency (EPA) has responsibility to regulate pesticides whether applied to, or produced by a plant. Their involvement with GMPs stems from the development of plants that expressed pesticidal proteins such as *Bacillus thuringiensis* proteins (*B.t.*). All three agencies have separate legal responsibilities and statutory jurisdiction entitled by different acts of law. The general philosophy of regulation among all agencies has been to use a science-based approach on a case-by-case basis. It is accepted that the method of transformation has little impact on the safety of the product. Each agency has recently issued its proposal on how it will regulate GMPs, their food products and pesticides produced by the plant.

The Animal and Plant Health Inspection Service (APHIS), a division of the USDA, regulates the movement and release of engineered plants and their parts under authority granted by the Plant Pest and Quarantine Act. APHIS permits are required for any importation and interstate movement, or release of a plant obtained by recombinant DNA techniques (Animal and Plant Health Inspection Service, 1987, 1991) (Figure 10.2.). Permits are not required for export. When plants contain genetic material originating from a plant pest, they are considered regulated articles. (The only exception to date is modified *Arabidopsis thaliana*, which has been exempted from regulation.) Data required to obtain a release permit include: donor organism, vector description, expression of trait, purpose of the release, sites, protocols, number of introductions, containment measures and disposal. A limited interstate movement or importation permit has different data requirements which include specification of the regulated article to be moved, origins and destination of all shipments, description of containment facilities and a description of the containers to be used during shipment. Individual states are involved with the permit process. Permit applications for GMPs are coordinated by a Deputy Director with the help of the Biotechnology Safety Officer and are reviewed by staff biotechnologists of the Plants Branch of APHIS (Figure 10.2.).

Before commercialization, the USDA/APHIS must determine that the GMP is not a pest and therefore is not subject to regulation under 7 CFR 340 (June 16, 1987). Information recommended to be submitted for this determination was published in draft form (Animal and Plant Health Inspection Service, 1992a) and includes:

1. The rationale for the development of the genetically engineered plant.
2. A description of the genetics, taxonomy, pollination characteristics and weediness of the crop.
3. A description of the transformation system and regulatory sequences used to generate the plant.

4. A description of the donor genes.
5. Genetic analysis and agronomic performance.
6. Environmental consequences of introduction of the GMPs.
7. Any unfavorable information that would impact the determination.

As of January 1993, two companies have submitted a petition for determination of non-regulated status of GMPs. Calgene's submission has been approved for their Flavr Savr® tomato, while Asgrow's petition for virus resistant squash is pending (Animal and Plant Health Inspection Service, 1992b,c).

Hundreds of applications are being submitted to the USDA requesting permission to conduct field tests with GMPs. Each year this number has been increasing at a rate such that resources at the APHIS are being strained because the present permit process demands much attention. Consequently, the USDA has issued a proposal (Animal and Plant Health Inspection Service, 1992a) based on their years of experience in granting field release permits that, if accepted, would significantly reduce the amount of oversight and paperwork prior to a field test. They propose a notification process prior to a release of a GMPs provided that plant meets certain criteria. First, the plant must be one of the following species: corn (*Zea mays* L.), cotton (*G. hirsutum* L.), potato (*Solanum tuberosum* L.), soybean (*Glycine*

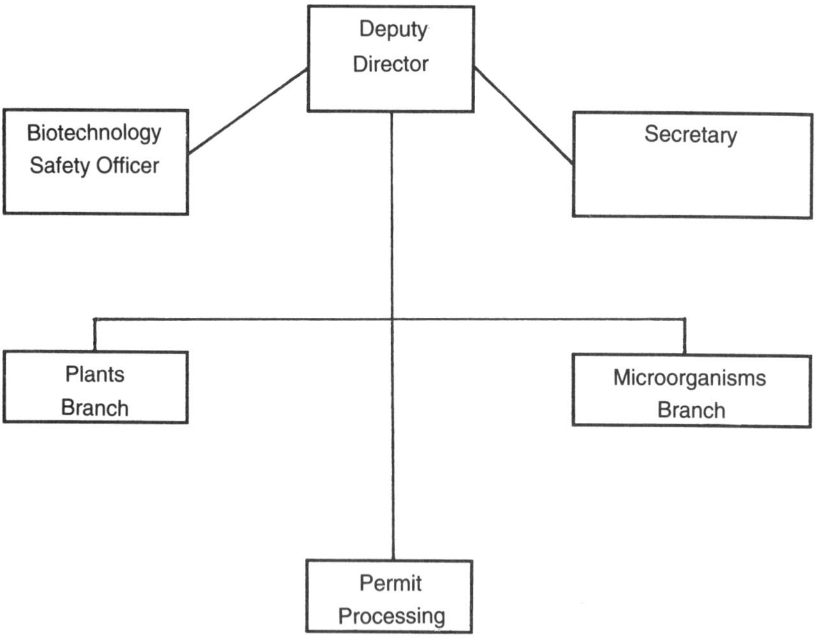

Figure 10.2 Organization/co-ordination chart for APHIS permits.

max L.), tobacco (*Nicotiana tabacum* L.), tomato (*Lycopersicon esculentum* L.) or 'any additional plant species that BBEP (Biotechnology, Biologicals and Environmental Protection) determines may be safely introduced' (Animal and Plant Health Inspection Service, 1992a). Other conditions are outlined in the document that must be met and would require extensive characterization of the GMPs. The procedure for conducting a field test would then be to inform the APHIS in writing prior to planting. This proposal is a significant change from the current system and would relieve, what APHIS calls, 'unnecessary restrictions on the introduction of regulated articles based on experience'. On March 31, 1993, the final policy was published (Animal and Plant Health Inspection Service, 1993) with some modification to the notification period. Notification will be required 30 days prior to release and 10 days prior to movement of genetically-modified seed.

The FDA is involved in regulation of GMP food products intended to enter commerce, and is not directly concerned with field tests. A discussion of the FDA is included here to show the consistency among US regulatory agencies. The FDA recently published a proposed set of guidelines which are science-based for the regulatory 'approval' of foods and feeds derived from GMPs (Food and Drug Administration, 1992). The FDA proposes to use a product-based approach, that emphasizes demonstration of substantial equivalence of the food derived from a GMPs compared to that from a non-modified plant. Like the other proposed guidelines from the US agencies, there was a 90 day discussion and comment period open to the public. The FDA is presently considering these comments and changes to their proposal.

The third governmental agency involved in oversight of field tests of GMPs in the US is the EPA. Currently, the EPA regulates pesticides as defined under the Federal Insecticide, Fungicide and Rodenticide Act (FIFRA) as 'any substance or mixture of substances intended for preventing, destroying, repelling, or mitigating any pest' (Environmental Protection Agency, 1992). They have further determined that certain GMPs fit this definition and thus should be regulated under FIFRA as plant-pesticides. A plant-pesticide is defined as: 'Pesticidal substances that are produced in the plant and the genetic material necessary for the production of those substances.' (Environmental Protection Agency, 1992). Additionally, the EPA has determined that selectable markers will be regulated as inert components. Thus, cotton modified to produce *B.t.* protein and the selection marker neomycin phosphotransferase II (*npt*II), would come under the EPA regulation. The *B.t.* protein would be viewed as the pesticidal component and *npt*II as an inert. Excluded from the EPA's authority, are GMPs that have altered nutritional composition, tolerance to chemical herbicides and altered flavour or texture of the foods. However, the herbicides and other chemical pesticides used with GMPs will continue to be regulated by the EPA.

Expanded field tests of greater than 10 acres (4 ha) involving plant pesticides, require that the EPA be petitioned for an experimental use permit (EUP). The data required for an EUP are contingent on the type requested. Experiments can be carried out under 'crop destruct', if the test is purely experimental and the product will not enter commerce. But, if the product obtained were to be used in commerce, a temporary or full tolerance could be requested as well as a temporary or complete exemption from the requirement of tolerance. In considering a petition for an EUP, the EPA would review the application in light of potential adverse affects to the environment. Information requested for an EUP involves:

1. A rationale that includes the ultimate product and why one wishes to do the test.
2. A complete description of the plant, its taxonomy, etc.
3. The molecular biology and biochemistry, expression of the new protein and characterization of the gene and its organism of origin.
4. The vector and transformation method.
5. Size of the test and safeguards for the test and movement of the materials.
6. Other safeguards such as storage, handling and method of destruction.

Once the EPA is satisfied that the pesticidal plant will pose no significant adverse effect on the environment, the petition is granted.

Like the USDA and FDA, the EPA has recently proposed guidelines for the regulation of certain GMPs referred to as plant-pesticides (Environmental Protection Agency, 1992). The agency 'believes that the majority of plant-pesticides pose minimal potential for unreasonable adverse effects on humans and the environment' (Environmental Protection Agency, 1992) and their proposal clarifies the EPA's position to regulate plant-pesticides posing the greatest risks. These would have two general characteristics: '(1) higher potential to present unique exposures to non-target organisms; and, (2) because of their mechanism of action, they have a higher potential for adversely affecting non-target organisms' (Environmental Protection Agency, 1992). An emphasis would be placed on regulating components produced by plants which have a toxic mode of action and create a new exposure to non-target organisms. Components typical of a genus would not be regulated. Also, their guidelines do not include regulating plants whose mode of protection is through mechanisms involving natural barriers, such as cuticle, and expression of normal components already present in the species at comparable levels. In the case of GMPs expressing viral coat proteins, the EPA is comfortable that no new exposure or risk will be present (Environmental Protection Agency, 1992).

10.2.4 Other nations

There is consensus in the European Community for the need to organize with respect to policy toward management of the risk associated with field testing GMPs. A positive step has been made by the publication of *Council*

Directive on the Deliberate Release to the Environment of Genetically Modified Organisms (EEC, 1990) which includes GMPs. Though in proposal form, the Council Directive requests a unified EEC review process. This means that all EEC nations will require the same information to support a proposed release. Furthermore, these requests will be in one of two forms depending on the purpose of the test. The first is an experimental request and the second is for commercial utility. It is accepted that experimental releases will be significantly smaller and manageable, with lower risk and no trade issues. Thus, approval for such a release involves submission of a technical dossier to a country where the test is planned. The Competent Authority of the country where the test will be conducted will distribute the complete dossier to the full European Commission who will advise and recommend. However, the Competent Authority is empowered to make the final judgement concerning the release. Because trade issues are involved with commercial releases, the same initial process for an experimental release is involved; except now the final approval must come from the Commission. Relying first on the input of Competent Authorities who evaluate the scientific basis of the risk, the Commission seeks a consensus. As may likely be the case, unanimity might not be achieved among the Competent Authorities. Then, the decision would become the responsibility of the Commission. For a region to reject a decision of the full Commission, they must present a detailed scientific explanation supporting their position. The proposed directive also has temporal requirements imposed on the process. After submission, the Competent Authority has 90 days to either approve or request more information on an experimental release. In the case of a commercial release, one must add 3 months for the Commission to review the dossier and reach agreement. While there is consistency in policy and procedure, the release of GMPs into field tests will be subject to regional attitudes.

Japan is regarded as being cautious in its approach to regulating field tests and appears to be lagging in development of regulatory policy regarding GMPs. At the present, obtaining a field test in Japan requires data from a completely contained growth chamber study, followed by a greenhouse experiment. These experiments must be performed in Japan and the data submitted and reviewed before one could proceed with a carefully controlled field release. This first field test requires that the plot(s) be completely contained to minimize environmental risks. Pending the results of this test, one could then proceed, with approval, to a more conventional field release. The Japanese are currently drafting guidelines since they are aware that products from GMPs could soon be available and they do not want to create any trade barriers with exporting nations.

10.3 ENVIRONMENTAL ASPECTS

10.3.1 Overview

Attitudes concerning acceptance of field testing of GMPs are varied. From the discussion above, it is apparent that, while these attitudes may vary, the types of questions and scientific basis for making decisions involve answering the same fundamental questions. In all documentation supporting GMPs, three issues are discussed:

1. The rationale for developing the product, for performing the test and the potential benefits.
2. The environmental impact.
3. The impact on human and animal safety.

To comment on the environmental impact of a GMP, an environmental assessment is performed which requires information concerning:

1. The movement of the genes (pollen dispersal and outcrossing potential).
2. The trait and its effect on the plants ecology (effect on fitness in the environment, invasiveness or weediness potential, and secondary effects which are defined as unpredicted effects of the genetic engineering such as altering levels of secondary metabolites).
3. The safety of gene products toward non-target organisms and uses.
4. The potential for other untoward environmental problems (gene movement by mechanisms other than intraspecific).

This section will focus on the environmental aspects of two GMPs which have been tested for several years in field trials at many locations. The first is insect resistant cotton (IRC), and the second is a canola (rapeseed) that is tolerant to the herbicide glyphosate, glyphosate-tolerant canola (GTCanola). Each discussion will address the risk of performing field tests. Also included will be a discussion of benefits since these must be weighed into any environmental assessment.

Before discussing the two examples, it is worth mentioning three topics that are broadly related to the environmental aspects of GMPs. First, the possibility that genes can move throughout the biota in a random, interspecific manner has been a source of lively debate (point 4 of the environmental assessment). While it is possible that in the course of evolution this direction of gene movement may have occurred, all evidence indicates that movement of genes from plants to microorganisms or to unrelated species of plants is improbable (Goodman and Newell, 1985). Many putative examples of gene flow even between compatible species can be explained by convergent evolution. This random gene movement remains an intellectual curiosity until experimental evidence can prove otherwise. Second, most genetic changes reduce rather than increase fitness (Crawley, 1992). The wild-type genotypes are usually more fit, better competitors, and hence, are more weedy. Since most genetic changes compromise the plant in some environmentally significant way, it is not expected that a GMP will have a greater potential of being a pest than any commercial cultivar obtained through traditional breeding and selection. Last, there has never been evidence of crown gall in plants derived from disarmed *Agrobacterium tumefaciens* mediated transformations (IRC and GTCanola) (Huttner *et al.*, 1992). This indicates that the organism is efficiently eliminated in tissue culture and is not propagated through the seed.

10.3.2 IRC

Much literature is available on the taxonomy, genetics, mode-of-reproduction and outcrossing potential for cotton. Commercially cultivated cotton (*G. hirsutum* and *G. barbadense*) is an allotetraploid unable to outcross to the two wild relatives of cotton that occur in the US, *G. thurberi* in Arizona and *G. tomentosum* in Hawaii. Furthermore, *G. tomentosum* is morphologically and temporally incompatible with commercial cotton varieties. The data strongly support that there is no reasonable mechanism for outcrossing of IRC genes into wild cotton relatives.

Weediness characteristics are well accepted to be multi-genic traits (Keeler, 1989). Attributes of the seed such as quantity, dispersal, dormancy, long soil persistence and ease of germination are not characteristic of cotton. The contrast is further enhanced by the fact that cotton has slower vegetative growth and a longer life cycle than one would associate with weediness. On the other hand it is evident that IRC could be more competitive than a non-modified cotton due to its resistance to damaging insects. This trait would not be expected to enhance weediness since, as noted above, cotton does not volunteer successfully in most of the cotton growing regions, outcrossing to weedy relatives is not a realistic possibility and cotton is not currently considered a weed problem in other crops.

Secondary effects of the genetic engineering have been assessed and IRC plants are unchanged relative to the control plants in terms of natural components such as gossypol expressed in the foliage. A change in the levels of such components could affect the susceptibility of IRC to plant pests. In addition, field observations have shown no differences in susceptibility to common fungal and viral diseases. An interesting benefit of this technology was the reduction in aflatoxin levels as a consequence of controlling *Pectinophora gossypiella* (pink bollworm), one of the target insect species and a vector for *Aspergillus flavus*.

Cotton has been modified to express the well characterized *Bacillus thuringiensis* ssp. *kurstaki* (*B.t.k.*) protein which is specific for control of lepidopteran pests such as cotton bollworm and pink bollworm. It has been shown that the protein expressed in the plant and that found in the commercial formulation Dipel® are indistinguishable based on specificity and selectivity (Fuchs *et al.*, 1993).

Like Dipel®, IRC does not harm beneficial insects and has shown no negative effects in feeding studies both with pure protein (gavage) and IRC seed or meal incorporated into diets. Experiments were also performed where the protein in a plant powder was added to soil and a degradation half-life was measured. The values obtained were comparable to published values for the microbial product; thus demonstrating that the gene product will not persist in the soil. Unlike Dipel®, IRC offers the benefit of season long control without repeated applications and reduced use of insecticides. It appears to have an excellent fit for Integrated Pest Management (IPM) systems.

The data for IRC were presented to the EPA in a request for an EUP which was granted (Federal Register, 1991).

10.3.3 GTCanola

The term Canola is a registered trademark in Canada and refers to rapeseed varieties and their processed components that meet standards for low levels of erucic acid and glucosinolates. Two members of the *Brassica* family (*B. napus,* also known as Argentine rapeseed, *B. campestris* a.k.a. *rapa,* also referred to as Polish canola) have been cultivated in Canada since the 1940s for their oil, though the defatted meal is used as an animal feed and a fertilizer. A third species, *B. juncea,* is being bred for lines with acceptable levels of erucic acid and glucosinolates. In particular, GTCanola is derived from a *B. napus* variety, Westar.

Many species of *Brassica* grow over a wide geographical range which extends across Europe and in Canada from central Ontario to northwestern Alberta. (Downey *et al.*, 1991) have reported results of extensive experiments designed to assess the potential of gene transfer among the important weeds and species involved in canola production. Controlled and natural interspecific crosses were performed among four *Brassica* species (*B. napus, B. campestris, B. juncea* and the weedy *B. nigra*) and *Sinapis arvensis* (*B. kaber*). The last species was of particular importance since it is a wild mustard that is recognized as a weed problem in canola. When *S. arvensis* was reciprocally crossed with *B. napus* and *B. campestris* under ideal conditions, no hybrid seed were obtained. Equally interesting was that only one seed was produced after performing 1881 backcrosses to *S. arvensis* and the plant produced was sterile. When reciprocal crosses were performed with *B. juncea* and *S. arvensis,* 2.5 hybrid seeds were obtained per 100 buds pollinated, though the cross *S. arvensis* × *B. juncea* produced no hybrid seed. These authors concluded that 'gene transfer from the three major oil crop species *B. napus, B. campestris,* and *B. juncea,* [to *S. arvensis*] was not achieved even under the most favorable conditions, and no hybrids were identified from natural crossing of these species when they were co-cultivated in field plots over a three year period (Downey *et al.*, 1991). An extremely remote possibility of a bridging mechanism was identified. Crosses of *B. juncea* to *B. nigra* were found to be 0.5% under controlled conditions. Also under these conditions, *B. nigra* was determined to cross to *S. arvensis* with a compatibility of 7%. This would give rise to a mathematical possibility of outcrossing. However, in observing 150 *B. nigra* plants under natural conditions, Downey saw no evidence of hybrid production with *B. juncea*. Given also that *B. nigra* is not typically found in cultivated areas and is not widely distributed in the major canola producing regions, it is reasonable to conclude that outcrossing possibilities are essentially zero. Downey concluded their study with the following: 'It is concluded from these controlled and natural crossing experiments that although gene transfer among the oilseed *Brassicas* under natural conditions can and probably does occur, the natural barriers for such gene flow to the weedy species, *B. nigra* and *S. arvensis*, is formidable and would not occur' (Downey *et al.*, 1991).

Downey has also studied the movement of pollen from *B. napus* and *B. rapa* under natural and artificial conditions over a two year period (Downey, 1992). Using a recessive mutant strain as an indicator, he was able to determine outcrossing as a function of distance. The results showed that *B. napus* outcrosses at 2.1, 1.1 and 0.6% at distances of 46, 137 and 366 m, respectively, while the self-incompatible *B. rapa* was noted to be 8.5, 5.8 and 3.7% at the same distances. Furthermore, their results indicate that wind does not effectively transfer pollen from either *Brassica* species. These facts, taken into account with the interspecific compatibility information above, do not support any contention that traits are easily transferred into weedy relatives and present a significant risk to the environment. Additional valuable information related to the pollen movement of herbicide tolerant rapeseed and other GMPs will be coming from the Planned Release of Selected and Modified Organisms (PROSAMO) program in the UK in 1993.

The second point to address in the environmental assessment is whether the trait will confer a special advantage to a plant that might give rise to a weediness problem. Crawley *et al.* (1993) have recently addressed this matter in a study sponsored by PROSAMO. They compared the demography of kanamycin and glufosinate-tolerant rapeseeds (*B. napus* subsp. *olifera* var. Westar) with non-modified rapeseed to assess the effect of genetic engineering on invasiveness potential. Separate experiments were conducted to determine seed dormancy, germination, plant survival and fecundity over 3 years at sites in Great Britain in 12 different habitats. The results of the experiments were then used to calculate a finite rate of increase (λ) of a plant population. Based on their analysis, Crawley concluded that 'there is no evidence that rape is invasive of undisturbed natural habitats, and no evidence that transgenic lines of rape are more invasive of, or more persistent in, disturbed habitats than their conventional counterparts' (Crawley *et al.*, 1993). The authors did caution that their data are specific to herbicide tolerance in rapeseed (canola) and that the risks of other traits and constructs must be based on a case-by-case basis.

From an agronomically practical perspective, canola is susceptible to many registered herbicides. Current practice to remove volunteers from wheat or barley in subsequent rotations is 2,4-dichlorophenoxyacetic acid application. The presence of the glyphosate-tolerance trait will not interfere with nor abrogate the current practices in controlling volunteers and should not afford a competitive advantage in terms of weediness over non-modified canola.

Secondary effects for GTCanola have been addressed through proximate and amino acid analyses, measurements of the levels of glucosinolates and fatty acid composition. All studies completed to date have shown that the GMP is substantially equivalent to the commercial cultivar Westar. All the data are within the acceptable ranges for plants derived from this variety. Furthermore, when evaluating lines with commercial levels of tolerance, no evidence of phenotypic variation indicative of pleiotropic or secondary effects was seen. Data on flowering and yields are all indicative of the fact that a single trait has been conferred and stably integrated.

At this time several studies need to be completed that will ensure the safety of the gene and transformation products. Feeding studies to assess the safety of the proteins are being completed as well as an wholesomeness assessment of the seed from a GMP relative to a non-modified canola seed.

10.4 CONCLUSIONS AND FUTURE PROSPECTS

Significant progress is being made towards harmonization of regulatory oversight of GMPs. Agencies throughout the world are aligning with a case-by-case regulation of GMPs using a science-based approach to granting approvals for field tests. Industry, academia, government and public interest groups are committed to ensuring the human, animal and environmental safety of products from biotechnology. Furthermore, the science-based, risk-based approach is the accepted method to assess these plants and their products. Based on a rapidly increasing database of successful field releases, some conclusions and recommendations are clear. Firstly, GMPs behave substantially the same as non-modified parental lines in terms of ecology in over 400 field trials worldwide. All the evidence indicates that species, not previously inclined toward weediness, will not develop into pests due to the introduction of traits such as herbicide tolerance or insect resistance. Secondly, the various safeguards used by plant breeding programs, protecting against unwanted gene flow and the current risk assessment methods are sufficient. Thus, all evidence to date clearly supports the contention that GMPs, exemplified by IRC and GTCanola will be safe for release and field testing.

ACKNOWLEDGEMENTS

We are thankful for the help and critical input from Robert Ingratta, and Drs Frank Serdy, Steve Rogers and Robb Fraley.

REFERENCES

Animal and Plant Health Inspection Service (1987) Plant Pests: Introduction of Genetically Engineered Organisms or Products; Final Rule. *Federal Register*, **52**, 22892–915.

Animal and Plant Health Inspection Service (1991) *Questions and Answers on Biotechnology Permits for Genetically Engineered Plants and Microorganisms*, United States Department of Agriculture, APHIS 21-35-001, pp. 1–5.

Animal and Plant Health Inspection Service. (1992a) Genetically Engineered Organisms and Products; Notification Procedures for the Introduction of Certain Regulated Articles; and Petition for Nonregulated Status. *Federal Register*, **57**, 53036–43.

Animal and Plant Health Inspection Service (1992b) Proposed Interpretive Ruling in Connection with Calgene, Inc. Petition for Determination of Regulatory Status of Flavr Savr® Tomato. *Federal Register*, **57**, 31170–1.

Animal and Plant Health Inspection Service (1992c) Proposed Interpretive Ruling in Connection with Upjohn Company Petition for Determination of Regulatory Status of ZW020 Virus Resistant Squash. *Federal Register*, **57**, 40632–3.

Animal and Plant Health Inspection Service (1993) Genetically Engineered Organisms and Products: Notification Procedures for the Introduction of Certain Regulated Articles; and Petition for Nonregulated Status; Final Rule., *Federal Register*, **58**, 17044–59.

Bureau of National Affairs (1992) OECD Group of National Experts, *World Food Regulation Review*, Reach Agreement on Food Safety in Biotechnology Report. ©The Bureau of Nation Affairs, *World Food Regulation Review*, 11–2.

Casper, R. and Landsmann, J. (eds) (1992) *Proceeedings of the 2nd International Symposium on The Biosafety Results of Field Tests of Genetically Modified Plants and Organisms*, Biologische Bundesanstalt fur Land- und Forstwirtschaft, Braunschweig, Germany, pp. 1-296.

Crawley, M. (1992) The Comparative Ecology of Transgenic and Conventional Crops, in *Proceedings of the 2nd International Symposium of The Biosafety Results of Field Tests of Genetically Modified Plants and Organisms*, (eds R. Casper and J. Landsmann), Biologische Bundesanstalt fur Land- und Forstwirtschaft, Braunschweig, Germany, pp. 43–52.

Crawley, M.J., Hails, R.S., Rees, M., Kohn, D. and Buxton, J. (1993) Ecology of transgenic oilseed rape in natural habitats. *Nature*, **363**, 620–3.

Downey, R. K., Bing, D. J. and Rakow, G. F. W. (1991) Potential of gene transfer among oilseed *Brassica* and their weedy relatives, in *Proceedings of the 8th International Rapeseed Congress*, Saskatoon, Canada, pp. 1022–7.

Downey, R.K. (1992) Biosafety of transgenic oilseed *Brassica* species, in Proceedings of the 2nd *International Symposium on The Biosafety Results of Field Tests of Genetically Modified Plants and Organisms*, (eds R. Casper and J. Landsmann), Biologische Bundesanstalt fur Land- und Forstwirtschaft, Braunschweig, Germany, pp. 17–21.

EEC (1990) Council Directive of 23 April 1990 on the Deliberate Release into the Environment of Genetically Modified Organisms (90/220/EEC), *Official Journal of the European Communities*, **L 117**, 15–27, 8.5.90.

Environmental Protection Agency (1992) EPA Proposal to Clarify the Regulatory Status of Plant-Pesticides. *Federal Regulations* **57**, 55531–59.

Federal Register (1991) Notification of EUP for Insect Resistant Cotton. *Federal Register*, **56**, 65073–4.

Food and Drug Administration (1992) Foods Derived from New Plant Varieties. *Federal Register* **57**, 22984–23007.

Fuchs, R.L., Berberich, S.A. and Serdy, F.S. (1993). Safety evaluation of genetically engineered plants and plant products, in *Biotechnology and Toxicology Assessment*, (eds J.A. Thomas and L.A. Myers), Raven Press, New York, pp. 199–212.

Gasser, C.S. and Fraley, R.T.(1989) Genetically engineering plants for crop improvement. *Science*, **244**, 1293–9.

Ginzburg, L.R. (ed.) (1991) *Assessing Ecological Risks of Biotechnology*. Butterworth-Heinemann, Oxford, pp. 1–379.

Goodman, R.M., and Newell, N. (1985) Genetic Engineering of Plants for Herbicide Resistance: Status and Prospects, in *Engineered Organisms in the Environment: Scientific Issues*, (eds H.O. Halvorson, D. Pramer and M. Rogul), American Society for Microbiology, Washington, DC, pp. 47–55.

Halvorson, H.O., Pramer, D. and Rogul, M. (eds) (1985) *Engineered Organisms in the Environment: Scientific Issues*. American Society for Microbiology, Washington, D. C. pp. 1–239.

Health and Welfare Canada (1992) *Information Letter, Health Protection Branch*, IL No. 806, August 5, pp. 1–9.

Huttner, S.L., Arntzen, C., Beachy, R., Breuning, G., De Francesco, L., Nester, E., Qualset, C. and Vidaver, A. (1992) Revising oversight of genetically modified plants, *Bio/Technology*, **10**, 967–71.

Kalous, M.J. and Duke, L.H. (1989) *The Regulation of Plant Biotechnology in Canada; Part 2, The Environmental Release of Genetically Altered Plant Material*, Agriculture Canada, Seed Division, Ottawa, Canada, pp. 1–63.

Keeler, K.H. (1989) Can genetically engineered crops become weeds? *Bio/Technology*, **7**, 1134–9.

National Research Council (1989) *Field Testing Genetically Modified Organism*, National Academy Press, Washington, DC, USA, pp. 16–76.

Environmental and regulatory aspects of using genetically-modified microorganisms in the field

L.A. Glover

11.1 INTRODUCTION

Advances in the techniques of molecular biology over the last two decades are such that genetically-modified microorganisms (GMMs), a category which includes viruses, bacteria and lower eukaryotes, are being used commercially in contained systems and are also being developed for intentional release into the environment. There are numerous potential applications of GMMs in the environment and these include crop protection (Blakeman and Fokkema, 1982; Lindow, 1983; Lindow *et al.*, 1983), biological control (Schippers, 1988; Bishop *et al.*, 1988), bioremediation (degradation of chemical wastes) (Jain and Sayler, 1987) and extraction of metal from ores (Lindow *et al.*, 1989). Similarly, genetically-modified plants are being developed for agricultural applications (see Chapters 5, 6 and 7). However, the deliberate release of GMMs differs fundamentally from these other cases as it is much more difficult to gauge their likely environmental impact because of the problems inherent in monitoring the organisms in the environment and assessing their interaction with other soil or aquatic biota. Indeed, if an unforeseen hazard was realized after the deliberate release of a GMM, it would be a major and perhaps impossible task to implement a recovery operation whereas this would be a feasible option in the case of genetically-modified plants or animals although dissemination of recombinant DNA via the pollen from transgenic plants may pose similar

Molecular Biology in Crop Protection. Edited by George Marshall and Dale Walters. Published in 1994 by Chapman & Hall, London. ISBN 0 412 54390 7(HB) 0 412 54400 8(PB).

difficulties (see Chapter 10). This fundamental difference has had important implications for the formulation of guidelines by the regulatory authorities responsible for granting licences for deliberate releases of GMMs.

In order that the risks of releasing GMMs into the environment can be assessed we need to be able to address some fundamental questions about the ecology of microorganisms in the environment. We need to know what factors affect their growth, survival, activity and movement through the environment. Information is also necessary on the impact the introduced GMMs will have on the indigenous biota in terms of populations and community structures, the persistence of the recombinant DNA and its transfer to the indigenous micro flora. For any of these questions to be rigorously addressed, it is necessary to be able to detect and quantify reliably concentrations of total, viable, culturable and non-culturable GMMs, to measure their actual and potential activity and to determine the presence and expression of recombinant DNA both in the introduced microorganism and in the indigenous population. The key to obtaining this information has been the development of techniques which allow the detection of the introduced GMMs in the environment. Such has been the importance of developing this technology that there have been a number of international initiatives directed at developing enabling technologies to allow the detection and risk assessment of GMMs deliberately released into the environment. One of the most prominent of these was the UK based PROSAMO Initiative (Programmed Release of Selected and Modified Organisms) which was jointly sponsored by UK government departments, research councils and multinational companies and promoted research into detection and risk assessment of the deliberate release of genetically-modified bacteria and plants (Killham, 1992). The experimental data from this and similar research has contributed to the formulation of effective guidelines for the use of recombinant organisms in the environment.

11.2 DETECTION OF INTRODUCED BACTERIA

There is no microbial detection technique which is perfectly suited to comprehensive risk assessment studies and so the combined use of traditional and molecular techniques is essential.

Traditional microbial detection techniques, which mainly require cultivation on laboratory media, are inadequate as they cannot detect non-culturable cells and are very selective depending on the choice of culture media and growth conditions. In addition, the efficiency of extraction of GMMs from environmental samples varies between organisms and sample type and the harsh methods required often reduce cell viability. Similarly, traditional immunological techniques are able to detect microorganisms but do not offer any information on their viability or activity. These deficiencies have stimulated the development of new methodologies, in particular molecular techniques, which offer the sensitivity and selectivity required to monitor GMMs and their recombinant DNA in the environment.

These techniques fall into three distinct groups: immunological methods, nucleic acid probing and molecular markers. Traditional immunological techniques have been enhanced by improvements in production of polyclonal and monoclonal antibodies, development of enzyme-linked immunosorbent assays (ELISA) and the potential application of flow cytometry which can distinguish and sort cells on the basis of their interaction with specific antibodies labelled with different fluorescent dyes. These immunological methods can provide information on total cell numbers but still cannot be used to assess viability or activity. Probing techniques involve the detection of specific DNA or RNA sequences and sensitivity has been greatly increased with the application of the polymerase chain reaction (PCR) to amplify the target sequences (Steffan and Atlas, 1991). The PCR has also been used more recently to attempt to quantify the persistence of introduced genetic material in soil (Romanowski *et al.*, 1993). This study suggested that genetic material may persist in soil for weeks and even months after its release from cells. In addition, this DNA in plasmid form was able to enter viable bacteria and be expressed indicating that it was still biologically active. Nucleic acid probing is very sensitive but tracks the DNA and not necessarily the GMM inoculum. The application of molecular markers has provided significant advantages for risk assessment studies in comparison with traditional methodologies and other molecular techniques. Such marker systems must incorporate several requirements:

1. The marker gene should not display any significant expression in the indigenous population.
2. It should be stably maintained in the GMM host (this usually means chromosomally located).
3. Its expression should not impose any significant metabolic burden on the host

A number of suitable marker genes have been employed although only a few have been shown to be appropriate for detection in the environment. These include resistance to antibiotics, heavy metals, *lac*ZY (encoding β-galactosidase), *xyl*E (encoding catechol-2,3-dioxygenase) and *lux* (encoding luciferase), each of which offer different possibilities for detection. The use of other marker genes such as *gus* (encoding β-glucuronidase) (Jefferson, 1989) have limited application for environmental detection due to the unacceptable background of indigenous *gus*$^+$ microorganisms. The use of *lux* genes has proved to have particular advantages in that, unlike the other marker systems, it is possible to monitor bioluminescence from environmental samples without prior extraction of the bacteria (Rattray *et al.*, 1990) and the sensitivity of bioluminescence detection allows the monitoring of single cells in soil (as illustrated on the front cover, after Silcock *et al.*, 1992) and microcolony detection on plant leaves (Waterhouse *et al.*, 1993). In addition, samples can be monitored non-destructively allowing time courses to be studied.

A variety of methodologies have been employed to introduce marker genes into the host GMM. The most straightforward is introduction of marker genes on a plasmid, e.g. the introduction of *lac*ZY genes into fluorescent pseudomonads (Drahos *et al.*, 1986), which can then be introduced into the host by a variety of methods including, transformation, electroporation and conjugation. Multicopy plasmids can be used which offer higher sensitivity of detection. However, this system has several drawbacks such as the imposition of a heavy genetic burden on the host cell thereby reducing its competitiveness in the environment, loss of the plasmid in the absence of a positive selection pressure or variation in plasmid copy number in response to changing environmental parameters both making quantification difficult and transfer of the plasmid to the indigenous population. This method of genetically marking a microorganism is of greatest use in studies on factors affecting transfer of DNA in the environment.

A more stable alternative for the introduction of marker genes is to incorporate them onto the chromosome of the host cell so that they are replicated with the host DNA in the absence of selection pressure. This can be done using dedicated integration vectors which cannot replicate in the intended host and either introduce the reporter genes to a specific site in the chromosome via a mechanism of homologous recombination (Cook *et al.*, 1993) or exploit suitably disabled transposon sequences designed to mediate stable integration into the chromosome (Herrero *et al.*, 1990; de Lorenzo *et al.*, 1990). In these cases, the marker genes are normally introduced as a single copy and this can reduce the sensitivity of detection. However, this small disadvantage is more than compensated by the stability of the cloned DNA.

The use of antibiotic resistance markers has been very common as they allow direct selection of the marked population by plating on antibiotic-containing medium which will select against background populations. This is particularly useful for detection and enumeration of marked organisms present at relatively low concentrations. However, there is a general move away from the use of antibiotic resistance genes as concern was voiced over the introduction of antibiotic resistance genes into the environment. The basis for this concern is not clear as there is a considerable background of antibiotic resistant bacteria present in the environment which can at times limit their usefulness for direct selection.

All of these molecular markers have been used successfully to monitor the presence of GMMs in environmental samples and details of each system can be found in some excellent review articles (Pickup, 1991; Prosser, 1994).

11.3 ECOLOGICAL IMPACT OF GMM RELEASES

There are a number of ecological concerns raised when considering the proposed release of a GMM. These include whether the microbial community structure and function will be disturbed by the introduced inoculum and whether the recombinant DNA will be transferred into the indigenous community.

Our knowledge of microbial ecology is limited although improving all the time and part of the impetus for this increased level of interest arises from the proposed deliberate release of genetically-modified or selected organisms into the environment. With the molecular tools described above in place, meaningful studies can be made on the impact an introduced inoculum will have on the indigenous microbial community. This has been tackled both by the use of contained microcosm and greenhouse experiments which can be conducted under controlled conditions and limited field introductions, both of which can be designed to answer specific questions about microbial ecology. Such field release experiments have been relatively few in number and this perhaps reflects the cautious approach to this subject taken by the regulatory authorities worldwide. As an example, in the US, by 1992, there had been only a handful of GMM releases including *Pseudomonas aureofaciens* modified to express the *Escherichia coli lac*ZY genes (Drahos *et al.*, 1986; Kleupfel *et al.*, 1991), isolates of *P. syringae* and *P. fluorescens* with specific deletions of the ice nucleation gene referred to as ice-minus (Lindow and Panopoulos, 1988; Supkoff *et al.*, 1988), a strain of *Clavibacter xyli* modified to express the *Bacillus thuringiensis* δ-endotoxin gene (Kostka *et al.*, 1988) and bioluminescent variants of *Xanthomonas campestris* pv. *campestris* (Shaw *et al.*, 1991, 1992a). Many of these releases were designed to add to our knowledge about the ecology of bacteria, e.g. the use of *lac*ZY genes to allow the environmental detection and monitoring of *P. aureofaciens*, and the use of *lux*-marked *X. campestris* to allow an investigation of the spread of a plant pathogen. Some involved the small-scale release of bacteria designed to protect crop plants such as potatoes and strawberries against damage from frost such as the release of *P. syringae* and *P. fluorescens* which had been geneticially modified to remove their genes for ice-nucleation. These modified strains of ice-nucleation-defective bacteria could then colonize plants and prevent the growth of ice-nucleation$^+$ bacteria which would cause frost damage on crops. Similarly, the *B. thuringiensis* δ-endotoxin gene was introduced into *C. xyli*, an endophyte of grasses, to act as a biocontrol agent against the European corn borer (see Chapter 7).

In contrast, during this period, over 190 different field tests of genetically-modified plants had been performed. A similar situation exists in the UK where there have been approximately 50 release experiments from 1986 to 1993 and again, the majority of these are of transgenic plants.

11.4 REGULATIONS GOVERNING GMM RELEASES

In general terms, the regulatory authorities worldwide have adopted a case-by-case approach to each application for permission for a release although the regulations remain different for individual countries. The main safety considerations are different from those already in place to deal with the contained use of genetically-modified organisms as, by their very nature, deliberately released microorganisms will be designed to persist in the environment. There have been a number of international conferences held to address the issues surrounding deliberate release such as REGEM 1 (Release of Genetically Engineered Micro-organisms) in 1988 (Sussman *et al.*, 1988) and REGEM 2 in 1991 (Stewart-Tull and Sussman, 1992) as well as a number of research programmes directed at biotechnology and risk assessment such as the EC Biotechnology Action Programme (BAP) (Economidis, 1990) and the PROSAMO Initiative (Killham, 1992). The decision of regulatory authorities to adopt a cautious approach to deliberate release has been due in part to the inability of scientists to predict the exact outcome of a release experiment and is also a reflection of a general public concern that the release of GMMs may have long-term environmental or health risks (Kemp, 1992). However, in both the UK (Steele, 1993) and the US (Shaw *et al.*, 1992b), the legislation governing the release into the environment of GMMs has recently been updated and simplified.

11.4.1 Current United Kingdom regulations

It is not within the scope of this chapter to address current legislation worldwide on the deliberate release of GMMs and so it is most useful to address one or two specific examples.

In the UK, prior to release, permission must be sought from the 'Secretary of State' (Secretary of State for the Environment in England, Secretary of State for Scotland or Secretary of State for Wales). In practice, the 'Secretary of State' is advised by an independent committee, the Advisory Committee on Releases to the Environment (ACRE) which comprises of scientific experts and representatives of various interests, such as industry and environmental groups, e.g. the Countryside Commission. The Department of the Environment (DoE), are now responsible for coordinating the consideration of all release applications, and they are responsible for liasing with other involved government departments such as the Health and Safety Executive (HSE) and the Ministry of Agriculture, Fisheries and Food (MAFF). A channel of communication has also been established between the DoE and the Commission of European Communities. A long-term aim is to establish common guidelines and regulations for deliberate release within Europe. This would be a rational progression as microorganisms do not observe national boundaries during release experiments and it would also have obvious benefits for commercial releases by trans-European companies.

Application is now made on a standard form and the amount of information requested about the organism and the recombinant DNA being used varies depending upon the degree of risk associated with the release and the purpose of the release. For example, no release permit would be granted if the release organism was to be *Escherichia coli* genetically-modified to express the toxin ricin (from plants), as ricin is lethal to humans and this would constitute an unacceptable risk for human health. This would contrast with the proposed release of a strain of *Rhizobium* genetically-modified via a natural *Rhizobium* plasmid DNA molecule to be resistant to an antibiotic. The presence of the antibiotic resistance would allow the modified organism to be monitored in the environment and information could be gathered on the ecology of the bacterium and how this related to its ability to fix atmospheric nitrogen. A release such as this would pose no environmental or human health hazard and would provide valuable ecological information. The application for consent for release asks for a wide variety of information including the characteristics of the donor, parental or recipient organisms, any phenotypic or genetic markers, how the organism can be identified, how sensitive those detection techniques are, information on the survivability and pathogenicity of the organism and involvement of the organism in environmental processes such as nutrient turnover and decomposition of organic matter and respiration. The complete regulations can be found in the DoE booklet, *The Regulation & Control of the Deliberate Release of Genetically Modified Organisms* (DoE, 1993).

11.4.2 Regulations in Australia

Similar guidelines are in place for the deliberate release of GMMs in other countries, e.g. Australia (*Genetic Manipulation: the Threat or the Glory?*, 1992). In this Australian government publication, consideration has been given to the potential benefits of the release of GMMs into the environment. An example of such a release with a positive environmental impact was the proposal to genetically-modify a strain of the nitrogen-fixing bacterium, *Rhizobium*, to make it fungicide resistant. The effect of this was to protect these nitrogen-fixing bacteria from fungicide damage when crops were sprayed to control fungal root infections. The use of these bacteria would decrease the need for Australian farmers to use nitrogenous fertilizers (which can lead to high levels of nitrite and nitrate in the drainage water) on pastures (Friend, 1991).

The regulations themselves operate in a very similar way to those in the UK. For each proposed release of GMMs, a range of information is requested including a justification for the release outlining why other methods (especially those not involving a release) are inferior, details of the genetic modification and its stability, the known effects of the unmodified organism as well as the modified organism on human, animal and plant health, agricultural productivity and the environment, evidence of the persistence, viability and dispersal of the modified organism, details of the release experiment and monitoring arrangements, and what contingency plans were in place to deal with environmental extremes such as floods during the release and control methods to attempt to eliminate the organism *post* release. Additional information is sought depending upon the nature of the modified organism or its end use. Special categories include, live vaccines, microorganisms associated with plants, microorganisms associated with animal (e.g. ruminants), microorganisms to be used in biological or pollution control and microorganisms to be used in food. For each of these categories, the proposer has to answer 23 questions with two to 13 additional questions depending upon the category of organism being assessed. The proposal is then considered by the Institution's own biosafety safety committee (IBC) which then sends the endorsed proposal to the Genetic Manipulation Advisory Committee (GMAC). The GMAC, in contrast to ACRE in the UK, consists predominantly of scientific experts appointed by government for a three year period with no appointments of persons from groups with special interests such as environmentalists and industrialists. The GMAC considers the proposal bearing in mind the criteria listed in the GMAC guidelines and prepares a report which is then sent back to the IBC and on to the appropriate regulatory authority, e.g. the Department of Health. It is this regulatory authority which has final responsibility for approving the release and it may seek and consider the opinion of special groups at this point. Adherence to these guidelines is voluntary although sanctions, mostly of a financial nature, may be enforced following breach of the guidelines.

11.4.3 Regulations in other countries

Most developing countries do not have specific regulations covering the production or release of GMMs and there is at least one documented case, involving the release of a genetically-modified virus, where this has led to a great deal of controversy. In 1986, an agreement was reached between the Wistar Institute, Philadelphia, USA and the Pan American Health Organization (PAHO) to conduct an experiment designed to test a genetically-modified rabies vaccine in cattle at an experimental farm operated by the Pan American Zoonoses Centre (CEPANZO) in Argentina. This experiment was terminated in September 1986 by Argentina's sanitary authorities and they destroyed the animals involved in the experiment. The allegation was made that the experiment was undertaken without the permission and knowledge of the Argentine authorities or scientific community. This incident, involving the use of a genetically-modified vaccinia virus, was discussed at the REGEM 2 conference (Sussman *et al.*, 1988). The basis of the complaint was that the PAHO staff imported the recombinant virus into

Argentina in a diplomatic bag and thereby circumvented Argentine regulations about the import of exotic microorganisms. In addition, the experiment itself and how it was managed gave rise to grave concerns for the health and safety of the workers involved and the potential ecological impact of the vaccinia virus.

Given all the serious concerns surrounding this incident, it still remains the case that this experiment would not have been allowed to proceed in a country where there are established guidelines and regulations. There have been releases of recombinant viruses in the UK (see Chapter 3) and the regulations and guidelines for recombinant viral release are the same as for any other microorganism. There are additional questions relevant to the risk assessment of the deliberate release of a recombinant virus. For example, the addition of foreign DNA might affect the host range of the recombinant virus and there might be gene transfer of the foreign DNA via recombination between viruses in the environment resulting in a virus with a different host range which could disseminate the foreign DNA. These questions are all addressed by current regulations which have been discussed here.

11.5 CONCLUSIONS AND FUTURE PROSPECTS

In general, it is the industrial concerns worldwide which are the most anxious to have legally enforceable guidelines in place. The impetus for this comes from their interest in having a responsible public image and also having a predictable and efficient regulatory framework within which to invest in programmes for deliberate release of GMMs.

No one could doubt the value of having carefully considered guidelines for the release of GMMs into the environment. Current guidelines have been formulated with due regard to our increasing knowledge of the ecology of microorganisms in the environment and many people see deliberate releases of GMMs as having a positive rather than just a neutral impact on the environment. Concerns about the transfer of 'introduced' genetic material from the GMMs to the indigenous population are being addressed as scientists provide more data on events such as the frequency of gene transfer in the environment. It is worthwhile remembering in this context that the environment provides a great natural opportunity for the exchange of genetic information and only where genetic manipulation crosses the 'natural' barriers of this genetic transfer would it be realistic to assume that the introduction of recombinant genetic material via a GMM poses any greater risk. As our knowledge of microbial ecology is increased and refined, the guidelines and regulations can evolve to address the relevant areas.

The great issue still to be addressed is how to provide the public at large with enough information to ensure that they are satisfied with the safety and purpose of proposed deliberate releases. In many countries, political issues are a major determinant of whether permission is given for the deliberate release of GMMs. With the increasing popularity of environmental 'Green' parties much political lobbying is being targeted against the use of GMMs. It is important to stress here that a balanced view must be taken on the likely environmental impact of the deliberate release of GMMs. Many GMMs are being designed to replace chemical pesticides or to deal with environmental disasters such as oil spills using biological means instead of adding to the pollution and environmental damage by extensive use of chemical dispersants.

However, at the end of the day the burden of gaining favourable public opinion regarding the use of GMMs in the environment must lie with both the scientists and the legislators.

ACKNOWLEDGEMENT

I would like to thank Dr Firoz Amijee, Department of the Environment for many useful discussions regarding the regulations governing release of GMMs.

REFERENCES

Bishop, D.H.L., Entwistle, P.F., Cameron, I.R., Allen, C.J. and Posse, R.D. (1988) Field trials of genetically-engineered baculovirus insecticides, in *The Release of Genetically-Engineered Micro-organisms* (eds M. Sussman, C. Collins, F. Skinner and D. Stewart-Tull), Academic Press, London, pp. 143–79.

Blakeman, J. and Fokkema, N.J. (1982) Potential for biological control of plant diseases on the phylloplane. *Annual Review of Phytopathology*, **20**, 167–92.

Cook, N., Silcock, D.J., Waterhouse, R.N., Prosser, J.I., Glover, L.A. and Killham, K. (1993) Construction and detection of bioluminescent strains of *Bacillus subtilis*. *Journal of Applied Bacteriology*, **75**, 350–9.

de Lorenzo, V., Herrero, M., Jakubzik, U. and Timmis, K.M. (1990) Mini-Tn5 transposon derivatives for insertion mutagenesis, promoter probing, and chromosomal insertion of cloned DNA in Gram-negative bacteria. *Journal of Bacteriology*, **172**, 6568–72.

DoE (1993) *The Regulation & Control of the Deliberate Release of Genetically Modified Organisms*, Department of the Environment, Publication Sales Unit, Block **3**, Spur 2, Room1/2, Government Buildings, Lime Grove, Eastcote, UK.

Drahos, D.J., Hemming, B.C. and McPherson, S. (1986) Tracking recombinant organisms in the environment: β-galactosidase as a selectable non-antibiotic marker for fluorescent pseudomonads. *Bio/Technology*, **4**, 439–44.

Economidis, I. (1990) *Biotechnology R&D in the EC: Biotechnology Action Programme (BAP) Parts I&II*, Commission of the European Communities, Brussels.

Friend, J. (1991) Submission for a Microbial Release to the Australian Government, Submission 72, p. 6.

Genetic Manipulation: The Threat or the Glory? (1992) Report by the House of Representatives Standing Committee on Industry, Science and Technology, The

Parliament of the Commonwealth of Australia, Australian Government Publishing Service, Canberra.

Herrero, M., de Lorenzo, V. and Timmis, K. (1990) Transposon vectors containing non-antibiotic resistance selection markers for cloning and stable chromosomal insertion of foreign genes in Gram-negative bacteria. *Journal of Bacteriology*, **172**, 6557–67.

Jain, R.K. and Sayler, G.S. (1987) Problems and potential for *in situ* treatment of environmental pollutants by engineered microorganisms. *Microbiological Science*, **4**, 59–63.

Jefferson, R.A. (1989) The GUS gene reporter system. *Nature*, **342**, 837-38.

Kemp, R (1992) Social implications and public confidence: risk perception and communication, in *The Release of Genetically Modified Microorganisms – REGEM 2*, (eds D. Stewart-Tull and M. Sussman), FEMS Symposium **63**, Plenum, London, pp. 99–114.

Killham, K. (1992) Positively luminescent. *Bio/Technology*, **10**, 830–1.

Kluepfel, D.A., Kline, E.L., Skipper, H.D., Drahos, D.J., Barry, G.F., Hemming, B.C., Gooden, D.T., Hughes, T.A. and Brandt, E.J. (1991) The release and tracking of genetically engineered bacteria in the environment. *Phytopathology*, **81**, 348–52.

Kostka, S.J., Tomasino, S.F., Turner, J.T. and Reeser, P.W. (1988) Field release of a transformed strain of *Clavibacter xyli* subsp. *cynodontis* (CXC) containing a delta-endotoxin gene from *Bacillus thuringiensis* subsp. *Kurstaki*. *Phytopathology*, **78**, 1540.

Lindow, S.E. (1983) Methods of preventing frost injury caused by epiphytic ice-nucleation-active bacteria. *Plant Diseases*, **67**, 327–33.

Lindow, S.E., Arny, D.C. and Upper, C.D. (1983) Biological control of frost injury. II. Establishment and effects of an antagonistic *Erwinia herbicola* isolate on corn in the field. *Phytopathology*, **73**, 1102–6.

Lindow, S.E. and Panopoulos, N.J. (1988) Field tests of recombinant ice$^-$ *Pseudomonas syringae* for biological frost control in potato, in *The Release of Genetically-Engineered Micro-Organisms*, (eds M. Sussman, C. Collins, F. Skinner and D. Stewart-Tull), Academic Press, London, pp. 121–38.

Lindow, S.E., Panopoulos, N.J. and McFarland, B.L. (1989) Genetic engineering of bacteria from managed and natural habitats. *Science*, **244**, 1300–7.

Pickup, R.W. (1991) Development of molecular methods for the detection of specific bacteria in the environment. *Journal of General Microbiology*, **137**, 1009–19.

Prosser, J.I. (1994) Molecular marker systems for detection of genetically engineered microorganisms in the environment. *Microbiology*, **140**, 5–17.

Rattray, E.A.S., Prosser, J.I., Killham, K. and Glover, L.A. (1990) Luminescence-based nonextractive technique for *in situ* detection of *Escherichia coli* in soil. *Applied and Environmental Microbiology*, **56**, 3368–74.

Romanowski, G., Lorenz, M.G. and Wackernagel, W. (1993) Use of polymerase chain reaction and electroporation of *Escherichia coli* to monitor the persistence of extracellular plasmid DNA introduced into natural soils. *Applied and Environmental Microbiology*, **59**, 3438–46.

Schippers, B. (1988) Biological control of pathogens with rhizobacteria. *Philosophical Transactions of the Royal Society of London*, **B18**, 1377–83.

Shaw, J.J., Dane, F., Geiger, D. and Kloepper, J.W. (1991) Bioluminescence for the detection of genetically-engineered microbes released into the environment, in *Proceedings of the International Symposium on the Biosafety Results of Field Releases of Genetically Modified Plants and Microorganisms, Kiawah Island*, (eds D.R. MacKenzie and S. Henry), South Carolina Agricultural Research Institute Press, Washington, DC, pp. 229–36.

Shaw, J.J., Dane, F., Geiger, D. and Kloepper, J.W. (1992a) Use of bioluminescence for detection of genetically-engineered microbes released into the environment. *Applied and Environmental Microbiology*, **58**, 267–73.

Shaw, J.J., Beauchamp, C., Dane, F. and Kriel, R.J. (1992b) Securing a permit from the United States Department of Agriculture for field work with genetically engineered microbes: a non-prohibitory process. *Microbial Releases*, **1**, 51–3.

Silcock, D.J., Waterhouse, R.N., Glover, L.A., Prosser, J.I. and Killham, K. (1992) Detection of a single genetically modified bacterial cell in soil by using charge coupled device-enhanced microscopy. *Applied and Environmental Microbiology*, **58**, 2444–8.

Steele, D. (1993) The new UK rules on releasing genetically modified organisms. *Chemistry & Industry*, 19th July, 542–4.

Steffan, R.J. and Altas, R.M. (1991) Polymerase chain reaction: applications in environmental microbiology. *Annual Review of Microbiology*, **45**, 137–61.

Stewart-Tull, D.E. and Sussman, M. (1992) *The Release of Genetically Modified Microorganisms – REGEM 2*, FEMS Symposium **63**, Plenum, London.

Supkoff, D.M, Gezark, L.G. and Opgenorth, D. (1988) Monitoring of the winter 1987 field release of genetically engineered bacteria in Contra Costa County, *Report BC 88*-1, California Department of Food and Agriculture, Sacramento, CA.

Sussman, M., Collins, C.H., Skinner, F.A. and Stewart-Tull, D.E. (1988) *The Release of Genetically-Engineered Micro-Organisms*, Academic Press, London.

Waterhouse, R.N., Silcock, D.J., White, H.L., Buhariwalla, H.K. and Glover, L.A. (1993) The cloning and characterization of phage promoters, directing high expression of luciferase in *Pseudomonas syringae* pv. *phaseolicola*, allowing single cell and microcolony detection. *Molecular Ecology*, **2**, 285–94.

Index

advantageous possibilities which include:

a) the newcomer need not be a potential pest if it out-competes the resident species in some arena other than competition for food from the crop (e.g. on wild plants in the fallow season, or for pupation sites);
b) the newcomer, although a potential pest, is at least not a disease vector as the resident is, and a plant disease problem may therefore be more easily controlled;
c) the newcomer may be severely affected by adverse climate (e.g. it is non-diapausing in temperate conditions). The release of the newcomer, if it were able to displace the resident in one season, would then end in the disappearance of both species.

An example in the last category concerns the field cricket (*Teleogryllus commodus*) in Australia. The race of the cricket in the colder south produces diapausing eggs, but the race in the north does not. When southern females are mated to northern males they produce only non-diapausing eggs, and thus matings following the release of northern males into the south should result in many eggs laid by the southern females failing to survive the winter.

One problem with the concept of competitive displacement is that there is usually no reason why the competitor should out-compete the resident species. Especially where the introduced competitor would need to replace the resident species within one season, it may be necessary to give the introduced competitor some selective advantage. An obvious technique would be to breed insecticide resistance into the competitor, and then manipulate insecticide use to exploit this artificial advantage.

8.5 Internal growth regulators

These are chemicals which interfere in an adverse manner with the normal growth and development of insects (Fig. 8.2). The idea of using such products originated from an insect endocrinologist, Carroll Williams, who suggested that the hormones produced internally by insects to regulate their moulting and metamorphosis could be turned back on them as 'third-generation insecticides'. Original claims made for such third-generation insecticides were that they should be highly specific, and that resistance to them would be unlikely.

Insect hormones are relatively non-specific among insects, but certainly they are likely to be less damaging to other types of organisms in the environment than are insecticides.

However, resistance to hormones and 'hormone mimics' has occurred, presumably because insects already possess the compounds to maintain their own hormone titre internally.

8.5.1 Juvenile hormone analogues

These function in the same way as the juvenile hormone in the regulation of metamorphosis, and may or may not be similar in chemical structure to the

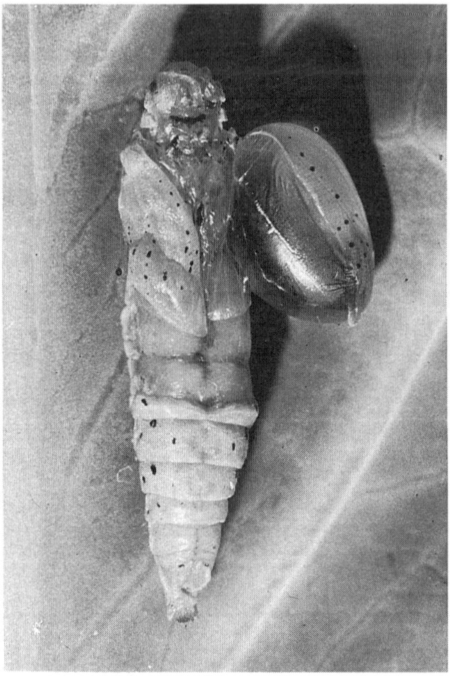

Fig. 8.2 Pupal deformity following application to large cabbage white butterfly (*Pieris brassicae*) caterpillars of a chemical with structural similarities to insect moulting hormone. Photograph courtesy of ICI Agrochemicals.

natural hormone. Two such analogues are methoprene and kinoprene. The effects of these compounds are usually seen during larval to pupal metamorphosis, and various degrees of incomplete metamorphosis become apparent. Larval-pupal mosaics may be produced, or strange deformations may appear on the pupal structure. Other uses of juvenile hormone analogues are in disrupting embryogenesis in the eggs and in preventing adult diapause. They have been extensively tested in public health and stored products work because of their relative safety to human beings.

8.5.2 Anti-juvenile hormones

A high titre of juvenile hormone in the insect maintains its larval characteristics, and if this effect can be counteracted in the early stages, then the larvae may metamorphose into miniature pupae or sterile adults. This is known as precocious development, and the name 'precocenes' has been given to a major group of anti-juvenile hormone compounds. None of the compounds developed so far have been sufficiently active for practical purposes, but there is still hope for the future.